Solid Mechanics and Its Applications

Volume 281

The fundamental questions arising in mechanics are: Why?, How?, and How much? The aim of this series is to provide lucid accounts written by authoritative researchers giving vision and insight in answering these questions on the subject of mechanics as it relates to solids. The scope of the series covers the entire spectrum of solid mechanics. Thus it includes the foundation of mechanics; variational formulations; computational mechanics; statics, kinematics and dynamics of rigid and elastic bodies; vibrations of solids and structures; dynamical systems and chaos; the theories of elasticity, plasticity and viscoelasticity; composite materials; rods, beams, shells and membranes; structural control and stability; soils, rocks and geomechanics; fracture; tribology; experimental mechanics; biomechanics and machine design. The median level of presentation is the first year graduate student. Some texts are monographs defining the current state of the field; others are accessible to final year undergraduates; but essentially the emphasis is on readability and clarity.

Springer and Professors Barber and Klarbring welcome book ideas from authors. Potential authors who wish to submit a book proposal should contact Dr. Mayra Castro, Senior Editor, Springer Heidelberg, Germany, email: mayra.castro@springer.com

Indexed by SCOPUS, Ei Compendex, EBSCO Discovery Service, OCLC, ProQuest Summon, Google Scholar and SpringerLink.

Jairo Useche Vivero

The Boundary Element Method for Engineering Analysis of Plates and Shells

Springer

Jairo Useche Vivero
Universidad Tecnológica de Bolívar
Cartagena, Colombia

ISSN 0925-0042 ISSN 2214-7764 (electronic)
Solid Mechanics and Its Applications
ISBN 978-3-031-93049-2 ISBN 978-3-031-93050-8 (eBook)
https://doi.org/10.1007/978-3-031-93050-8

This Springer imprint is published by the registered company Springer Nature Switzerland AG
The registered company address is: Gewerbestrasse 11, 6330 Cham, Switzerland

To my loved wife Johana
To my sons Nicolas and Julian

Preface

The structural and mechanical analysis of plates and shells remains a crucial field in engineering. The multiple applications of thin-walled structures in aeronautical, civil, mechanical, and oceanic engineering, among others, have required continuous research over the years. Much research is carried out on topics ranging from plate theories, new materials, nonlinear dynamic behavior, and applications at high temperatures. The mathematical complexity inherent in the governing equations proposed by the different existing theories of plates and shells makes it necessary to use computational methods for their solution, which is a very active research topic. In recent decades, advances in research and the application of reduced mesh numerical methods, such as the Boundary Element Method (BEM), have allowed their practical use in engineering, reducing computational costs compared to traditional domain discretization methods, such as the finite element method.

Although numerous works have been published on plate and shell analysis using BEM, this topic is very specialized and requires many hours of study to master the subject. The study sources generally could be more friendly for those who come to the subject for the first time. The novice who wants to enter this world requires studying many different bibliographic sources, which are easier to understand with an appropriate guide. The BEM literature for plates and shells is spread across numerous research documents, such as journal articles and master's and doctoral theses. Few books treat the subject in a complete, detailed, and didactic way. Currently, no books are focused on the BEM analysis of plates and shells in an understandable manner that covers the topic broadly and with a practical approach. The application of the BEM in the analysis of plates and shells still has much to explore and study and much to exploit in terms of its application in engineering, among other reasons, due to the little diffusion that the method has had beyond the academic field.

The primary aim of this book is to provide a comprehensive introduction to the BEM and its practical application in the analysis of bending problems in plates and shells in engineering. It covers a wide range of practical issues, including static problems, transient dynamics, modal analysis, harmonic, as well as problems of buckling and fracture mechanics. The book presents both the classical theory of plates and shells and the theory of thick plates and shells. The topics are developed sequentially, starting with a thorough introduction to the necessary mathematical concepts and then moving on to the application of the BEM to problems of plane elasticity, thin and thick plates and shells, for both isotropic materials and laminated composites. The book also delves into the analysis of fractured plates and shells, providing a detailed explanation of the application of the dual boundary element method in plates and shells and the calculation of stress intensification factors, in both static and dynamic regimes.

A library of computational codes developed in MATLAB for BEM analysis of plates and shells, following the formulations presented in this book, has been made available for free download at the link: https://github.com/jfuseche/BEM_Plate_Shell_Book_Juseche.git

Cartagena, Colombia Jairo Useche Vivero

Competing Interests The author has no competing interests to declare that are relevant to the content of this manuscript.

About This Book

The specialized literature does not currently offer a similar book with the proposed scope, which broadly and clearly and with an engineering approach, shows the application of the BEM in static, dynamic, fracture, buckling, and nonlinear problems in plates, shells, and thin-walled space structures. Likewise, few open computer programs are available for BEM analysis in solid mechanics, and the codes available for studying plates, shells, and thin-walled structures, as offered in this book, are practically nonexistent. The open source offered fulfills the double purpose of allowing the computational implementation of the theory presented throughout the book to be studied and used as an analysis tool for practical use in engineering.

The book is aimed at final-year students, graduate students, and practicing engineers and is structured as follows:

Chapter 1 describes basic concepts in vector and tensor calculus and interpolation functions. An important aspect is found in the numerical integration schemes used in the Boundary Element Method (BEM), such as the Gauss-Legendre integration method and logarithmic integration, which are presented in this chapter.

Chapter 2 presents an introduction to the BEM. This chapter uses a simple example to demonstrate the application of the method in its various steps to solve a problem posed from its governing differential equation to its numerical solution using BEM. The concepts of Sominglian identity, fundamental solutions, the boundary integral formulation, and the BEM application procedure are presented.

The plate and shell theories used throughout the book are presented in Chap. 3. This chapter presents in detail the thin-plate (Kirchhoff-Love) and thick-plate (Reissner-Midlin) theories, as well as their extension to the case of laminated materials and shallow shells. For each case, the deformation displacement relationships, constitutive relationships, and equilibrium equations are presented.

The integral formulations for thin isotropic plates in Chap. 4. Fundamental solutions for isotropic thin plates are presented and the deduction of the boundary integral equations. The treatment of quasi-singular and singular contour integrals are explained in detail. Also, the cell integration method is presented for treating domain integrals. Application examples related with plates under static loads.

Chapter 5 presents the boundary element analysis of thick isotropic plates. The fundamental isotropic solutions for Reissner's plate equations are offered. The boundary integrals equations and their discretization through the BEM are developed. The treatment of singular integrals is presented in detail, as well as domain integrals through the dual reciprocity boundary element method. Application examples related with static, dynamic, and modal response of plates are presented.

Chapter 6 presents the integral displacement representation of the governing equation for thick shallow shells. The isotropic fundamental solutions are introduced, and the boundary integral equations are derived. The evaluation of singular integrals is presented in detail. Treatment of domain integrals using the dual reciprocity BEM for the transformation of domain integrals to boundary integrals. Numerical examples are given for static and dynamic loads.

Chapter 7 presents the application of the BEM to the analysis of thin laminated plates and shells based on the classical laminated plate theory. Fundamental solutions for symmetric laminates and recently proposed fundamental solutions for unsymmetric laminates are presented. For both cases, the application of BEM to the analysis of static and dynamic problems is presented.

The BEM analysis of thick plates based on first-order plate theory is presented in Chap. 8. Fundamental solutions for thick plates with symmetric laminates are also presented, as are recent fundamental solutions for unsymmetric laminates. The development of the BEM formulation is presented throughout the chapter, and its application is demonstrated through examples of applications to static and dynamic cases. Similarly, in Chap. 9 the application of these formulations to the analysis of thick laminated shells is presented.

Chapter 10 presents the fracture mechanics analysis for plates and thick shells. For this, the dual reciprocity boundary element method is presented for modeling of cracks in isotropic thick plates and shallow shells. Crack modeling in laminated materials is aborted using the domain sub-division technique. The stress intensification factors for bending and plane stress are defined. The evaluation of stress intensity factors using the Crack Tip Opening Displacement method and the J-integral are presented. Static and dynamic application examples are presented.

Chapter 11 analyzes thin-walled spatial structures for both isotropic and laminated cases. In the first part of the chapter, a multi-region formulation is developed to model this type of problems. The compatibility relationships between translations and rotations and the equilibrium equations for membrane traction, bending, and shear are established to do this. A special case is presented in this chapter related to the application of the multi-region technique for the analysis of thick plates fractured and repaired with patches of laminated composite materials.

Chapter 12 presents the buckling analysis of isotropic and laminated plates and shells. For this, the boundary integral equations for displacements and tractions for plates and shells are presented with the domain integrals corresponding to the coupling of the internal membrane forces and the rotations and deflections of the plate/shells. The solution strategy is presented using the BEM. The application of the dual reciprocity method and the radial integration method for the treatment of contour integrals is presented. At the end of the chapter, examples of applications in engineering are presented.

Finally, Chap. 13 presents advanced topics related to the BEM analysis of nonlinear problems involving large deflections, for which boundary integrals for isotropic plates and shells are proposed. The chapter also covers the elastoplastic analysis of plates and shells using boundary elements.

Contents

About the Author

Prof. Jairo Useche Vivero Ph.D., M.Sc., is a Mechanical Engineer from the Technological University of Bolívar—UTB (located in Cartagena, Colombia) with a Master's degree in Mechanical Engineering in solid mechanics from the Universidad de los Andes—Colombia. He has Ph.D. in Mechanical Engineering in computational solid mechanics and structures from the State University of Campinas, São Paulo—Brazil. He is a Titular Professor in the Department of Mechanical Engineering at UTB. Professor Useche was Dean of the Engineering School at the UTB from 2011 until 2017. He has been the Director of the Research Office at UTB since 2017.

He teaches various engineering courses at both the undergraduate and graduate levels. He has developed and taught several solid mechanics engineering courses, including Mechanics of Materials, Advanced Numerical Modeling, Structural Vibrations, Fracture Mechanics, Finite Element Analysis, and Boundary Element Method in Solid and Structural Mechanics. He has also developed and taught professional short courses for industry. Their research and consulting expertise includes structural mechanics, structural reliability, mechanics of oil and gas pipelines, fitness for service analysis, structural panels and pipes under blast loads, piping stress analysis, composite mechanics, and composite patch mechanics in fracture plates. Dr. Useche's research has been funded by the Ministerio de Ciencia y Tecnologìa de Colombia, Corporación para el Desarrollo de la Industria Naval, Maritima y Fluvial—COTECMAR, and industries and consulting firms in Colombia.

Dr. Useche currently serves on the SHIP Science and Technology Journal editorial board. He is organizing the Continuum Mechanics and Structures Symposium NLCMS/SMEC 2008, 2009, 2011, 2014, and 2016. He is on the scientific committee of the following congresses: VIII Colombian Congress in Numerical Methods—CCMN 2011; IX Colombian Congress in Numerical Methods—CCMN 2013; VI International Congress of Materials—CIM 2011; II International Ship Design and Naval Engineering Congress—CIDIN 2011; III International Ship Design and Naval Engineering Congress—CIDIN 2013, IV International Ship Design and Naval Engineering Congress—CIDIN 2015, V International Ship Design and Naval Engineering Congress—CIDIN 2017 and VI International Ship Design and Naval Engineering Congress—CIDIN 2019. He has been a member of the Science and Technology Bureau of Cotecmar since 2013. He was appointed by the President Office of the Universidad Tecnológica de Bolívar as a member of the Board of Directors of Cotecmar since 2013 until now, Delegate member to OCAD CTi, Technical Secretary of Codecti—Bolívar 2018 until 2023. Technical Secretary of Comité Universidad-Empresa-Estado-CUEE Cartagena, 2018 until now.

He has delivered over 60 papers and presentations at national and international workshops, conferences, and journals. He has received several awards and honors for excellence in teaching, research, and professional service, including the Universidad Tecnológica Award for Excellence in Research 2010 and 2015 UTB Quality Assessment and the Cotecmar Medal for Outstanding Contributions, 2018. From 2016 until now, Prof. Useche has been recognized as a Senior Researcher, the highest category granted to scientific researchers in Colombia.

Also, he is a consulting engineer in Colombia and worked as a mechanical engineer in the industry for over 20 years. He has participated in projects related to numerical modeling in structural mechanics, composite materials, vibrations in structures, and evaluating the structural reliability of pipelines. His area of expertise includes Advanced Numerical Analysis in Structural Mechanics Analysis, Structural Reliability, Residual Strength and Damage, Impact and Dynamic Load Assessment, Pipeline Structural Mechanics, Fitness for Service (API 579/ASME FFS), fatigue and fracture propagation analysis, computational code development for advanced engineering analysis using the Finite Element Method, the Boundary Element Method Smoothed Particle Hydrodynamics Method, and Finite Volume Method. As a specialist in computational mechanics of solids and structures, he has developed commercial computational tools based on the Finite Element Method and Boundary Element Method for advanced structural analysis for the oil and gas industry.

List of Figures

Chapter 1
Mathematical Preliminaries

Abstract This chapter presents the fundamentals used in the mathematical basis of the Finite Element Method. An essential aspect of this scaffolding is the index notation or Einstein notation, which allows the writing of complex and large matrix and tensor expressions in a compact form. Another fundamental aspect, which is at the heart of the method, is the divergence theorems. In its most general form, Green's Theorem allows transforming domain integrals to boundary integrals, the basis of the boundary element method. Likewise, this chapter deals with the concept of interpolation functions, which allow the approximation of field variables within elements and cells in the domain, using transformations of affine coordinates for this purpose. Related to this are the numerical integration methods of Gauss-Legrende and logarithmic integration, which are fundamental for the numerical approximation of the integrals that appear in the formulation of the method.

1.1 Introduction

This chapter presents the fundamentals used in the mathematical basis of the Finite Element Method. An essential aspect of this scaffolding is the index notation or Einstein notation, which allows the writing of complex and large matrix and tensor expressions in a compact form. Another fundamental aspect, which is at the heart of the method, is the divergence theorems. In its most general form, Green's Theorem allows transforming domain integrals to boundary integrals, the basis of the boundary element method.

Likewise, this chapter deals with the concept of interpolation functions, which allow the approximation of field variables within elements and cells in the domain, using transformations of affine coordinates for this purpose. Related to this are the numerical integration methods of Gauss-Legrende and logarithmic integration, which are fundamental for the numerical approximation of the integrals that appear in the formulation of the method.

J. Useche Vivero, *The Boundary Element Method for Engineering Analysis of Plates and Shells*, Solid Mechanics and Its Applications 281,
https://doi.org/10.1007/978-3-031-93050-8_1

1.2 Vectors and Tensors

One of the most critical applications of vectors and matrices is in the solution of systems of equations, written in compact form as [3]

$$\mathbf{A} \cdot \mathbf{x} = \mathbf{b} \tag{1.1}$$

where $\mathbf{A}$ is the coefficient matrix, $\mathbf{x}$ is the vector of unknowns, and $\mathbf{b}$ is the vector of independent terms. In index notation:

$$A_{ij}x_j = b_i, \qquad i = 1\ldots, m, \quad j = 1\ldots, n \tag{1.2}$$

A matrix is a rectangular array of m rows and n columns of numbers or quantities. A matrix consisting only of elements in a single row is called a *row vector*. Similarly, if the elements are located in a single column, it is a *column vector*. To represent vectors in the form of components, we introduce in our physical space a right-handed system of rectangular Cartesian axes $Ox_1x_2x_3$, and identify with these axes the triad of unit vectors of basis $\hat{\mathbf{e}}_1$, $\hat{\mathbf{e}}_2$ y $\hat{\mathbf{e}}_3$ (Fig. 1.1). In this book, we work with straight orthogonal coordinate systems, so these basis vectors form an orthogonal basis. Any vector $\mathbf{v}$ is expressed in its basis components, $\hat{\mathbf{e}}_1$, $\hat{\mathbf{e}}_2$ y $\hat{\mathbf{e}}_3$, as

$$\mathbf{v} = v_1\hat{\mathbf{e}}_1 + v_2\hat{\mathbf{e}}_2 + v_3\hat{\mathbf{e}}_3 = \sum_{i=1}^{3} v_i\hat{\mathbf{e}}_i \tag{1.3}$$

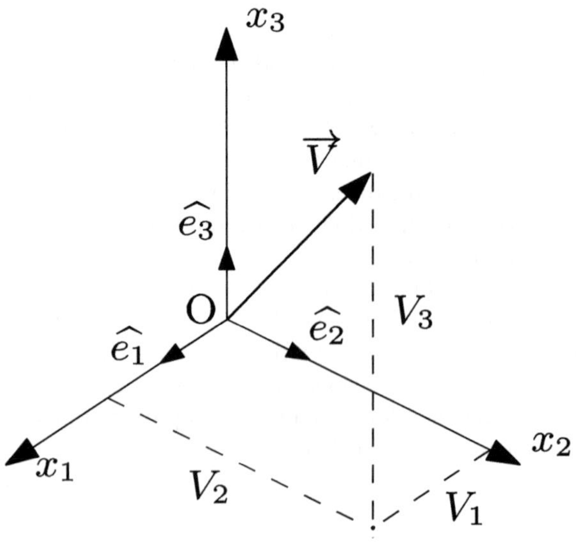

Fig. 1.1 Vector in a rectangular system

Throughout this book, vectors will be represented in bold lowercase letters, and their magnitude will be presented by: ($\|\cdot\|$), which means the vector's norm.

The trace of a matrix of size $n \times m$ is defined as the sum of the terms of its main diagonal:

$$tr(\mathbf{A}) = A_{11} + A_{22} + A_{33} \cdots + A_{nn} \tag{1.4}$$

On the other hand, we define $\mathbf{F} = diag[\cdot]$ as a block diagonal matrix consisting of several matrices. That is, consider $\mathbf{A}_1, \mathbf{A}_2, \ldots \mathbf{A}_m$ matrices (not necessarily square), the matrix $\mathbf{F} = diag[\mathbf{A}_1, \mathbf{A}_2, \ldots \mathbf{A}_m]$ is a block diagonal matrix formed by the matrices, with the diagonal of $\mathbf{A}_n,\ n = 1, 2 \ldots, m$ located on the diagonal of $\mathbf{F}$.

On the other hand, tensors are mathematical objects that are used to describe physical quantities. Tensors generalize scalars and vectors; a scalar is a tensor of rank zero, and a vector is a tensor of first rank. The rank of a tensor is defined by the number of directions needed to describe it. For example, properties that require one direction (first rank) can be fully described by a 3×1 column vector, and properties that require two directions (second rank tensors) can be described by 9 components, such as a 3×3 matrix. As such, in general, a tensor of rank n can be described by $3n$ coefficients. This book will present the tensors in capital letters and subindex. For example, the tensor A of rank 2 is written as: A_{ij}, of rank 3: A_{ijk} and of rank 4: A_{ijkl}, where $i, j, k, l = 1, \ldots, n$.

1.3 Kronecker Delta

The Kronecker Delta function is widely used in the mathematical architecture of the boundary element method to represent mathematical operations using indexical notation (see Sect. 1.7). The Kronecker Delta function is defined as

$$\delta_{ij} = \begin{cases} 1 & \text{if } i = j \\ 0 & \text{if } i \neq j \end{cases} \tag{1.5}$$

we see that

$$\hat{\mathbf{e}}_i \cdot \hat{\mathbf{e}}_j = \delta_{ij} \tag{1.6}$$

Also, note that by the summation convention:

$$\delta_{ii} = \delta_{11} + \delta_{22} + \delta_{33} = 1 + 1 + 1 = 3 \tag{1.7}$$

For example, using index notation:

$$\delta_{ij}\hat{\mathbf{e}}_j = \delta_{11}\hat{\mathbf{e}}_1 + \delta_{12}\hat{\mathbf{e}}_2 + \delta_{13}\hat{\mathbf{e}}_3 = \hat{\mathbf{e}}_1. \tag{1.8}$$

1.4 Permutation Symbol

The permutation symbol ϵ_{ijk} is defined as

$$\epsilon_{ijk} = \begin{cases} 1 \text{ if } ijk \text{ appear as in the sequence: } & 1-2-3-1-2 \\ -1 \text{ if } ijk \text{ appear as in the sequence: } & 3-2-1-3-2 \\ 0 \text{ if } ijk \text{ appear in any other sequence} & \end{cases} \tag{1.9}$$

The interchange of any two subscripts in ϵ_{ijk} causes a change of sign so that

$$\epsilon_{ijk} = -\epsilon_{kji} = \epsilon_{kij} = -\epsilon_{ikj} \tag{1.10}$$

For repeated subscripts: $\epsilon_{ijk} = 0$.

1.5 $\epsilon - \delta$ Identity

The product of the permutation symbols $\epsilon_{miq}\epsilon_{jkq}$ can be expressed in terms of Kronecker deltas by the identity:

$$\epsilon_{miq}\epsilon_{jkq} = \delta_{mj}\delta_{ik} - \delta_{mk}\delta_{ij} \tag{1.11}$$

as may be proven by direct expansion.

1.6 Dirac Delta Function

The Dirac Delta function, represented as $\delta(x)$, has a definitive application in the boundary element method by allowing the definition of local perturbations, represented in the case of plate and shell problems, forces and/or specific moments applied in the determination of the **fundamental solutions** of the governing equations of the problem.

The Dirac Delta Function is defined as a function that takes the value of zero except at the origin, where it takes an infinite value, such that its integral in the domain $x \in]-\infty, +\infty[$:

$$\int_{-\infty}^{+\infty} \delta(x)dx = 1 \tag{1.12}$$

From this definition, the function has the following properties:

1. Sampling property:

$$\int_{-\infty}^{+\infty} f(x)\delta(x-a)dx = f(a) \tag{1.13}$$

2. Scaling property (if $a \neq 0$):

$$\delta(ax) = \frac{1}{|a|}\delta(x) \tag{1.14}$$

3. Derivative property:

$$\int_{-\infty}^{+\infty} f(x)\delta'(x)dx = -f'(0). \tag{1.15}$$

1.7 Indicial Notation

The initial notation allows tensor operations of addition, multiplication, differentiation, etc., without the use or appearance of the basis vectors $\hat{\mathbf{e}}_i$ in the equations. We agree that the number of *free*, indicates a term's tensor rank (order), That is, non-repeating subscripts appearing in that term. Consequently, a term without free indices represents a scalar, a term with one free index, a vector, a term having two free indices, a second-order tensor, etc. In particular, a vector is represented as v_i, a second-order tensor is defined as T_{ij}, a third-order tensor as S_{ijk} and a fourth-order tensor as D_{ijkl}, etc.

For tensors defined in a three-dimensional space, the free indices successively take the values 1–3, and we say that these indices have a rank of three. If N is the number of free indices in a tensor, that tensor has $3N$ components in three spaces. We should note that exactly two types of subscripts appear in index notation:

- "Free" indices are represented by letters that appear only once in a given term.
- "Dummy" indexes are represented by letters appearing twice in a given term.

In addition, each term of a valid equation must have the same letter subscripts for the free indices. No letter subscript may appear more than twice in a given term. In Table 1.1, where its application in different operations is shown. The index notation will be used throughout this book. Subindices in Greek letters vary from 1 to 2, and subindices in Roman letters from 1 to 3.

Table 1.1 Examples of index notation

Operation	Indicial notation
Dot product	$\mathbf{x} \cdot \mathbf{y} = x_1 y_1 + x_2 y_2 + \ldots x_n y_n = x_i y_i$
Matrix dot product	$\mathbf{b} = \mathbf{A} \cdot \mathbf{x} = A_{ij} x_j = b_i$
Matrix product	$\mathbf{C} = \mathbf{A} \cdot \mathbf{B} = A_{ik} B_{kl} = C_{il}$
Derivatives	$\frac{\partial f}{\partial x_1} = f_{,1}, \ \frac{\partial^2 f}{\partial x_1^2} = f_{,11}, \frac{\partial^2 f}{\partial x_1 \partial x_2} = f_{,12}$
Divergence	$\nabla \phi = \frac{\partial \phi}{\partial x_1} + \frac{\partial \phi}{\partial x_2} = \phi_{,i} \delta_{ii}$
Laplacian	$\nabla^2 \phi = \frac{\partial^2 \phi}{\partial x_1^2} + \frac{\partial^2 \phi}{\partial x_2^2} = \phi_{,ii}$
Gradient	$\nabla \phi = \frac{\partial \phi}{\partial x_1} \hat{\mathbf{e}}_1 + \frac{\partial \phi}{\partial x_2} \hat{\mathbf{e}}_2 = \phi_{,i} \delta_{ii} \hat{\mathbf{e}}_i$
Tensor product	$C_{ijkl} = T_{ij} L_{kl}$

1.8 Gradient and Divergence Theorem

Let's represent by ∇ y ∇^2 the gradient and Laplacian operators, respectively. In Cartesian rectangular coordinates, these operators are defined as [4]

$$\nabla = \left\{ \frac{\partial}{\partial x}, \frac{\partial}{\partial x} \right\}, \quad \nabla^2 = \nabla \cdot \nabla = \left\{ \frac{\partial^2}{\partial x}, \frac{\partial^2}{\partial x} \right\} \tag{1.16}$$

The Gauss divergence theorem establishes a relation between the surface integral having a tensor of order-n $G_{ij\ldots k}$ as the integrand to the integral volume for which a coordinate derivative of $G_{ij\ldots k}$ is the integrand. Specifically for a second-order tensor G_{ij}:

$$\int_{\Omega} \nabla G_{ij} \, dxdy = \oint_{\Omega} \boldsymbol{n} G_{ij} \, ds \tag{1.17}$$

where $\boldsymbol{n}$ represents a unit vector normal to the surface Γ of the domain Ω, n_x y n_y are the rectangular components of the normal vector. As a special case, this theorem applied to the case of a vector field $\mathbf{t}$, we obtain the Grient Theorem

$$\int_{\Omega} \nabla \cdot \boldsymbol{t} \, dxdy = \oint_{\Gamma} \boldsymbol{n} \cdot \boldsymbol{t} \, ds \tag{1.18}$$

The following identities are derived from the gradient and divergence theorems. Let w and G be scalar functions defined on the two-dimensional domain Ω bounded by the boundary Γ, one has then

$$\int_{\Omega} (\nabla G) w dx dy = -\int_{\Omega} (\nabla w) G dx dy + \oint_{\Gamma} \boldsymbol{n} w G ds \tag{1.19}$$

In Cartesian components:

$$\int_{\Omega} w \frac{\partial G}{\partial x} dx dy = -\int_{\Omega} \frac{\partial w}{\partial x} G dx dy + \oint_{\Gamma} n_x w G ds$$

$$\int_{\Omega} w \frac{\partial G}{\partial y} dx dy = -\int_{\Omega} \frac{\partial w}{\partial y} G dx dy + \oint_{\Gamma} n_y w G ds \tag{1.20}$$

Likewise, it must be noted that

$$-\int_{\Omega} (\nabla^2 G) w dx dy = \int_{\Omega} (\nabla w) \cdot (\nabla G) dx dy - \oint_{\Gamma} \frac{\partial G}{\partial n} w ds \tag{1.21}$$

where $\partial/\partial n = \boldsymbol{n} \cdot \nabla^T()$.

1.9 Geometric Transformations

If we consider localized points x_i in a Cartesian space of dimension n, X^n, $(n = 1, 2, 3)$ together with their corresponding curvilinear coordinates ξ_i in the space C^n, $(n = 1, 2, 3)$, x_i can be expressed as a function of the normalized coordinates $1 \geq \xi \leq 1$ as: $x_i = f(\xi)$ and conversely also: $\xi = g(x_1, x_2, x_3) = \xi(x_i)$ (see Fig. 1.2). Thus, the relationship between line differential elements and X and Z are related as [6])

$$dx_i = \frac{\partial x_i}{\partial \xi} d\xi = x_{i,i} d\xi \tag{1.22}$$

where $\boldsymbol{J}$ is the Jacobian matrix of the transformation, defined as

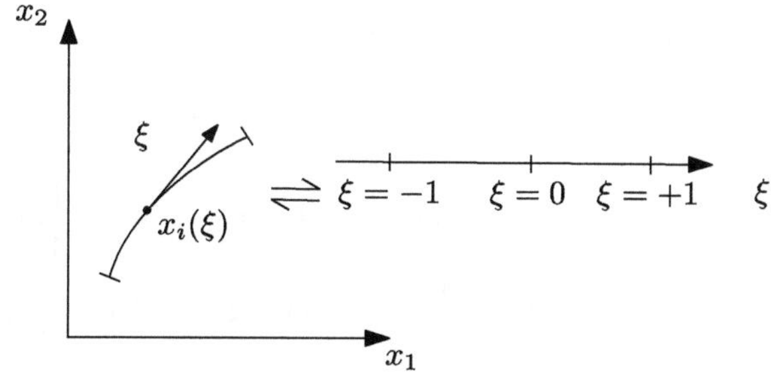

Fig. 1.2 Interpolation functions

$$J = \left[\frac{\partial x_i}{\partial \xi}\right] \tag{1.23}$$

with its inverse: $d\xi = J^{-1}dx$, where $J^{-1} = \partial\xi/\partial x_i$.

From the above expressions, the transformation from X to ξ or a general line element of length $dl(x)$ yields that

$$dl(x) = J_l d\eta \tag{1.24}$$

where the Jacobian J_l is given by

$$J_l = \left[\left(\frac{\partial x_1}{\partial \xi}\right)^2 + \left(\frac{\partial x_2}{\partial \xi}\right)^2\right]^{1/2}. \tag{1.25}$$

1.10 Interpolation Functions

Consider a four-sided cell, as presented in Fig. 1.4, whose points x_i are defined in the Cartesian space X_i and the coordinates ξ of a normalized space Z_i are related. If the values of a certain function ψ defined at each of the nodes $i = 1, 2, 3, 4$ in X, one can define a polynomial of order n $\hat{\psi}(\xi_1, \xi_2)$ that approximates ψ over the cell:

$$\psi(\xi_1, \xi_2) = \sum_{i=0}^{n} c_i \left[\sum_{j=0}^{n} c_j \xi_1^j\right] \xi_2^i \tag{1.26}$$

where c_i are the coefficients of the polynomial.

As a special case, consider the polynomial of degree two defined in the region Ω_c, as shown in Fig. 1.3:

$$\psi \approx \hat{\psi}(\xi_1, \xi_2) = c_0 + c_1\xi_1 + c_2\xi_2 + c_3\xi_1\xi_2 \tag{1.27}$$

Knowing the values of ψ at points 1–4, the following interpolation function is obtained

$$\hat{\psi}(\xi_1, \xi_2) = N_1(\xi_1, \xi_2)\psi_1 + N_2(\xi_1, \xi_2)\psi_2 + N_3(\xi_1, \xi_2)\psi_3 + N_4(\xi_1, \xi_2)\psi_4 = N_i\psi_i \tag{1.28}$$

where the *shape functions* N_i are given by

$$N_1 = \frac{1}{4}(1 - \xi_1)(1 - \xi_2), \quad N_2 = \frac{1}{4}(1 + \xi_1)(1 - \xi_2),$$
$$N_3 = \frac{1}{4}(1 + \xi_1)(1 + \xi_2), \quad N_4 = \frac{1}{4}(1 - \xi_1)(1 + \xi_2) \tag{1.29}$$

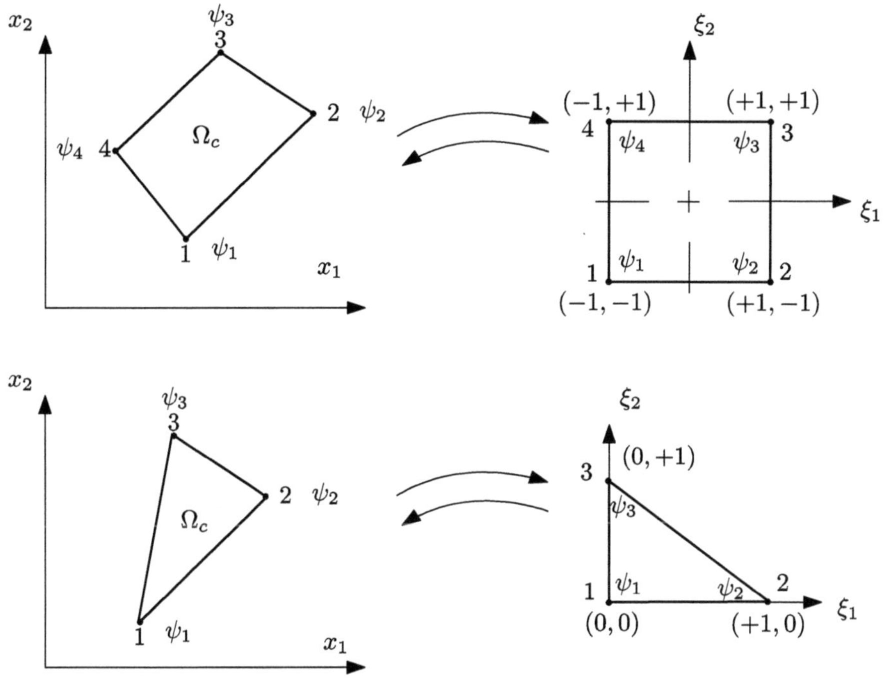

Fig. 1.3 Interpolation over four-node quadrilater (above) and three-node triangle (below)

Fig. 1.4 Interpolation functions

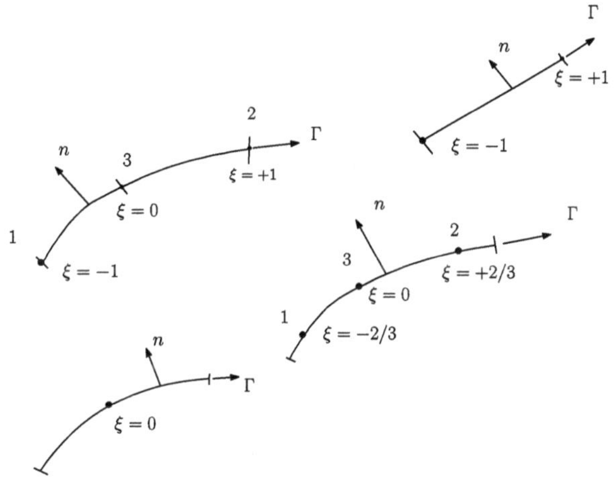

Similarly, for a region defined by three nodes, we have (see Fig. 1.3)

$$\hat{\psi}(\xi_1, \xi_2) = N_1(\xi_1, \xi_2)\psi_1 + N_2(\xi_1, \xi_2)\psi_2 + N_3(\xi_1, \xi_2)\psi_3 \tag{1.30}$$

In this case, the shape functions are given by

$$N_1 = 1 - \xi_1 - \xi_2, \quad N_2 = \xi_1, \quad N_3 = \xi_2 \tag{1.31}$$

For functions defined on line segments (see Fig. 1.3), where the value of the ψ function at two end nodes is known, the linear interpolation function is obtained

$$\hat{\psi}(\xi_1) = N_1(\xi_1)\psi_1 + N_2(\xi_1)\psi_2 \tag{1.32}$$

where:

$$N_1 = 1/2(1 - \xi_1), \quad N_2 = 1/2(1 + \xi_1) \tag{1.33}$$

Similarly, for a line segment with three nodes (quadratic interpolation):

$$\hat{\psi}(\xi_1) = N_1(\xi_1)\psi_1 + N_2(\xi_1)\psi_2 + N_3(\xi_1)\psi_3 \tag{1.34}$$

where:

$$N_1 = \frac{1}{2}\xi_1(1 + \xi_1), \quad N_2 = -\frac{1}{2}\xi_1(1 - \xi_1), \quad N_3 = 1 - \xi_1^2 \tag{1.35}$$

Finally, for line segments with end nodes located 2/3 from the center of node 2 (Fig. 1.4), we have (discontinuous quadratic interpolation)

$$N_1 = \frac{3}{4}\xi_1\left(\frac{3}{2}\xi_1 - 1\right), \quad N_2 = \left(1 - \frac{3}{2}\xi_1\right)\left(1 + \frac{3}{2}\xi_1\right), \quad N_3 = \frac{3}{4}\xi_1\left(\frac{3}{2}\xi_1 + 1\right). \tag{1.36}$$

1.11 Coordinate Transformation

The coordinates x_i defining a line can be described equally by linear, quadratic, or higher-order interpolation polynomials using the approximation polynomials presented in the previous section. For a straight-line element, we have the following coordinate transformation:

$$x_i = N_1(\xi_1)x_i^1 + N_2(\xi_1)x_i^2 = N_k(\xi_1)x_i^k \tag{1.37}$$

where the shape functions are given by the equations (1.33), (1.35) and (1.36). For this element, the determinant of the Jacobian of the transformation $dl(x) = det(J)d\xi$ is given by

$$J_l = \left[\left(\frac{\partial x_1}{\partial \xi_1}\right)^2 + \left(\frac{\partial x_2}{\partial \xi_1}\right)^2\right]^{1/2} \tag{1.38}$$

For this case:

$$\frac{\partial x_1}{\partial \xi_1} = \frac{\partial N_1}{\partial \xi_1} x_1^1 + \frac{\partial N_2}{\partial \xi_1} x_1^2 = -\frac{1}{2} x_1^1 + \frac{1}{2} x_1^2 = -\frac{1}{2}(x_1^2 - x_1^1)$$

$$\frac{\partial x_2}{\partial \xi_1} = \frac{\partial N_1}{\partial \xi_1} x_2^1 + \frac{\partial N_2}{\partial \xi_1} x_2^2 = -\frac{1}{2} x_2^1 + \frac{1}{2} x_2^2 = -\frac{1}{2}(x_2^2 - x_2^1) \tag{1.39}$$

Therefore:

$$det(J) = \frac{1}{2}\left[(x_1^2 - x_1^1)^2 + (x_2^2 - x_2^1)^2\right]^{1/2} = \frac{1}{2} l_e \tag{1.40}$$

where l_e is the element length. For elements with quadratic interpolation, a similar expression is obtained.

On the other hand, the coordinates x_i for three and four-node regions defined in the $x_1 - x_2$, plane, the coordinate transformation function is given by

$$x_i = N_1(\xi_1, \xi_2)x_i^1 + N_2(\xi_1, \xi_2)x_i^2 + N_3(\xi_1, \xi_2)x_i^3 + N_4(\xi_1, \xi_2)x_i^4 = N_k(\xi_1, \xi_2)x_i^k \tag{1.41}$$

with the N_k shape functions given by the expressions (1.29) and (1.31), respectively. In this case, the determinant of the Jacobian in the transformation $dx_1 dx_2 = det(J) d\xi_1 d\xi_2$ is given by

$$det(J) = det\left(\left[\frac{\partial x_i}{\partial \xi_i}\right]\right) = \begin{vmatrix} \frac{\partial x_1}{\partial \xi_1} & \frac{\partial x_1}{\partial \xi_2} \\ \frac{\partial x_2}{\partial \xi_1} & \frac{\partial x_2}{\partial \xi_2} \end{vmatrix} \tag{1.42}$$

Similarly, the transformation is obtained for quadratic elements and three-node triangular elements.

1.11.1 Radial Basis Interpolation Functions

Radial Basis Functions (RBF) are widely used in the Boundary Element Method (BEM) to evaluate domain integrals since they depend on the distance between points, facilitating their evaluation in complex geometries. They are also very appropriate in the Radial Integration Method (RIM) and in the Dual Reciprocity Boundary Element Method (DRBEM) to transform domain integrals into boundary integrals.

A radial basis function has the general form [2, 5]:

$$\varphi(r) = \varphi(\| \mathbf{x} - \mathbf{x}' \|) \tag{1.43}$$

where $\mathbf{x}$ y $\mathbf{x}'$ are points in the domain; $r = \| \mathbf{x} - \mathbf{x}' \|$ the distance between the points; $\varphi(r)$ is a function that depends only on r. Although any functions can be used, $\varphi(r)$,

Table 1.2 Commonly used radial basis functions

$\varphi(r) = 1 + r$
$\varphi(r) = 1 + r + r^2$
$\varphi(r) = \sqrt{c^2 + r}$
$\varphi(r) = r^2 \ln(r)$

in the boundary element method, the functions listed in Table 1.2 are most commonly used. An interpolation polynomial $f(r)$ based on radial basis functions is written in the general form:

$$q(\mathbf{x}) \approx \sum_{i=1}^{n} \alpha_i \varphi_i(r) = \alpha_i \varphi_r \tag{1.44}$$

In this case, $q(\mathbf{x})$ is interpolated from the radial functions $\varphi_i(r)$ and the constants α_i, which are determined from the known values $q(\mathbf{x})$ at n interpolation points in the domain (see Fig. 1.5). In this way, a system of n equations and n unknown values of α_i, is obtained, which is solved to obtain these coefficients:

$$\boldsymbol{\alpha} = \mathbf{F}^{-1} \cdot \mathbf{q} \tag{1.45}$$

where $\boldsymbol{\alpha} = \{\alpha_1, \alpha_2, \ldots, \alpha_n\}^T$, $\boldsymbol{q} = \{q_1, q_2, \ldots, q_n\}^T$ and $\mathbf{F}$ is a matrix whose rows are defined as: $\mathbf{F}_j = \{\varphi(r_{j1}), \varphi(r_{j2}), \ldots, \varphi(r_{ji})\}^T$, where $r_{ji} = \|\mathbf{x}_i - \mathbf{x}_j\|$. Thus, the function $q(\mathbf{x})$is approximated as

$$q(\mathbf{x}) \approx (\mathbf{F}^{-1} \cdot \mathbf{q})\varphi. \tag{1.46}$$

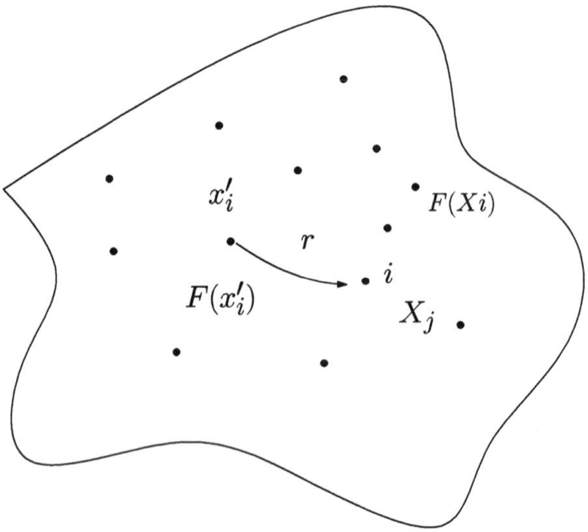

Fig. 1.5 Radial base functions

1.12 Numerical Integration

1.12.1 Gauss-Legrende Quadrature

Using numerical integration schemes is essential in evaluating integrals and calculating integrals in elements and cells. The standard practice has been the use of the Gauss-Legrende Quadrature, which is given by the general expression [1]:

$$\int_{-1}^{+1} f(\xi)d\xi \simeq \sum_{k=1}^{npg} w_k f(\xi_k) \tag{1.47}$$

where $f(x)$ s the function to integrate; $npg \geq 1$ represents the number of Gaussian integration points, w_k are integration weighting numbers, and x_k are the abscissae of the integration points in the interval $[-1, 1]$. Table 1.3 shows the coordinates of the integration points and the values of w_k for each point. The Gauss-Legendre quadrature with N integration points exactly integrates polynomials up to order equal to or less than $2N - 1$. Using this quadrature requires transforming the domain of integration to the normed space Z where $-1 \leq \xi \leq +1$. For that, we use the transformation functions seen in the previous sections. Thus, let $f(x_i)$ be a function defined in the space X, its transformation to the space Z is obtained by considering $x_i = x_i(\xi_i) = N_k(\xi_i)x_i^k$, such that: $f(x_i) = f(x_i(\xi_i))$, where x_i^k are the spatial coordinates of the nodes defining the line element. Thus, the line integral of the function $f(x_i)$ defined along the element is transformed as:

$$\int_{s_1}^{s_1} f(x_i)ds = \int_{-1}^{+1} f(x_i(\xi_1))det(J)d\xi \simeq \sum_{k=1}^{npg} w_k f(x_i(\xi_1^k))det(J). \tag{1.48}$$

Table 1.3 Abscissae and weight coefficients of the Gaussian formula

Puntos	$\pm\xi_i$	w_i
1	0.000000000000000	2.000000000000000
2	0.577350269189626	1.000000000000000
3	0.774596669241483	0.555555555555555
	0.000000000000000	0.888888888888888
4	0.861136311594053	0.347854845137454
	0.339981043584856	0.652145154862546
5	0.906179845938664	0.236926885056189
	0.538469310105683	0.478628670499366

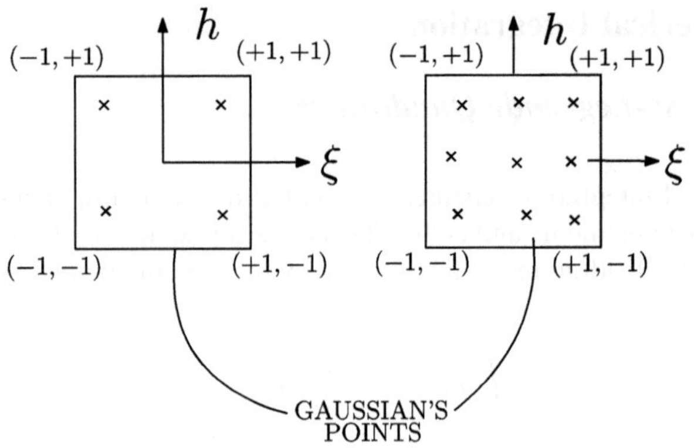

Fig. 1.6 Gaussian integration

Gaussian quadrature with two integration points

```
%-----------------------------------------------
%Gaussian quadrature with two integration points
%-----------------------------------------------
e=[-sqrt(1/3) sqrt(1/3)];
n=[-sqrt(1/3) sqrt(1/3)];
wi=[1 1 1 1];
wj=[1 1 1 1];
F=0;
for j=1:2
  f = 0;
  for i=1:2
    Fp=6*n(i)^2*e(i)^3+7*e(i)*n(i)^2-n(i)+2; %integrant
    f=f+wi(i)*Fp;
  end
  F=F+wj(j)*f;
end
F
```

Similarly, it has the following general expression for a function of two variables defined on the domain A (see Fig. 1.6):

Table 1.4 Abscissae and weight coefficients for integration in triangles

n	i	ξ_{i1}	ξ_{i2}	ξ_{i3}	$2w_i$
1 (linear)	1	1/3	1/3	1/3	1
3 (quadratic)	1	1/2	1/2	0	1/3
	2	0	1/2	1/2	1/3
	3	1/2	0	1/2	1/3
4 (cubic)	1	1/3	1/3	1/3	−9/16
	2	3/5	1/5	1/5	25/48
	3	1/5	3/5	1/5	25/48
	4	1/5	1/5	3/5	25/48
7 (quintic)	1	0.33333333	0.33333333	0.33333333	0.22500000
	2	0.79742699	0.10128651	0.10128651	0.12593918
	3	0.10128651	0.79742699	0.10128651	0.12593918
	4	0.10128651	0.10128651	0.79742699	0.12593918
	5	0.05971587	0.47014206	0.47014206	0.13239415
	6	0.47014206	0.05971587	0.47014206	0.13239415

$$\int_A f(\xi(\xi_1, \xi_2))dA = \int_{-1}^{+1}\int_{-1}^{+1} f(x_i(\xi_1, \xi_2))det(J)d\xi_1 d\xi_2$$
$$\approx \sum_{i=1}^{N_i} w_i \sum_{j=1}^{N_j} w_j f(x_i(\xi_1^k, \xi_2^k))det(J) \tag{1.49}$$

In the case of triangular domains, the Gaussian integration can be easily extended. For triangular domains, as presented in Fig. 1.3, the general expression Gaussian integration is given by

$$\int_A f(x_1, x_2)dA = \int_0^1 \left(\int_0^{1-\xi_2} f(\xi_1, \xi_2)d\xi_1 \right) \det(\xi_1, \xi_2)d\xi_2$$
$$\approx \frac{1}{2} \sum_{i=1}^{n} w_i f(\xi_1^i, \xi_2^i) \det(\xi_1^i, \xi_2^i) \tag{1.50}$$

The weights and integration points are presented in Table 1.4 for this case.

1.12.2 Logarithmic Gaussian Integration

Logarithmic Gaussian integration is a variant of Gaussian quadrature adapted to integrate functions with logarithmic terms. In the boundary element method, the integrals appearing in the formulation usually involve logarithmic type functions, which arise in the fundamental solution of plate problems. In this sense, logarithmic integrals of the general form [1, 5]:

$$I = \int_0^1 \ln\left(\frac{1}{\xi}\right) f(\xi)d\xi \approx \sum_{i=1}^{n} w_i f(\xi) \tag{1.51}$$

The weights and integration points used in this method are presented in Table 1.5. The logarithmic terms found in boundary integrals are generally evaluated in the domain $\xi \in [-1, +1]$. Since the logarithmic Gaussian integration is defined for $\xi \in [0, 1]$, the domain of integration, the transformation: $\xi(\chi) = 2\chi - 1$ y $d\xi = 2d\chi$, is employed, whereby an integral of the type:

$$I = \int_{-1}^{+1} \ln\left(\frac{1}{\xi}\right) f(\xi)d\xi. \tag{1.52}$$

is transformed as

$$I = \int_{-1}^{+1} \ln\left(\frac{1}{\xi}\right) f(\xi)d\xi = \int_0^1 2\ln\left(\frac{1}{2\chi - 1}\right) f(2\chi - 1)d\chi. \tag{1.53}$$

Table 1.5 Abscissae and weight coefficients of the Gaussian Logarithmic Formula

Puntos	$\pm\xi_i$	w_i
1	0.0090425944	0.1209547400
2	0.0539710540	0.1863631000
3	0.1353113400	0.1956606600
4	0.2470516900	0.1735772300
5	0.3802117100	0.1356959700
6	0.5237915900	0.0936470840
7	0.6657747200	0.0557879380
8	0.7941901900	0.0271598930
9	0.8981610200	0.0095151992
10	0.9688479800	0.0016381586

1.13 Summary

In this chapter, we present the fundamental mathematical concepts that support the formulation of the boundary element method. Their use will be discussed in the following chapters. The computational implementation of Gauss quadrature and logarithmic integration in MATLAB(R) has also been presented. Software for algebraic manipulation, such as Mathematica(R) and Maxima, are widely used as tools for developing the formulation of the method since they facilitate the task of algebraic manipulation of complex expressions that appear in the mathematical formulation of the boundary element method.

References

1. Brebbia CA, Wrobel LC (1996) Boundary element techniques in engineering. Springer
2. Gao X (2002) The radial integration method for evaluation of domain integrals with boundary only discretization. Eng Anal Bound Elem 26:905–916
3. Lay D (2015) Linear algebra and its applications, 5th edn. Pearson
4. Sahraee S, Wriggers P (2023) Tensor calculus and differential geometry for engineers: with solved exercises. Springer
5. Wrobel LC, Aliabadi MH (2002) The boundary element method. Volume 2: Applications in solid and structures. Wiley, New York
6. Zienkiewicz OC, Taylor L (1995) The finite element method, 3rd edn. Addison Wesley, Massachusetts

Chapter 2
A Brief Introduction to the Boundary Element Method

Abstract This chapter introduces the boundary element method by presenting its application to a problem such as bending of prestressed elastic membrane subjected to a distributed pressure. The Laplace equation governs this problem and, due to its simplicity, allows us to illustrate the application of the boundary element method step by step. The purpose of this chapter is to know the steps for applying the method to the solution of the governing equation that governs the problem. How the integral equations of this equation are obtained from this equation, how these equations are discretized using BEM, and how the numerical solution of the problem is reached. This chapter then presents the general methodology of analysis that will be used in the book.

2.1 Introduction

This chapter introduces the boundary element method by presenting its application to a problem such as bending a prestressed elastic membrane subjected to a distributed pressure. The Laplace equation governs this problem and, due to its simplicity, allows us to illustrate the application of the boundary element method step by step. The purpose of this chapter is to know the steps for applying the method to the solution of the governing equation that governs the problem. How the integral equations of this equation are obtained from this equation, how these equations are discretized using BEM, and how the numerical solution of the problem is reached. This chapter then presents the general methodology of analysis that will be used in the book.

2.2 Deflection of an Elastic Membrane

Let us consider the bending problem of an elastic membrane occupying the region Ω bounded by the curve Γ in the space $x_1 - x_2 - x_3$, as shown in Fig. 2.1. This membrane is prestressed with a constant stress value T throughout the domain and

J. Useche Vivero, *The Boundary Element Method for Engineering Analysis of Plates and Shells*, Solid Mechanics and Its Applications 281,
https://doi.org/10.1007/978-3-031-93050-8_2

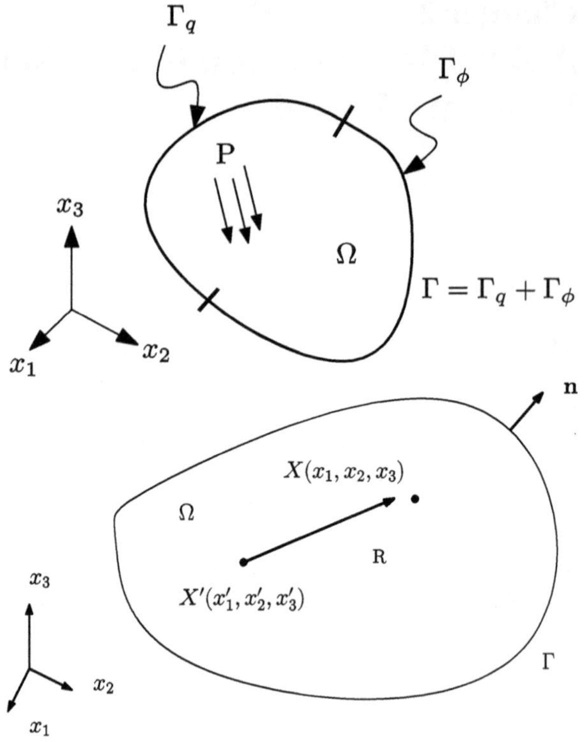

Fig. 2.1 Deflection of an elastic membrane. Top: Solution domain. Bottom: Definition of source and field point

subjected to a pressure force p. This problem is governed by the governing equation (Laplace equation) [1]:

$$\nabla^2 \frac{1}{T} \phi = p \tag{2.1}$$

With the boundary conditions:

$$\phi = \hat{\phi} \quad \text{en} \quad \Gamma_\phi$$

$$\phi_{,n} = \frac{1}{T} q_n \quad \text{en} \quad \Gamma_q$$

where ϕ represents the membrane deflection and q the distributed pressure per unit area acting on the membrane, Γ_ϕ is part of the boundary where the displacement is known (Dirichlet boundary condition), and Γ_q is part of the boundary where $\phi_{,n}$ is known (Neumann boundary condition).

2.2.1 Fundamental Solutions

Fundamental solutions are the basis of the boundary element method. A basic solution corresponds to the solution of a differential equation in an infinite domain when the source function is concentrated at a point (in this case, the pressure p). Thus, the fundamental solution to Laplace's equation must satisfy the following equation:

$$\nabla^2 \phi^*(\mathbf{x}', \mathbf{x}) = \frac{1}{k}\delta(\mathbf{x}', \mathbf{x}) \tag{2.2}$$

where ϕ^* is the fundamental solution, $\mathbf{x}' = \{x_1', x_2'\}$ are the coordinates of the point where the source function is located (placement point) and $\mathbf{x} = \{x_1, x_2\}$ are the field coordinates, that is, any point in the domain where ϕ^* (is calculated (field point) (Fig. 2.1).

A function that satisfies the previous equation for the two-dimensional case is given by

$$\phi^*(\mathbf{x}', \mathbf{x}) = Aln(r) \tag{2.3}$$

where r is the distance between the placement point and the field point:

$$r(\mathbf{x}', \mathbf{x}) = \|\mathbf{r}\| = \sqrt{(x_1 - x_1')^2 + (x_2 - x_2')^2} \tag{2.4}$$

The first derivative of the fundamental solution concerning x_i is

$$\frac{\partial \phi^*}{\partial x_i} = A\left(\frac{\partial \phi^*}{\partial r}\right)\left(\frac{\partial r}{\partial x_i}\right) = A\left(\frac{x_i - x_i'}{r^2}\right) \tag{2.5}$$

Similarly:

$$\frac{\partial \phi^*}{\partial x_2} = A\left(\frac{x_2 - x_2'}{r^2}\right) \tag{2.6}$$

The second order derivatives are given by

$$\frac{\partial^2 \phi^*}{\partial x_1^2} = \frac{r}{r^2} - \frac{2A(x_1 - x_1')^2}{r^4} \tag{2.7}$$

and,

$$\frac{\partial^2 \phi^*}{\partial x_2^2} = \frac{A}{r^2} - \frac{2A(x_2 - x_2')^2}{r^4} \tag{2.8}$$

Replacing in Eq. (2.2), we have

$$\frac{A}{r^2} - \frac{2A(x_1 - x_1')^2}{r^4} + \frac{A}{r^2} - \frac{2A(x_2 - x_2')^2}{r^4} = \frac{1}{T}\delta(\mathbf{x} - \mathbf{x}') \tag{2.9}$$

Rearranging terms:

$$\frac{1}{r^2} - \frac{(x_1 - x_1')^2 + (x_2 - x_2')^2}{r^4} = \frac{1}{2AT}\delta(\mathbf{x} - \mathbf{x}') \tag{2.10}$$

We observe that

$$\frac{1}{r^2} - \frac{(x_1 - x_1')^2 + (x_2 - x_2')^2}{r^4} = \delta(\mathbf{x} - \mathbf{x}') = \begin{cases} \infty \text{ si } \mathbf{x} = \mathbf{x}' \\ 0 \text{ si } \mathbf{x} \neq \mathbf{x}' \end{cases} \tag{2.11}$$

Then Eq. (2.3) is written as

$$\delta(\mathbf{x} - \mathbf{x}') = \frac{1}{2AT}\delta(\mathbf{x} - \mathbf{x}') \tag{2.12}$$

such that $A = 1/2T$. Replacing it in Eq. (2.3), we find the fundamental solution for displacement:

$$\phi^*(\mathbf{x}', \mathbf{x}) = \frac{1}{2T} ln(r) \tag{2.13}$$

In general, the following fundamental solution to Laplace's equation is used in BEM:

$$\phi^*(\mathbf{x}', \mathbf{x}) = \frac{1}{2\pi T} \ln(r) \tag{2.14}$$

2.2.2 Boundary Integral Equation

The boundary integral equation of the governing differential equation given by Eq. (2.1) is obtained by applying the Gradient Theorem, presented in Chap. 1 [1–3]:

$$\int\limits_{\Omega} \phi \left(\frac{\partial^2 \Psi}{\partial^2 x_1} + \frac{\partial^2 \Psi}{\partial^2 x_2} \right) d\Omega - \int\limits_{\Gamma} \phi \frac{\partial \Psi}{\partial n} d\Gamma + \int\limits_{\Gamma} \Psi \frac{\partial \phi}{\partial n} d\Gamma + \frac{1}{k} \int\limits_{\Omega} \Psi f d\Omega = 0 \tag{2.15}$$

where $n = \{n_1, n_2\}$ is the normal vector to the boundary. The function $G = \Psi$, has been taken in applying the Gradient Theorem, where Ψ is an arbitrary weighting function. To obtain an integral equation that does not have domain integrals, the weighting function Ψ must be chosen such that the domain integral disappears in Eq. (2.15). If the weighting function Ψ, is taken to be equal to the fundamental solution of the differential equation, ϕ^*, the first term of Eq. (2.15) can be written as

$$\int_{\Omega} \phi \left(\frac{\partial^2 \phi^*}{\partial^2 x_1} + \frac{\partial^2 \phi^*}{\partial^2 x_2} \right) d\Omega = \int_{\Omega} \phi \frac{1}{k} \delta(\mathbf{x}', \mathbf{x}) d\Omega = \frac{c}{k} \phi(\mathbf{x}') \tag{2.16}$$

given that

$$\nabla^2 \phi^*(\mathbf{x}', \mathbf{x}) = \frac{1}{k} \delta(\mathbf{x}', \mathbf{x}) \tag{2.17}$$

In Eq. (2.16), c represents a constant that depends on the location of the placement point (source point) if the point lies on the domain boundary $0 < c < 1$. If the placement point lies in the domain $c = 1$ and if it lies outside the domain $c = 0$. It can be shown that for a smooth domain (continuous curvature), $c = 1/2$. Thus, Eq. (2.16) can be written as

$$\frac{c}{k} \phi(\mathbf{x}') - \int_{\Gamma} \phi \frac{\partial \phi^*}{\partial n} d\Gamma + \int_{\Gamma} \phi^* \frac{\partial \phi}{\partial n} d\Gamma + \frac{1}{k} \int_{\Omega} \phi^* f d\Omega = 0 \tag{2.18}$$

Defining:

$$q = k \frac{\partial \phi}{\partial n}, \qquad q^* = k \frac{\partial \phi^*}{\partial n} \tag{2.19}$$

And replacing in (2.18), we obtain the boundary integral equation for the governing equation:

$$\frac{c}{k} \phi(\mathbf{x}') = \int_{\Gamma} \phi(\mathbf{x}) q^*(\mathbf{x}', \mathbf{x}) d\Gamma - \int_{\Gamma} \phi^*(\mathbf{x}', \mathbf{x}) q(\mathbf{x}) d\Gamma - \frac{1}{k} \int_{\Omega} \phi(\mathbf{x}', \mathbf{x}) f(\mathbf{x}) d\Omega \tag{2.20}$$

Based on the fundamental solution ϕ^*, the fundamental solution for flows $\phi_{,n}$ can be found as

$$q^*(\mathbf{x}', \mathbf{x}) = -k \frac{\partial \phi^*}{\partial n} = k \left(n_1 \frac{\partial \phi^*}{\partial x} + n_2 \frac{\partial \phi^*}{\partial y} \right) \tag{2.21}$$

Replacing the fundamental solution, we finally obtain

$$q^*(\mathbf{x}', \mathbf{x}) = -\frac{k}{2\pi r^2} \{(x_1 - x_1')n_1 + (x_2 - x_2')n_2\}. \tag{2.22}$$

2.2.3 *Discretization of the Boundary Integral Equation*

Discretization of the boundary integral equation Once the boundary integral equation has been obtained, the next step is to discretize this equation so that the integrals over the boundary are written as the sum of integrals over discrete sections of boundary *elements* $\Gamma_1, \Gamma_2, \Gamma_3, \ldots, \Gamma_n$, such that (see Fig. 2.2)

$$\Gamma = \sum_{j=1}^{n} \Gamma_j \tag{2.23}$$

In this way, Eq. (2.20) can be written in discrete form as

$$\frac{c}{k}\phi(\mathbf{x}') = \sum_{j=1}^{n}\left[\int_{\Gamma_j} \phi(\mathbf{x})q^*(\mathbf{x}',\mathbf{x})d\Gamma_j\right] - \sum_{j=1}^{n}\left[\int_{\Gamma_j} \phi^*(\mathbf{x}',\mathbf{x})q(\mathbf{x})d\Gamma_j\right] - \frac{1}{k}\int_{\Omega} \phi^* f(\mathbf{x})d\Omega \tag{2.24}$$

This equation represents the discretized form of the boundary integral. In the simplest case, let us consider the geometry of the segments $\Gamma_1, \Gamma_2, \Gamma_3, \ldots, \Gamma_n$ to be approximated by rectilinear elements, as shown in Fig. 2.2. It can be observed that

$$\lim_{i\to\infty} \sum_{j=1}^{n} \Gamma_j \to \Gamma \tag{2.25}$$

Let us also consider that the functions ϕ y q_n are constants within these elements. These functions are evaluated at node j, located at the midpoint of the element, as shown in Fig. 2.2. In this way, the expression (2.20) is written as

$$\frac{1}{2}\phi(\mathbf{x}') = \sum_{j=1}^{n}\left[\phi_j \int_{\Gamma_j} q^*(\mathbf{x}',\mathbf{x})d\Gamma_j\right] - \sum_{j=1}^{n}\left[q_j \int_{\Gamma_j} \phi^*(\mathbf{x}',\mathbf{x})d\Gamma_j\right] \tag{2.26}$$

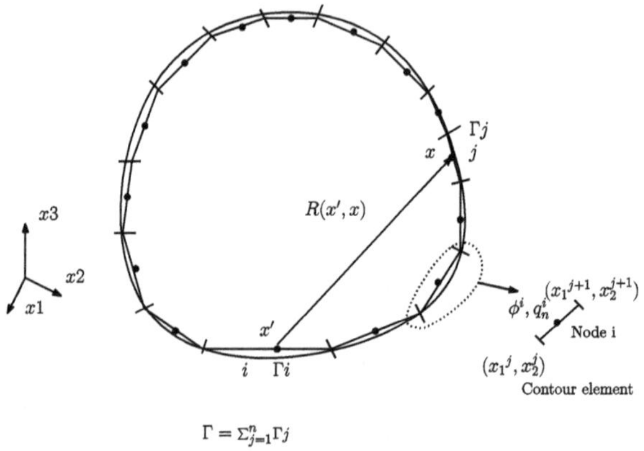

Fig. 2.2 Discretized boundary using linear elements and constant interpolation

The domain integral was not written for simplicity, as its treatment will be presented later. Considering that the source point is located at the collocation node i, the previous equation can be rewritten as

$$\frac{1}{2}\phi_i = \sum_{j=1}^{n}\left[\phi_j \int_{\Gamma_j} q_{ij}^* d\Gamma_j\right] - \sum_{j=1}^{n}\left[q_j \int_{\Gamma_j} \phi_{ij}^* d\Gamma_j\right] \tag{2.27}$$

Defining:

$$H_{ij} = \int_{\Gamma_j} q_{ij}^* d\Gamma_j si \qquad i \neq j \tag{2.28}$$

ó

$$H_{ij} = -\frac{1}{2} + \int_{\Gamma_j} q_{ij}^* d\Gamma_j si \qquad i = j \tag{2.29}$$

and,

$$G_{ij} = \int_{\Gamma_j} \phi_{ij}^* d\Gamma_j \tag{2.30}$$

Equation (2.27) can be written in a compact form as

$$\sum_{j=1}^{n} H_{ij}\phi_j = \sum_{j=1}^{n} G_{ij} q_j \tag{2.31}$$

This equation represents the discrete form of the integral equation (2.26). The matrix H_{ij} is the influence matrix for displacement, and the matrix G_{ij} is the stress influence matrix. Applying this equation to n placement points on the boundary, together with the boundary conditions of the problem, generates a linear system of simultaneous equations of size $n \times n$, whose solution allows calculating the unknown displacements and stresses on the membrane boundary. Once the unknowns on the boundary are known, the deflection of the membrane at specific points in the domain is calculated through Eq. (2.24):

$$\phi(\mathbf{x}') = \sum_{j=1}^{n}\left[\phi_j \int_{\Gamma_j} q^*(\mathbf{x}', \mathbf{x}) d\Gamma_j\right] - \sum_{j=1}^{n}\left[q_j \int_{\Gamma_j} \phi^*(\mathbf{x}', \mathbf{x}) d\Gamma_j\right] \tag{2.32}$$

The previous procedure represents the application of the boundary element method, which is shown in more detail in the following example.

2.3 Example of Application

Application Example

Let's consider a square elastic membrane with two edges restricted, as shown in Fig. 2.3. As an example of applying the boundary element method, let's assume the solution with four line elements with constant interpolation for deflections and boundary stress, as shown in the figure.

Taking node i as the source point (placement point), we have

$$H_{i1}\phi_1 + H_{i2}\phi_2 + H_{i3}\phi_3 + H_{i4}\phi_4 = G_{i1}q_1 + G_{i2}q_2 + G_{i3}q_3 + G_{i4}q_4 + p_i \tag{2.33}$$

The term p_i comes from the domain integral involving the distributed pressure in Eq. (2.24). We can construct a consistent system of equations by applying Eq. (2.33)

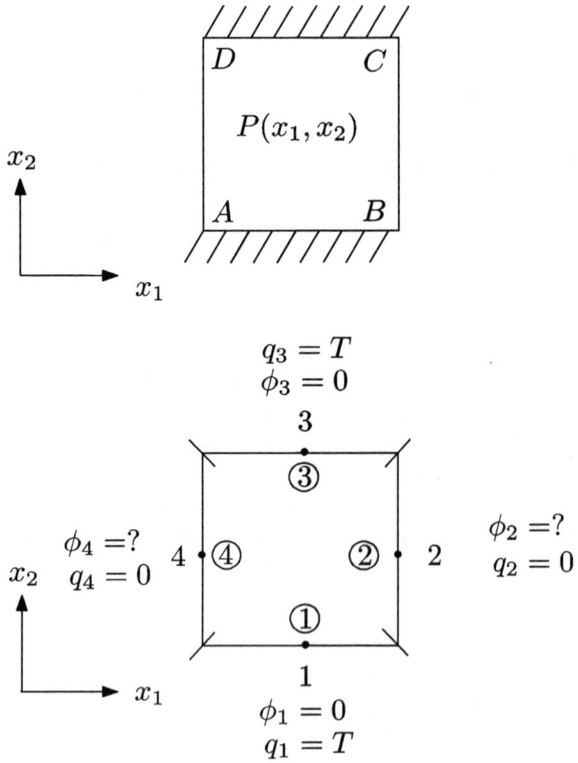

Fig. 2.3 Square elastic membrane with distributed and constrained pressure on all edges (top). Discretization with four line elements with constant interpolation (bottom)

using successively nodes 1–4 as source points. The complete system of equations is thus obtained

$$
\begin{aligned}
H_{11}\phi_1 + H_{12}\phi_2 + H_{13}\phi_3 + H_{14}\phi_4 &= G_{11}q_1 + G_{12}q_2 + G_{13}q_3 + G_{14}q_4 + p_1\\
H_{21}\phi_1 + H_{22}\phi_2 + H_{23}\phi_3 + H_{24}\phi_4 &= G_{21}q_1 + G_{22}q_2 + G_{23}q_3 + G_{24}q_4 + p_2\\
H_{31}\phi_1 + H_{32}\phi_2 + H_{33}\phi_3 + H_{34}\phi_4 &= G_{31}q_1 + G_{32}q_2 + G_{33}q_3 + G_{34}q_4 + p_3\\
H_{41}\phi_1 + H_{42}\phi_2 + H_{43}\phi_3 + H_{44}\phi_4 &= G_{41}q_1 + G_{42}q_2 + G_{43}q_3 + G_{44}q_4 + p_4
\end{aligned}
$$

Knowing the boundary conditions for this problem, we can reorganize the previous system of equations such that the unknown variables are located on the left side and the known terms on the right side. Thus, the terms involving q_1, ϕ_2, q_3, ϕ_4 known variables are placed on the right side, and the remaining ϕ_1, q_2, ϕ_3, q_4:

$$
\begin{aligned}
-G_{11}q_1 + H_{12}\phi_2 - G_{13}q_3 + H_{14}\phi_4 &= -H_{11}\phi_1 + G_{12}q_2 - H_{13}\phi_3 + G_{14}q_4 + p_1\\
-G_{21}q_1 + H_{22}\phi_2 - G_{23}q_3 + H_{24}\phi_4 &= -H_{21}\phi_1 + G_{22}q_2 - H_{23}\phi_3 + G_{24}q_4 + p_2\\
-G_{31}q_1 + H_{32}\phi_2 - G_{33}q_3 + H_{34}\phi_4 &= -H_{31}\phi_1 + G_{32}q_2 - H_{33}\phi_3 + G_{34}q_4 + p_3\\
-G_{41}q_1 + H_{42}\phi_2 - G_{43}q_3 + H_{44}\phi_4 &= -H_{41}\phi_1 + G_{42}q_2 - H_{43}\phi_3 + G_{44}q_4 + p_4
\end{aligned}
$$

In this way, the set of equations found can be written in a compact form as

$$
\begin{bmatrix} -G_{11} & H_{12} & -G_{13} & H_{14}\\ -G_{21} & H_{22} & -G_{23} & H_{24}\\ -G_{31} & H_{32} & -G_{33} & H_{34}\\ -G_{41} & H_{42} & -G_{43} & H_{44} \end{bmatrix}
\begin{Bmatrix} q_1\\ \phi_2\\ q_3\\ \phi_4 \end{Bmatrix}
=
\begin{bmatrix} -H_{11} & G_{12} & -H_{13} & G_{14}\\ -H_{21} & G_{22} & -H_{23} & G_{24}\\ -H_{31} & G_{32} & -H_{33} & G_{34}\\ -H_{41} & G_{42} & -H_{43} & G_{44} \end{bmatrix}
\begin{Bmatrix} \phi_1\\ q_2\\ \phi_3\\ q_4 \end{Bmatrix}
+
\begin{Bmatrix} p_1\\ p_2\\ p_3\\ p_4 \end{Bmatrix}
$$

Since the right-hand side of this system of equations is known, the following system of equations can be written:

$$\mathbf{Ax} = \mathbf{b} + \mathbf{p} \tag{2.34}$$

The solution to this system gives us the unknown variables.

Once the boundary variables are known, the calculation of ϕ in the domain is obtained by applying Eq. (2.32), defining the points where this variable is to be calculated

$$\phi_i = \sum_{j=1}^{n}\left[\phi_j \int_{\Gamma_j} q_{ij}^* d\Gamma_j\right] - \sum_{j=1}^{n}\left[q_j \int_{\Gamma_j} \phi_{ij}^* d\Gamma_j\right] \tag{2.35}$$

Based on the matrices H_{ij} and G_{ij}, we have

$$\phi_i = \sum_{j=1}^{n} H_{ij}\phi_j - \sum_{j=1}^{n} G_{ij}q_j \tag{2.36}$$

In this case, always $i \neq j$.

2.3.1 *Calculation of the Matrices H_{ij} and G_{ij}*

The boundary element method presents some singular integrals, that is, they contain integrands that tend to infinity when the independent variables of the same approach specific points ($\phi^* \rightarrow \infty$ y ($q_n^* \rightarrow \infty$ when $r \rightarrow 0$). In the case of the formulation that we have developed for two-dimensional problems, the singularities that appear are of two types: in the matrix G_{ij}, they are singularities of the type $ln(r)$ and are known as *weak singularities* or improper integrals. In the matrix H_{ij} there is an integral of the form $1/r$ called *strong singularity* (integral in the sense of the Cauchy principal value).

As will be seen later, the treatment of this type of singularity will not be necessary due to the characteristics of the matrix H_{ij}. In the case of weak singularity, there are numerical or analytical treatments. The analytical treatment is only possible with constant elements or linear interpolation. In the case of quadratic or higher-order elements, the resulting expressions are complicated or even impossible to obtain, in which case numerical integration must be used. Among the numerical methods used for numerical integration of integrals with weak singularity, Logarithmic Gauss Quadrature is among the most used.

Non-singular Integration

Calculate the matrices H_{ij} and G_{ij} when the source point is not located in the integration element, that is, when $i \neq j$. Let's start with the matrix G_{ij}; replacing (2.14) in (2.30):

$$G_{ij} = \int_{\Gamma_j} \phi_{ij}^* d\Gamma_j = -\frac{1}{2\pi k} \int_{\Gamma_j} ln(r) d\Gamma_j \tag{2.37}$$

where: $r = \sqrt{(x_1 - x_1')^2 + (x_2 - x_2')^2}$. In this case, the coordinates of the source point (x_1', x_2') are the coordinates of node i and the coordinates of field point (x_1, x_2) are the coordinates of element j at the Gauss integration points. Consider the element shown in Fig. 2.4.

As seen in Chap. 1, the geometric transformation of the element is obtained through the expressions:

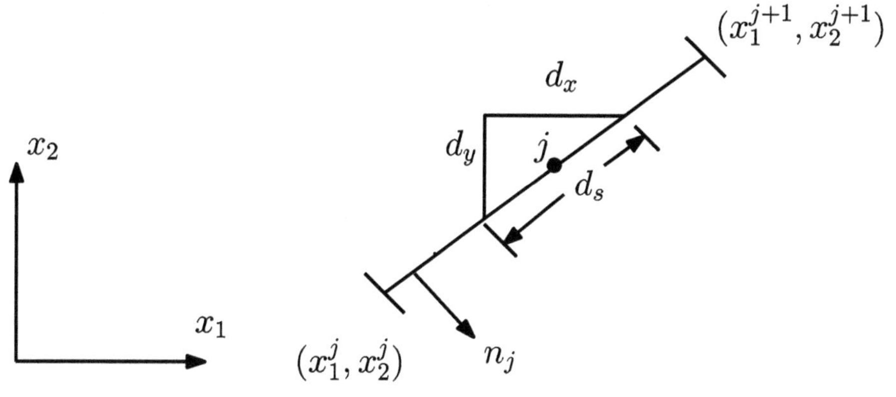

Fig. 2.4 Boundary element with constant interpolation

$$x_1(\xi) = \frac{1}{2}[(x_1^{j+1} - x_1^j)\xi + x_1^j + x_1^{j+1}] \tag{2.38}$$

$$x_2(\xi) = \frac{1}{2}[(x_2^{j+1} - x_2^j)\xi + x_2^j + x_2^{j+1}] \tag{2.39}$$

$$d\Gamma = \sqrt{dx_1^2 + dx_2^2} \tag{2.40}$$

with $\xi \in [-1, +1]$. Therefore, we have

$$dx_1 = \frac{1}{2}(x_1^{j+1} - x_1^j)d\xi \tag{2.41}$$

$$dx_2 = \frac{1}{2}(x_2^{j+1} - x_2^j)d\xi \tag{2.42}$$

And replacing the last of the expressions in Eq. (2.40):

$$d\Gamma_j = \frac{1}{2}\sqrt{(x_1^{j+1} - x_1^j)^2 + (x_2^{j+1} - x_2^j)^2}d\xi = \frac{1}{2}l_e \tag{2.43}$$

where l_e is the length of element j. Replacing it in Eq. (2.28), we have

$$G_{ij} = -\frac{l_e}{4\pi k}\int_{-1}^{+1} \ln(r)d\xi \tag{2.44}$$

Finally, using Gaussian quadrature:

$$G_{ij}^{ns} = -\frac{l_e}{4\pi k}\int_{-1}^{+1} \ln(r)d\xi \approx -\frac{l_e}{4\pi k}\sum_{i=1}^{npg} \ln[r(\xi_i)]w_i \tag{2.45}$$

Where the superscript ns is used to denote "non-singular." The matrix H_{ij} is calculated from Eq. (2.29). Replacing the fundamental solution given in Eq. (2.14) into this:

$$H_{ij} = \int_{\Gamma_j} q_{ij}^* d\Gamma_j = \frac{1}{2k\pi} \int_{\Gamma_j} \frac{1}{r^2} \{r_1 n_1 + r_2 n_2\} d\Gamma_j \tag{2.46}$$

where $r_1 = x_1^{j+1} - x_1^j$ the $r_2 = x_2^{j+1} - x_2^j$. Las componentes n_1 and n_2 components of the normal vector to the element can be calculated from dx_1 and dx_2 as follows. We will consider the unit vector ds along the element, which is given by

$$d\mathbf{s}_j = \{ds_1, ds_2\} = \frac{1}{l_e^j}\{dx_1, dx_2\} \tag{2.47}$$

Since $\mathbf{n}_j$ and $d\mathbf{s}_j$ are perpendicular vectors, we have that

$$\mathbf{n}_j \cdot d\mathbf{s}_j = n_1 ds_1 + n_2 ds_2 = 0 \tag{2.48}$$

This relation is satisfied for $n_{x1} = -ds_2$ and $n_{x2} = ds_1$, which gives

$$n_1 = \frac{1}{2}(x_2^{j+1} - x_2^j) \tag{2.49}$$

$$n_2 = -\frac{1}{2}(x_1^{j+1} - x_1^j) \tag{2.50}$$

Replacing $d\Gamma_j = \frac{1}{2} l_e$ in (2.29), we obtain

$$H_{ij} = \frac{l_e}{4k\pi} \int_{-1}^{+1} \frac{1}{R^2} \{R_{x1} n_{x1} + R_{x2} n_{x2}\} d\xi \tag{2.51}$$

Finally, using Gaussian integration of npg points, we have

$$H_{ij}^{ns} = \frac{l_e}{4k\pi} \int_{-1}^{+1} \frac{1}{R^2} \{R_{x1} n_{x1} + R_{x2} n_{x2}\} d\xi \approx$$

$$-\frac{l_e}{4k\pi} \sum_{i=1}^{npg} \frac{1}{R(\xi_i)^2} \{R_{x1}(\xi_i) n_{x1} + R_{x2}(\xi_i) n_{x2}\} w_i. \tag{2.52}$$

Singular Integration

On the other hand, the calculation of the singular terms of the matrix H_{ij} (those when the placement point coincides with one of the element's nodes) is performed considering that the expression (2.29) can be rewritten as

$$\frac{1}{r^2}\{r_1 n_1 + r_2 n_2\} = \frac{1}{r^2}(\mathbf{r} \cdot \mathbf{n}_j) = 0 \tag{2.53}$$

since the vectors **r** and $\mathbf{n}_j$ are parallel. Therefore $H_{ii}^s = -1/2$, where the superindex s means singular. Finally, since the singular integral in G_{ij} can be integrated numerically over the element, the expression (2.30) reduces to

$$G_{ij}^s = -\frac{l_e}{4\pi k} \int_{-l_e/2}^{+l_e/2} \ln(R) d\Gamma_j = \frac{l_e}{2\pi k}\left[1 - \ln\left(\frac{l_e}{w}\right)\right]. \tag{2.54}$$

Domain Integral Calculation

One of the boundary element method's central aspects is evaluating domain integrals. Several methods have been proposed for this purpose, among which we can highlight the most widely used in plate and shell analysis: the cell integration method, the radial integration method, and the dual reciprocity boundary element method. Each of these methods will be presented and used throughout the book.

To show how the vector **p** is obtained from the system of equations found previously, we will use the cell method, the simplest of all. In this method, the solution domain is divided into 4 cells, as presented in Fig. 2.4, such that the domain integral, for a given placement point, is calculated by Gaussian integration over a triangle, as presented in Chap. 1. In this way, the domain integral that appears in Eq. (2.24) for the placement point i on the boundary is approximated as (see Fig. 2.5)

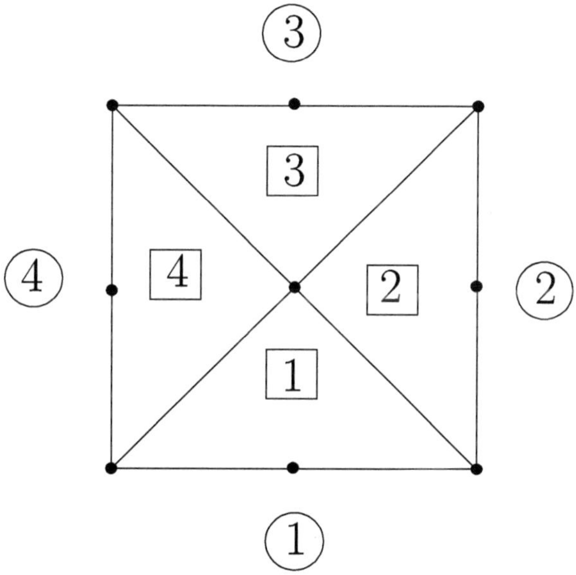

Fig. 2.5 Membrane domain discretization using four cells

$$
\begin{aligned}
p_i &= \frac{1}{T}\int_{\Omega} \phi^*(\mathbf{x}_i, \mathbf{X})p(\mathbf{X})d\Omega \\
&\approx \sum_{k=1}^{4}\left[\frac{1}{2T}\sum_{i=1}^{3} w_i\phi^*(\xi_1^i, \xi_2^i)p(\mathbf{X}(\xi_1^i, \xi_2^i))\det(\xi_1^i, \xi_2^i)\right]_k .
\end{aligned} \tag{2.55}
$$

2.4 Summary

The methodology presented in this chapter applies to solving the plate and shell problems discussed throughout this book. While this implementation is simple, it demonstrates the various components standard to the computational implementation of the method for plate-shell problems.

References

1. Brebbia CA, Wrobel LC (1996) Boundary element techniques in engineering. Springer
2. Hughes TJR (2000) The finite element method: linear static and dynamic finite element analysis. Dover Publications, New York
3. Wrobel LC, Aliabadi MH (2002) The boundary element method. Volume 2: Applications in solid and structures. Wiley, New York

Chapter 3
Plate and Shell Theories

Abstract This chapter presents the theories used throughout the book for isotropic and laminated materials for plate and shell analysis, both thin and thick, using the boundary element method. For thin plates, the classical theory of linear elastic plates based on Kirchhoff's hypotheses is presented, along with its extension to the case of thin shells. In both cases, the mechanical hypotheses governing their behavior are presented, and their governing equations are developed. Likewise, the thick plate theory, or Reissner-Midlin theory, is presented in the case of isotropic elastic thick plates. The extension of this theory to the case of laminated plates is given through the first-order shear theory. Their extension to developing the governing equations for isotropic thick shells and laminated material is presented for both cases. The governing equations presented here are again presented in their corresponding chapter on the application of the method so that each chapter is self-contained as much as possible.

3.1 Introduction

This chapter presents the theories used throughout the book for isotropic materials and laminated materials for plate and shell analysis, both thin and thick, using the boundary element method.

For thin plates, the classical theory of linear elastic plates based on Kirchhoff's hypotheses is presented, along with its extension to the case of thin shells. In both cases, the mechanical hypotheses governing their behavior are presented, and their governing equations are developed. Likewise, the thick plate theory, or Reissner-Midlin theory, is presented in the case of isotropic elastic thick plates. The extension of this theory to the case of laminated plates is given through the first-order shear theory. Their extension to developing the governing equations for isotropic thick shells and laminated material is presented for both cases.

The governing equations presented here are again presented in their corresponding chapter on the application of the method so that each chapter is self-contained as much as possible.

J. Useche Vivero, *The Boundary Element Method for Engineering Analysis of Plates and Shells*, Solid Mechanics and Its Applications 281,
https://doi.org/10.1007/978-3-031-93050-8_3

3.2 Isotropic Thin Plates

Classical thin-plate theory (Kirchhoff plate theory) is a mathematical model describing the behavior of thin plates under loading [2]. It is based on the assumption that cross sections remain plane and perpendicular to the midsurface after deformation. Traditionally, this theory helps analyze plates subjected to bending, such as ship hull and aerospace structures. It is applicable when the plate thickness is small compared to its plate dimensions and when shear effects are negligible, that is, in thin plates and moderate deformations.

3.2.1 Basic Assumptions

The linear theory of thin elastic plates with arbitrary midsurface shape is derived from the Kirchhoff-Love assumptions in developing the plate bending theory presented in Part I. These assumptions are formulated for the linear theory of thin plates of arbitrary shape as follows [6, 7]:

- The normal to the undeformed midsurface remain straight and normal to the deformed midsurface and do not undergo extension. This assumption implies that all strain components (normal and shear) in the direction of the surface normal to the midsurface vanish.
- The transverse normal stress is small compared to other normal stress components and can be neglected.
- The plate material is linearly elastic, isotropic, and homogeneous.
- The plate thickness is uniform in Ω.
- The strains and displacements are small.
- The shear deformations at the plate surfaces are zero.

The thickness of the plate is assumed to be small compared to other dimensions—for example, the smallest radius of the plate's midsurface. We also think that the displacements of an arbitrary point on a plate are small compared to its thickness. As a consequence of this assumption, the products of the displacements and their partial derivatives will be neglected as second-order quantities of smallness. Furthermore, we can refer all calculations to the original configuration of the plate and ensure that the differential equations are linear. From now on, we assume that the shell material is homogeneous, isotropic, and obeys Hooke's Law.

3.2.2 Strain-Displacement Relationships

From the previous kinematic considerations raised, the displacement components u_α, w_3 of any point x_i located at a distance x_3 from the middle plane as [2, 5]

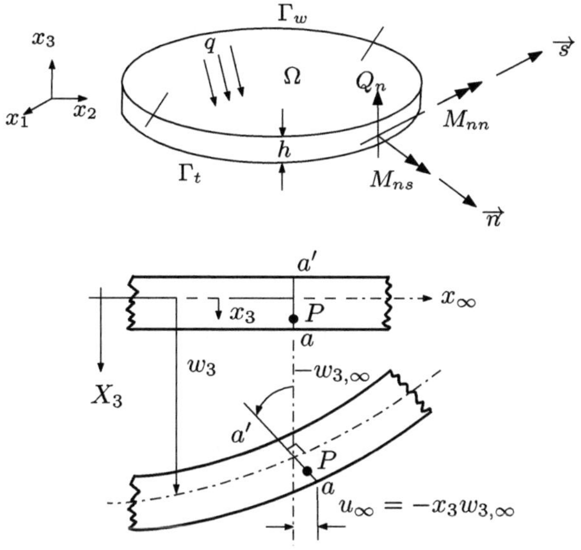

Fig. 3.1 Thin-plate geometry and strain definition

$$
\begin{aligned}
u_\alpha(x_i, t) &= u^0_\alpha(x_\alpha, t) - x_3 w^0_\alpha(x_\alpha, t) \\
w_3(x_i, t) &= w^0_3(x_\alpha, t)
\end{aligned} \tag{3.1}
$$

where u^0_α, w^0_3 are the displacement components in the midplane of the plate (see Fig. 3.1). Therefore, the strain components through the Cauchy tensor are

$$
\varepsilon_{\alpha\beta} = \frac{1}{2}(u_{\alpha,\beta} + u_{\beta,\alpha}) \tag{3.2}
$$

Based on the displacement components defined above, the relationship between unit strain and displacement is defined as

$$
\varepsilon_{\alpha\beta} = \frac{1}{2}(u^0_{\alpha,\beta} - x_3 w^0_{\alpha,\beta} + u^0_{\beta,\alpha} - x_3 w^0_{\beta,\alpha}) = \varepsilon^o_{\alpha\beta} + x_3 \kappa_{\alpha\beta} \tag{3.3}
$$

where $\varepsilon^o_{\alpha\beta}$ is the membrane strain and $\kappa_{\alpha\beta}$ are the angular deformation components due to curvature.

3.2.3 Plane Stress Constitutive Relationships

For a plate in the $x_1 - x_2$ plane, the transverse stress components σ_{33}, σ_{13} and σ_{23} are considered negligible compared to the stress components acting in the plane of the plate. If the material is assumed to be linearly elastic, the stress-strain relationship for a plane stress state is obtained from Hooke's law:

$$\begin{Bmatrix} \sigma_{11} \\ \sigma_{22} \\ \sigma_{12} \end{Bmatrix} = \frac{E}{1-\nu^2} \begin{bmatrix} 1 & \nu & 0 \\ \nu & 1 & 0 \\ 0 & 0 & 1-\nu \end{bmatrix} \begin{Bmatrix} \varepsilon_{11} \\ \varepsilon_{22} \\ \varepsilon_{12} \end{Bmatrix} \tag{3.4}$$

$$\sigma_{\alpha\beta} = \frac{E}{1-\nu^2} \left\{ (1+\nu)\varepsilon_{\theta\theta} - \nu\varepsilon_{\alpha\gamma}\delta_{\alpha\gamma} \right\} \delta_{\alpha\beta}$$

$$\sigma_{\alpha\beta} = \frac{E}{2(1+\nu)} = G \tag{3.5}$$

where E is Young's modulus and G is the rigidity modulus of the plate material.

3.2.4 Equations of Motion

Consider a plate element, as shown in Fig. 3.2. Let σ_{11} and σ_{22} be the regular forces, acting on the faces of the component, σ_{12} and σ_{21} are shear forces on these faces, acting parallel to the midsurface, and σ_{13} and σ_{23} are transverse shear forces acting normal to the midsurface. The resultant forces $N_{\alpha\beta}$ and resultant bending moments $M_{\alpha\beta}$ per unit length, due to the forces acting on the plate thickness, are given by

$$N_{\alpha\beta} = \int_{-h/2}^{h/2} \sigma_{\alpha\beta} dx_3$$

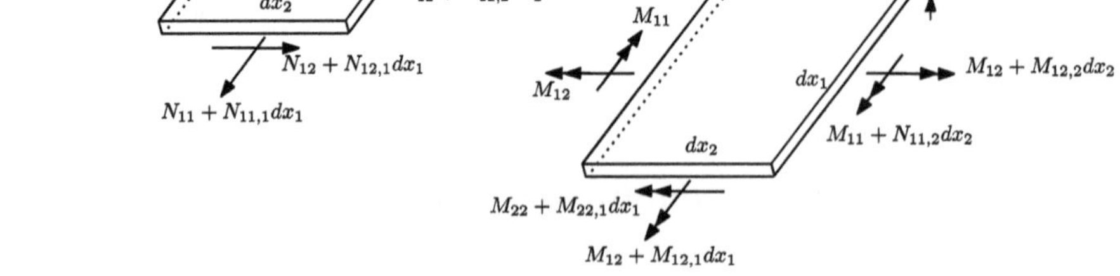

Fig. 3.2 Differential element for thin plate

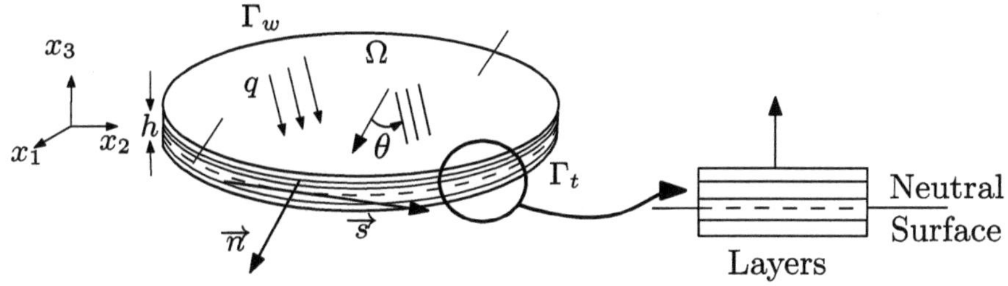

Fig. 3.3 Laminated thin plate

$$M_{\alpha\beta} = \int_{-h/2}^{h/2} \sigma_{\alpha\beta} x_3 dx_3 \tag{3.6}$$

On the other hand, the mass moments of inertia (I_0, I_1, I_2) are given by

$$\begin{aligned} I_0 &= \int_{-h/2}^{h/2} \rho dx_3 = \rho h \\ I_1 &= \int_{-h/2}^{h/2} \rho x_3 dx_3 = \frac{\rho h^2}{8} \\ I_2 &= \int_{-h/2}^{h/2} \rho x_3^2 dx_3 = \frac{\rho h^3}{12} \end{aligned} \tag{3.7}$$

With the help of Fig. 3.3, and taking into account the effect of the domain charges qi acting on the whole mean surface, the equilibrium of forces along the axes x_1, x_2, and x_3 is [2]

$$\begin{aligned} \frac{\partial N_{11}}{\partial x_1} + \frac{\partial N_{12}}{\partial x_2} &= I_0 \frac{\partial^2 u_1^0}{\partial t^2} - I_1 \frac{\partial^2 w_1^0}{\partial t^2} \\ \frac{\partial N_{12}}{\partial x_1} + \frac{\partial N_{22}}{\partial x_2} &= I_0 \frac{\partial^2 u_2^0}{\partial t^2} - I_1 \frac{\partial^2 w_2^0}{\partial t^2} \\ \frac{\partial^2 M_{11}}{\partial x_1^2} + 2\frac{\partial^2 M_{12}}{\partial x_1 \partial x_2} + \frac{\partial^2 M_{22}}{\partial x_2^2} + q &= I_0 \frac{\partial^2 w_3^0}{\partial t^2} - I_2 \frac{\partial^2}{\partial t^2}\left(\frac{\partial w_1^0}{\partial x_1} + \frac{\partial w_2^0}{\partial x_2}\right) \\ &\quad + I_1 \frac{\partial^2}{\partial t^2}\left(\frac{\partial u_1^0}{\partial x_1} + \frac{\partial u_2^0}{\partial x_2}\right) \end{aligned} \tag{3.8}$$

where q is the distributed force per unit area of the plate surface, in index notation, these equations can be written as

$$\begin{aligned} N_{\alpha\beta,\beta} &= I_0 u^0_{\alpha,tt} - I_1 w^0_{\alpha,tt} \\ M_{\alpha\beta,\alpha\beta} + q &= I_0 w^0_{3,tt} - I_2 (w^0_{\alpha,\alpha})_{,tt} + I_1 (u^0_{\alpha,\alpha})_{,tt} \end{aligned} \tag{3.9}$$

Replacing N and M, we obtain the equations of motion

$$\begin{aligned} B\frac{\partial^2 u_1}{\partial x_1^2} + B\nu\frac{\partial^2 u_2}{\partial x_1 \partial x_2} + G_{12}h\left(\frac{\partial^2 u_1}{\partial x_2^2} + \frac{\partial^2 u_2}{\partial x_1 \partial x_2}\right) &= I_0\frac{\partial^2 u_1^0}{\partial t^2} - I_1\frac{\partial^2 w_1^0}{\partial t^2} \\ G_{12}h\left(\frac{\partial^2 u_2}{\partial x_1^2} + \frac{\partial^2 u_1}{\partial x_1 \partial x_2}\right) + B\frac{\partial^2 u_2}{\partial x_2^2} + B\nu\frac{\partial^2 u_1}{\partial x_1 \partial x_2} &= I_0\frac{\partial^2 u_2^0}{\partial t^2} - I_1\frac{\partial^2 w_2^0}{\partial t^2} \\ D\frac{\partial^4 w_3}{\partial x_1^4} + 2D\frac{\partial^4 w_3}{\partial x_1^2 \partial x_2^2} + D\frac{\partial^4 w_3}{\partial x_2^4} - q(x_1, x_2) &= I_0\frac{\partial^2 w_3^0}{\partial t^2} \end{aligned} \tag{3.10}$$

where $B = Eh/(1-\nu^2)$ is the in-plane stiffness factor and $D = 2h^3E/(3(1-v^2))$ is the bending stiffness factor. The Dirichlet and Newmann boundary conditions for classical thin theory are

$$\begin{aligned} u_n = u_n^*, \quad u_s = u_s^*, \quad w_3 = w_3^*, \quad & in \quad \Gamma_w \\ V_n = Q_n + M_{ns,s} = V_n^*, \quad M_{nn} = M_{nn}^*, \quad & in \quad \Gamma_t, \end{aligned} \tag{3.11}$$

where $N_{\alpha\beta}$, $M_{\alpha\beta}$, and Q_α are related with N_s, N_n, and Q_n as

$$\begin{Bmatrix} N_{nn} \\ N_{ns} \end{Bmatrix} = \begin{bmatrix} n_1^2 & n_2^2 & 2n_1n_2 \\ -n_1n_2 & n_1n_2 & n_1^2 - n_2^2 \end{bmatrix} \begin{Bmatrix} N_{11} \\ N_{22} \\ N_{12} \end{Bmatrix} \tag{3.12}$$

and,

$$\begin{Bmatrix} M_{nn} \\ M_{ns} \end{Bmatrix} = \begin{bmatrix} n_1^2 & n_2^2 & 2n_1n_2 \\ -n_1n_2 & n_1n_2 & n_1^2 - n_2^2 \end{bmatrix} \begin{Bmatrix} M_{11} \\ M_{22} \\ M_{12} \end{Bmatrix} \tag{3.13}$$

$$Q_n = Q_1 n_1 + Q_2 n_2 \tag{3.14}$$

The initial conditions at $t = 0$:

$$u_n = u_n^0, \quad u_s = u_s^0, \quad w_3 = w_3^0, \tag{3.15}$$

$$\dot{u}_n = u_n^0, \quad \dot{u}_s = u_s^0, \quad \dot{w}_3 = w_3^0, \quad in \;\; \Omega \tag{3.16}$$

with: $u_1 = u_n n_1 - u_s n_2$ and $u_2 = u_n n_2 + u_s n_1$.

3.3 Laminated Thin Plates

Laminated plates are structural materials formed by multiple layers of different materials or by a single metallic sheet that has undergone a lamination process to improve its mechanical properties. In plates made of composite materials, each laminate is formed by a base material (matrix) and embedded fibers oriented in it, thus obtaining a laminate of orthotropic material whose properties are spatially defined by the orientation of its fibers (see Fig. 3.3).

The hypotheses governing the basic theory of laminated thin plates are the same hypotheses of Kirchhoff's theory given in Sect. 3.2.1, with the difference that the plate is made up of sheets of orthotropic stacked material firmly adhered to each other in such a way that the continuity of displacement and therefore of unit deformations in the thickness is assured. Thus, the kinematic equations for thin laminated plates are the same as those in Sect. 3.2.2.

3.3.1 Laminated Constitutive Equations

For a laminate in the $x_1 - x_2$ plane, the transverse stress components σ_{33} are considered negligible compared to the stress components acting in the plane of the laminate. It is assumed that each layer is orthotropic in terms of the material axes and obeys Hooke's Law. Thus, the stress components for an orthotropic laminate in the material axis system $x_1' - x_2'$ are given for the kth layer (see Fig. 3.3):

$$\left\{ \begin{matrix} \sigma_1 \\ \sigma_2 \\ \sigma_6 \end{matrix} \right\}^k = \begin{bmatrix} Q_{11} & Q_{12} & 0 \\ Q_{21} & Q_{22} & 0 \\ 0 & 0 & \bar{Q}_{66} \end{bmatrix}^k \left\{ \begin{matrix} \varepsilon_1 \\ \varepsilon_2 \\ \varepsilon_6 \end{matrix} \right\}^k \tag{3.17}$$

where $\bar{Q}_{ij}^k$ are the reduced in-plane stress components, which are related to the material elastic constants E_1, E_2, G_{12} through the expressions:

$$Q_{11}^k = \frac{E_1^k}{1 - \nu_{12}^k \nu_{21}^k}, \quad Q_{12}^k = \frac{E_2^k \nu_{12}^k}{1 - \nu_{12}^k \nu_{21}^k}, \quad Q_{22}^k = \frac{E_2^k}{1 - \nu_{12}^k \nu_{21}^k}, \quad Q_{66}^k = G_{12}^k \tag{3.18}$$

The stress components referred to a global system $x_1 - x_2$ for an orthotropic laminate in a plane state of stress are given by (see Fig. 3.3)

$$\begin{Bmatrix} \sigma_{11} \\ \sigma_{22} \\ \sigma_{12} \end{Bmatrix}^k = \begin{bmatrix} \bar{Q}_{11} & \bar{Q}_{12} & 0 \\ \bar{Q}_{21} & \bar{Q}_{22} & 0 \\ 0 & 0 & \bar{Q}_{66} \end{bmatrix}^k \begin{Bmatrix} \varepsilon_{11} \\ \varepsilon_{22} \\ \varepsilon_{12} \end{Bmatrix}^k \tag{3.19}$$

where the components $\bar{Q}_{ij}$ re explicitly given by

$$\begin{aligned}
\bar{Q}_{11} &= Q_{11}\cos^4\theta + 2(Q_{12} + 2Q_{66})\sin^2\theta\cos^2\theta + Q_{22}\sin^4\theta \\
\bar{Q}_{12} &= (Q_{11} + Q_{22} - 4Q_{66})\sin^2\theta\cos^2\theta + Q_{12}(\sin^4\theta + \cos^4\theta) \\
\bar{Q}_{22} &= Q_{11}\sin^4\theta + 2(Q_{12} + 2Q_{66})\sin^2\theta\cos^2\theta + Q_{22}\cos^4\theta \\
\bar{Q}_{16} &= (Q_{11} - Q_{12} - 2Q_{66})\sin\theta\cos^3\theta + (Q_{12} - Q_{22} + 2Q_{66})\sin^3\theta + \cos\theta) \\
\bar{Q}_{26} &= (Q_{11} - Q_{12} - 2Q_{66})\sin^3\theta\cos\theta + (Q_{12} - Q_{22} + 2Q_{66})\sin\theta + \cos^3\theta) \\
\bar{Q}_{66} &= (Q_{11} + Q_{22} - 2Q_{12} - 2Q_{66})\sin^2\theta\cos^2\theta + Q_{66}(\sin^4\theta + \cos^4\theta)
\end{aligned} \tag{3.20}$$

where θ is the orientation angle of the material in the local system x_i' (see Fig. 3.3). Since the stresses are variable in each layer in direction x_3, the resultant forces $N_{\alpha\beta}$ and resultant bending moments $M_{\alpha\beta}$, given by Eq. (3.6) are now given by the following expressions:

$$\begin{Bmatrix} N_{11} \\ N_{22} \\ N_{12} \end{Bmatrix} = \sum_{k=1}^{N} \int_{x_3^k}^{x_3^{k+1}} \begin{Bmatrix} \sigma_{11} \\ \sigma_{22} \\ \sigma_{12} \end{Bmatrix}^k dx_3 = \sum_{k=1}^{N} \int_{x_3^k}^{x_3^{k+1}} \begin{bmatrix} \bar{Q}_{11} & \bar{Q}_{12} & \bar{Q}_{16} \\ \bar{Q}_{21} & \bar{Q}_{22} & \bar{Q}_{26} \\ \bar{Q}_{16} & \bar{Q}_{26} & \bar{Q}_{66} \end{bmatrix}^k \begin{Bmatrix} \varepsilon_{11}^0 + z\varepsilon_{11}^1 \\ \varepsilon_{22}^0 + z\varepsilon_{22}^1 \\ \varepsilon_{12}^0 + z\varepsilon_{12}^1 \end{Bmatrix} dx_3 =$$

$$\begin{Bmatrix} N_{11} \\ N_{22} \\ N_{12} \end{Bmatrix} = \begin{bmatrix} A_{11} & A_{12} & A_{16} \\ A_{21} & A_{22} & A_{26} \\ A_{16} & A_{26} & A_{66} \end{bmatrix} \begin{Bmatrix} \varepsilon_{11}^0 \\ \varepsilon_{22}^0 \\ \varepsilon_{12}^0 \end{Bmatrix} + \begin{bmatrix} B_{11} & B_{12} & B_{16} \\ B_{21} & B_{22} & B_{26} \\ B_{16} & B_{26} & B_{66} \end{bmatrix} \begin{Bmatrix} \varepsilon_{11}^1 \\ \varepsilon_{22}^1 \\ \varepsilon_{12}^1 \end{Bmatrix} \tag{3.21}$$

$$\begin{Bmatrix} M_{11} \\ M_{22} \\ M_{12} \end{Bmatrix} = \sum_{k=1}^{N} \int_{x_3^k}^{x_3^{k+1}} \begin{Bmatrix} \sigma_{11} \\ \sigma_{22} \\ \sigma_{12} \end{Bmatrix}^k z\,dx_3 = \sum_{k=1}^{N} \int_{x_3^k}^{x_3^{k+1}} \begin{bmatrix} \bar{Q}_{11} & \bar{Q}_{12} & \bar{Q}_{16} \\ \bar{Q}_{21} & \bar{Q}_{22} & \bar{Q}_{26} \\ \bar{Q}_{16} & \bar{Q}_{26} & \bar{Q}_{66} \end{bmatrix}^k \begin{Bmatrix} \varepsilon_{11}^0 + z\varepsilon_{11}^1 \\ \varepsilon_{22}^0 + z\varepsilon_{22}^1 \\ \varepsilon_{12}^0 + z\varepsilon_{12}^1 \end{Bmatrix} z\,dx_3$$

$$\begin{Bmatrix} M_{11} \\ M_{22} \\ M_{12} \end{Bmatrix} = \begin{bmatrix} B_{11} & B_{12} & B_{16} \\ B_{21} & B_{22} & B_{26} \\ B_{16} & B_{26} & B_{66} \end{bmatrix} \begin{Bmatrix} \varepsilon_{11}^0 \\ \varepsilon_{22}^0 \\ \varepsilon_{12}^0 \end{Bmatrix} + \begin{bmatrix} D_{11} & D_{12} & D_{16} \\ D_{21} & D_{22} & D_{26} \\ D_{16} & D_{26} & D_{66} \end{bmatrix} \begin{Bmatrix} \varepsilon_{11}^1 \\ \varepsilon_{22}^1 \\ \varepsilon_{12}^1 \end{Bmatrix} \tag{3.22}$$

where N is the number of layers in the laminate. In the above expressions the coupling matrices A_{ij}, B_{ij} and D_{ij} (with $i, j = 1, 2, 6$) are defined as

$$A_{ij}=\sum_{k=1}^{N}\bar{Q}_{ij}^{k}(x_{3(k+1)}-x_{3(k)}),\quad B_{ij}=\frac{1}{2}\sum_{k=1}^{N}\bar{Q}_{ij}^{k}(x_{3(k+1)}^{2}-x_{3(k)}^{2}),$$

$$D_{ij}=\frac{1}{3}\sum_{k=1}^{N}\bar{Q}_{ij}^{k}(x_{3(k+1)}^{3}-x_{3(k)}^{3}).\tag{3.23}$$

3.3.2 Equilibrium Equations in Terms of Displacements

Applying force and moment equilibrium on the coordinate axes using the differential element shown in Fig. 3.2, the force and moment equilibrium equations presented in Sect. 3.2.4 considering the force the equations of motion for a thin laminated plate are obtained [2]

$$A_{11}\frac{\partial^2 u_1}{\partial x_1^2}+A_{12}\frac{\partial^2 u_2}{\partial x_1\partial x_2}+A_{66}\left(\frac{\partial^2 u_1}{\partial x_2^2}+\frac{\partial^2 u_2}{\partial x_1\partial x_2}\right)+-B_{11}\frac{\partial^3 w_3}{\partial x_1^3}$$
$$-B_{12}\frac{\partial^3 w_3}{\partial x_1\partial x_2^2}-2B_{16}\frac{\partial^3 w_3}{\partial x_2^3}$$
$$A_{16}\frac{\partial^2 u_1}{\partial x_1\partial x_2}+A_{26}\frac{\partial^2 u_2}{\partial x_2}+A_{66}\left(\frac{\partial^3 u_1}{\partial x_1\partial x_2}+\frac{\partial^2 u_2}{\partial x_2^2}\right)$$
$$-B_{12}\frac{\partial^3 w_3}{\partial x_1\partial x_2^2}-B_{22}\frac{\partial^3 w_3}{\partial x_2^3}-2B_{26}\frac{\partial^3 w_3}{\partial x_1\partial x_2^2}=I_0\frac{\partial^2 u_1^0}{\partial t^2}-I_1\frac{\partial^2 w_1^0}{\partial t^2}$$

$$A_{66}\left(\frac{\partial^2 u_2}{\partial x_1^2}+\frac{\partial^2 u_1}{\partial x_1\partial x_2}\right)+A_{22}\frac{\partial^2 u_2}{\partial x_2^2}+A_{12}\frac{\partial^2 u_1}{\partial x_1\partial x_2}-B_{16}\frac{\partial^3 w_3}{\partial x_1^3}$$
$$-B_{26}\frac{\partial^3 w_3}{\partial x_1\partial x_2^2}-2B_{66}\frac{\partial^3 w_3}{\partial x_2^3}$$
$$+A_{12}\frac{\partial^2 u_1}{\partial x_1\partial x_2}+A_{22}\frac{\partial^2 u_2}{\partial x_2}+A_{26}\left(\frac{\partial^3 u_2}{\partial x_1\partial x_2}+\frac{\partial^2 u_1}{\partial x_2^2}\right)$$
$$-B_{12}\frac{\partial^3 w_3}{\partial x_1^3}-B_{22}\frac{\partial^3 w_3}{\partial x_2^3}-2B_{26}\frac{\partial^3 w_3}{\partial x_1\partial x_2^2}=I_0\frac{\partial^2 u_2^0}{\partial t^2}-I_1\frac{\partial^2 w_2^0}{\partial t^2}$$

$$B_{11}\frac{\partial^3 u_1}{\partial x_1^3}+B_{12}\frac{\partial^3 u_2}{\partial x_1^2\partial x_2}+B_{16}\left(\frac{\partial^3 u_2}{\partial x_2^2\partial x_1}+\frac{\partial^3 u_2}{\partial x_1^3}\right)-D_{11}\frac{\partial^4 w_3}{\partial x_1^4}$$
$$-2D_{12}\frac{\partial^4 w_3}{\partial x_1^2\partial x_2^2}-4D_{16}\frac{\partial^4 w_3}{\partial x_1^3\partial x_2}$$

$$+ 2B_{16}\frac{\partial^3 u_1}{\partial x_1^2 \partial x_2} + 2B_{26}\frac{\partial^3 u_2}{\partial x_1 \partial x_2} + 2B_{66}\left(\frac{\partial^3 u_1}{\partial x_1 \partial x_2^2} + \frac{\partial^3 u_2}{\partial x_1^2 \partial x_2}\right)$$

$$- 2D_{26}\frac{\partial^4 w_3}{\partial x_1 \partial x_2^3} - 4D_{66}\frac{\partial^4 w_3}{\partial x_1^2 \partial x_2^2}$$

$$+ B_{12}\frac{\partial^3 u_1}{\partial x_1 \partial x_2^2} + B_{22}\frac{\partial^3 u_2}{\partial x_3} + B_{26}\frac{\partial^3 u_1}{\partial x_2^3} - D_{22}\frac{\partial^4 w_3}{\partial x_2^4} - 2D_{26}\frac{\partial^4 w_3}{\partial x_1 \partial x_2^3} + q(x_1, x_2) =$$

$$- I_0\frac{\partial^2 w_3^0}{\partial t^2} + I_2\frac{\partial^2}{\partial t^2}\left(\frac{\partial w_1^0}{\partial x_1} - \frac{\partial w_2^0}{\partial x_2}\right) - I_1\frac{\partial^2}{\partial t^2}\left(\frac{\partial u_1^0}{\partial x_1} - \frac{\partial u_2^0}{\partial x_2}\right) \tag{3.24}$$

The boundary and initial conditions for above equations are the same given for thin plate in Sect. 3.2.4.

3.4 Isotropic Thick Plates

The Reissner-Mindlin theory is an extension of the classical Kirchhoff plate theory, including transverse shear effects (see [4]). It allows the analysis of thick and moderately thick plates, where shear stresses cannot be neglected. Unlike the classical plate theory, here the cross sections can rotate independently of the normal to the mean surface, which better captures the real deformation. It is used in structures subjected to high loads or where bending and shear have a significant influence, such as in thick plates, composite structures, ship hull, and aerospace elements.

Consider an isotropic plate of thickness h with its Ω midplane located at $x_1 - x_2$, as presented in Fig. 3.4. The x_3-axis is taken positive downward from the midplane. The upper and lower surfaces of the plate are located at $z_3 = -h/2$ and $z_3 = h/2$, respectively. The boundary of the plate is defined by $\Gamma = \Gamma_u + \Gamma_t$, where Γ_u is subject to definite displacements and Γ_t to imposed shear forces and bending moments. Unlike classical plate theory, in Reissner's theory, the out-of-plane shear deformation components, $\gamma_{\alpha 3}$ and $\gamma_{\alpha 3}$ are not negligible and cause the perpendicular planes to remain flat but not perpendicular to the midplane of the plate. The following hypotheses support the thick plate theory, also known as the Reissner-Midlin theory (see [2, 4]):

- The plate material is isotropic elastic-linear and very homogeneous.
- The plate thickness is uniform in Ω.
- Unit deformations and displacements are small.
  Shear deformations at the plate surfaces are zero.

Unlike in classical plate theory, in Reissner's theory, the out-of-plane shear deformation components, $\gamma_{\alpha 3}$ and $\gamma_{\alpha 3}$ are not negligible and cause the perpendicular planes to remain flat but not perpendicular to the midplane of the plate.

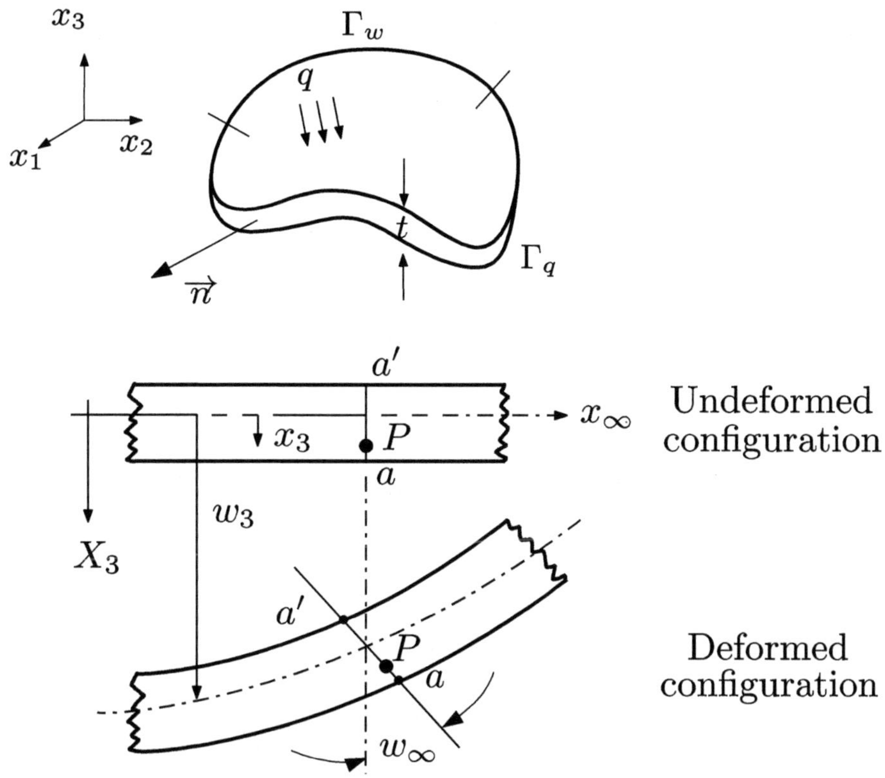

Fig. 3.4 Laminated thick plate

3.4.1 Strain-Displacement Relationships

In addition to the unit strain components given by the Reissner's theory, the shear unit strain components $\gamma_{\alpha 3}$ and $\gamma_{\alpha 3}$ are given by [2]

$$\gamma_{\alpha 3} = w_\alpha + w_{3,\alpha} \tag{3.25}$$

and the curvature relationships or the flexural strain is

$$\kappa_{\alpha\beta} = w_{\alpha,\beta} + w_{\beta,\alpha}. \tag{3.26}$$

3.4.2 Constitutive Equations

The resultant forces $N_{\alpha\beta}$ and resultant bending moments $M_{\alpha\beta}$ due to the forces are given by

$$N_{\alpha\beta} = \int_{-h/2}^{h/2} \sigma_{\alpha\beta} dx_3, \quad M_{\alpha\beta} = \int_{-h/2}^{h/2} \sigma_{\alpha\beta} x_3 dx_3, \quad Q_\alpha = \int_{-h/2}^{h/2} \sigma_{\alpha 3} dx_3 \tag{3.27}$$

The stress resultant-displacement relationships are given by

$$\begin{aligned} N_{\alpha\beta} &= B\frac{1-\nu}{2}\left(u_{\alpha,\beta} + u_{\beta,\alpha} + \frac{2\nu}{1-\nu}u_{\gamma,\gamma}\delta_{\alpha\beta}\right) \\ M_{\alpha\beta} &= D\frac{1-\nu}{2}\left(w_{\alpha,\beta} + w_{\beta,\alpha} + \frac{2\nu}{1-\nu}w_{\gamma,\gamma}\delta_{\alpha\beta}\right) \\ Q_\alpha &= C(w_\alpha + w_{3,\alpha}) \end{aligned} \tag{3.28}$$

where $C = E(1-\nu)\lambda^2/2$ is the shear stiffness and $\lambda = \sqrt(10)/h$ is called the shear factor. Therefore, the thick isotropic plate equilibrium equations are [2]

$$\begin{aligned} &\frac{\partial N_{11}}{\partial x_1} + \frac{\partial N_{12}}{\partial x_2} + q_1 = I_0\frac{\partial^2 u_1}{\partial t^2} + I_1\frac{\partial^2 w_1}{\partial t^2} \\ &\frac{\partial N_{12}}{\partial x_1} + \frac{\partial N_{22}}{\partial x_2} + q_2 = I_0\frac{\partial^2 u_1}{\partial t^2} + I_1\frac{\partial^2 w_2}{\partial t^2} \\ &\frac{\partial M_{11}}{\partial x_1} + \frac{\partial M_{21}}{\partial x_2} - Q_1 = I_2\frac{\partial^2 w_1}{\partial t^2} + I_1\frac{\partial^2 u_1}{\partial t^2} \\ &\frac{\partial M_{12}}{\partial x_1} + \frac{\partial M_{22}}{\partial x_2} - Q_2 = I_2\frac{\partial^2 w_2}{\partial t^2} + I_1\frac{\partial^2 u_2}{\partial t^2} \\ &\frac{\partial Q_1}{\partial x_1} + \frac{\partial Q_2}{\partial x_2} + q_3 = I_0\frac{\partial^2 w_3}{\partial t^2} \end{aligned} \tag{3.29}$$

As a function of the displacements, these last equations are given by

$$B\nabla^2 u_1 + \frac{B}{2}(1+\nu)\frac{\partial}{\partial x_2}\left(-\frac{\partial u_1}{\partial x_2} + \frac{\partial u_2}{\partial x_1}\right) + q_1 = I_0\frac{\partial^2 u_1}{\partial t^2} + I_1\frac{\partial^2 w_1}{\partial t^2}$$

$$\frac{B}{2}(1+\nu)\frac{\partial}{\partial x_1}\left(\frac{\partial u_1}{\partial x_2} - \frac{\partial u_2}{\partial x_1}\right) + B\nabla^2 u_2 + q_2 = I_0\frac{\partial^2 u_1}{\partial t^2} + I_1\frac{\partial^2 w_2}{\partial t^2}$$

$$D\nabla^2 w_1 + \frac{D}{2}(1+\nu)\frac{\partial}{\partial x_2}\left(-\frac{\partial w_1}{\partial x_2} + \frac{\partial w_2}{\partial x_1}\right) - Cw_1 - C\frac{\partial w_3}{\partial x_1} = I_2\frac{\partial^2 w_1}{\partial t^2} + I_1\frac{\partial^2 u_1}{\partial t^2}$$

$$\frac{D}{2}(1+\nu)\frac{\partial}{\partial x_1}\left(\frac{\partial w_1}{\partial x_2}-\frac{\partial w_2}{\partial x_1}\right)+D\nabla^2 w_2 - C w_2 - C\frac{\partial w_3}{\partial x_2} = I_2\frac{\partial^2 w_2}{\partial t^2}+I_1\frac{\partial^2 u_2}{\partial t^2}$$

$$C\nabla^2 w_3 + C\frac{\partial w_1}{\partial x_1}+C\frac{\partial w_2}{\partial x_2}+q_3 = I_0\frac{\partial^2 w_3}{\partial t^2} \tag{3.30}$$

The essential and natural boundary conditions are given by

$$\begin{aligned} u_n = u_n^*, \quad u_s = u_s^*, \quad w_3 = w_3^*, \quad w_n = w_n^*, \quad w_s = w_s^*, \quad \text{in } \Gamma_w \\ N_{nn} = N_{nn}^*, \quad N_{ns} = N_{ns}^*, \quad Q_n = Q_n^*, \quad M_{nn} = M_{nn}^*, \quad M_{ns} = M_{ns}^*, \quad \text{in } \Gamma_t \end{aligned} \tag{3.31}$$

where super-index "*" intends a known quantity and the initial conditions at $t = 0$:

$$u_n = u_n^0, \quad u_s = u_s^0, \quad w_3 = w_3^0, \quad w_n = w_n^0, \quad w_s = w_s^0, \tag{3.32}$$

$$\dot{u}_n = u_n^0, \quad \dot{u}_s = u_s^0, \quad \dot{w}_3 = w_3^0, \quad \dot{w}_n = w_n^0, \quad \dot{w}_s = w_s^0 \text{ in } \Omega. \tag{3.33}$$

3.4.3 *Equations for Large Deflection*

Large deflections (geometric nonlinearity) in plates significantly impact their structural behavior, generating nonlinear effects that must be considered in the analysis. As the deflection increases, the plate develops in-plane (membrane) stresses, contributing to structural stiffness. At large deflections (put as a condition for this type of problem: $u_{\alpha,\beta} \sim w_\alpha \sim \left(w_{3,\alpha}\right)^2 \ll 1$), the plate strength comes not only from bending but also from these membrane stresses, which can make the plate stronger than predicted by linear models.

From a mechanical point of view, large deflections and nonlinear terms appear in the unit strain tensor related to curvature and deformations at the mean surface, describing the physical coupling between membrane effects (in-plane response) and bending effects.

For this type of problem, the unit strain tensor for the plate is written as [2, 9]

$$\epsilon_{\alpha\beta} = \frac{1}{2}\left(u_{\alpha,\beta}+u_{\beta,\alpha}+w_{3,\alpha}w_{3,\beta}\right)+\frac{1}{2}x_3\left(w_{\alpha,\beta}+w_{\beta,\alpha}\right) \tag{3.34}$$

where $w_\alpha = u_{\alpha,3}$ are rotations. The nonlinear term $w_{3,\alpha}w_{3,\beta}/2$ in Eq. (3.1) is considered comparable in magnitude to the linear terms $u_{\alpha,\beta}$ and w_α. The stress resultants and moments resolved in the coordinate directions are

$$N_{\alpha\beta} = \int_{-h/2}^{+h/2} \sigma_{\alpha\beta} \mathrm{d}x_3, \quad Q_\alpha = \int_{-h/2}^{+h/2} \left(\sigma_{\alpha 3} + \sigma_{\alpha\beta} w_{3,\beta}\right) \mathrm{d}x_3 \tag{3.35}$$

$$M_{\alpha\beta} = \int_{-h/2}^{+h/2} \sigma_{\alpha\beta} x_3 \, \mathrm{d}x_3 \tag{3.36}$$

The nonlinear term $\sigma_{\alpha\beta} w_{3,\beta}$ for the shear strain Q_α acting in the x_3 direction arises from the realignment of the membrane tensions $\sigma_{\alpha\beta}$ as the plate deforms. Using Hooke's law, the stress resultants and moments can be expressed in terms of displacements and rotations:

$$\begin{aligned} N_{\alpha\beta} &= \frac{B}{2}\left((1-v)\left(u_{\alpha,\beta} + u_{\beta,\alpha}\right) + 2vu_{\gamma,\gamma}\delta_{\alpha,\beta}\right) \\ &+ \frac{B}{2}\left((1-v)w_{3,\alpha}w_{3,\beta} + vw_{3,\gamma}w_{3,\gamma}\delta_{\alpha,\beta}\right) \\ &= N^l_{\alpha,\beta} + N^n_{\alpha,\beta} \end{aligned} \tag{3.37}$$

Based on Reissner's theory, the resulting stress-strain relationships of the shear-deformable plate for large deflection can be written as

$$Q_\alpha = \frac{5}{6} Dh\left(w_\alpha + w_{3,\alpha}\right) + N_{\alpha\beta} w_{3,\beta} = Q^l_\alpha + Q^n_\alpha \tag{3.38}$$

$$M_{\alpha\beta} = \frac{D}{2}\left((1-v)\left(w_{\alpha,\beta} + w_{\beta,\alpha}\right) + 2\nu w_{\gamma,\gamma}\delta_{\alpha,\beta}\right) \tag{3.39}$$

The nonlinear terms $N^n_{\alpha\beta}$ and Q^n_α incorporate the effects of the large deflection (w_3) of the plate and result in a coupling between the equations describing the loading and bending of the membrane.

The equations of equilibrium for plane tensions in terms of displacements u_α are

$$L^s_{\alpha\beta} u_\alpha + f_\alpha = 0 \tag{3.40}$$

where $L^s_{\alpha\beta}$ is the linear operator:

$$L^s_{\alpha\beta} = \frac{B}{2}\left[(1-v)\delta_{\alpha\beta}\nabla^2 + (1+v)\frac{\partial}{\partial x_\alpha}\frac{\partial}{\partial x_\beta}\right] \tag{3.41}$$

And equivalent body forces:

$$f_\alpha = f^0_\alpha\left(x_1, x_2\right) + N^n_{\alpha\beta,\beta} \tag{3.42}$$

The equations of equilibrium for the resultants of shear stress and bending moments in terms of the rotation w_α and deflection w_3 are

$$L_{ij}^{p} w_j + q_i = 0 \tag{3.43}$$

where L_{ij}^{p} is the differential operator for plate bending problems:

$$L_{\alpha\beta}^{p} = \frac{D}{2}\left[(1-\upsilon)\left(\nabla^2 - \lambda^2\right)\delta_{\alpha\beta} + (1+\upsilon)\frac{\partial}{\partial x_\alpha}\frac{\partial}{\partial x_\beta}\right] \tag{3.44}$$

$$L_{\alpha 3}^{p} = -L_{3\alpha}^{p} = -\frac{(1-\upsilon)D}{2}\lambda^2\frac{\partial}{\partial x_\alpha} \tag{3.45}$$

$$L_{33}^{p} = -\frac{(1-\upsilon)D}{2}\nabla^2. \tag{3.46}$$

3.5 Isotropic Thick Shell

Shell theory with shear deformation is applied in various areas of engineering where thin walled structures resistant to complex loads are required. This theory is an extension of classical shell theory, incorporating the effects of transverse shear, which makes it more accurate for thick shells or shells with anisotropic behavior. Some prominent applications include aircraft fuselages and wings, rocket decks, thick pipes and pressure vessels, and ship and submarine hulls.

A shell is called thin if: $\max(h/R) \ll 1/20$; otherwise, it will be considered a thick shell [2, 6]. On the other hand, a shell is considered shallow, if the ratio between the height h and its projected length (see Fig. 3.4) is $<1/5$ according to. In the theory of shallow shells, the rectangular coordinates $x_1 - x_2$ (projected on the plane tangent to the mean surface) of the points on the mean surface defined from a curvilinear coordinate system $x_1 - x_2$ of the form [8]:

$$x_1 = \xi_1, \quad x_2 = \xi_2, \quad x_3 = f(\xi_1, \xi_2) \tag{3.47}$$

where $f(\xi_1, \xi_2)$ is the equation of the average surface area of the shell. The general shallow shell theory is based on the following hypotheses:

- The order-2 terms of the derivatives f_α are negligible with respect to unity, that is to say, $f_{,\alpha}^2 \ll 1$ and $(f_{,\alpha})(f_{,\beta}) \ll 1$.
- The effects of the shear stress resultants Q_α on the force balance equations in the direction of the x_1 and x_2 axes are negligible.
- The influence of tangent displacements u_α on the out-of-plane shear deformations $\gamma_{\alpha 3}$ is negligible.

3.5.1 Strain-Stress Relationships

The deformation-strain association for shallow shells involve a coupling of the transversal displacement w_3 with the axial strain components $\varepsilon_{\alpha\alpha}$ through the curvature terms (see [1]). Thus:

$$\varepsilon_{\alpha\alpha} = u_{\alpha,\alpha} + \kappa_{\alpha\alpha} w_3 \tag{3.48}$$

The shear strain components $\gamma_{\alpha\beta}$ are the same given by expressions for thick plates and re-written here:

$$\gamma_{\alpha\beta} = u_{\alpha,\beta} + u_{\beta,\alpha} \tag{3.49}$$

The curvature terms couples the tangential displacements u_α with the transversal shear strains $\gamma_{\alpha 3}$. However using the third hypotesis proposed for shallow shells this term can be neglected. Thus:

$$\gamma_{\alpha 3} = w_\alpha - \frac{u_\alpha}{R_\alpha} + w_{3,\alpha} \approx w_\alpha + w_{3,\alpha} \tag{3.50}$$

where $w_{3,\alpha}$ are rotations. Finally, the curvature are related with generalized displacements as

$$\kappa_{\alpha\beta} = w_{\alpha,\beta} + w_{\beta,\alpha} \tag{3.51}$$

On the other hand, the resulting membrane tension per unit length $N_{\alpha\beta}$ and the shear stress $Q\alpha$ are given as [2, 7]

$$N_{11} = \int_{-h/2}^{h/2} \left(1 + \frac{x_3}{R_2}\right) \sigma_{11} dx_3 \qquad N_{22} = \int_{-h/2}^{h/2} \left(1 + \frac{x_3}{R_1}\right) \sigma_{22} dx_3$$

$$N_{12} = \int_{-h/2}^{h/2} \left(1 + \frac{x_3}{R_2}\right) \sigma_{12} dx_3 \qquad N_{21} = \int_{-h/2}^{h/2} \left(1 + \frac{x_3}{R_1}\right) \sigma_{21} dx_3$$

$$Q_1 = \int_{-h/2}^{h/2} \left(1 + \frac{x_3}{R_2}\right) \sigma_{13} dx_3 \qquad Q_2 = \int_{-h/2}^{h/2} \left(1 + \frac{x_3}{R_1}\right) \sigma_{23} dx_3 \tag{3.52}$$

The resulting bending moments $M_{\alpha\beta}$, per unit length are given by

$$M_{11} = \int_{-h/2}^{h/2} \left(1 + \frac{x_3}{R_2} x_3\right) \sigma_{11} dx_3 \qquad M_{22} = \int_{-h/2}^{h/2} \left(1 + \frac{x_3}{R_1} x_3\right) \sigma_{22} dx_3$$

$$M_{12} = \int_{-h/2}^{h/2} \left(1 + \frac{x_3}{R_2} x_3\right) \sigma_{12} dx_3 \qquad M_{21} = \int_{-h/2}^{h/2} \left(1 + \frac{x_3}{R_1} x_3\right) \sigma_{21} dx_3 \tag{3.53}$$

In Reissner plate theory [3], it is assumed that stresses due to membrane forces are uniform, stresses due to bending and twisting moments vary linearly, and shear stresses vary parabolically over the thickness. Thus, the following expressions define these stress components:

$$\left(1 + \frac{x_3}{R_\gamma}\right) \sigma_{\alpha\beta} = \frac{1}{h} N_{\alpha\beta} + \frac{12 x_3}{h^3} M_{\alpha\beta} \tag{3.54}$$

with $\gamma = \beta$ if $\alpha \neq \beta$ and $\gamma \neq \beta$ if $\alpha = \beta$. And,

$$\left(1 + \frac{x_3}{R_\gamma}\right) \sigma_{3\beta} = \frac{3}{2h} \left[1 - \left(\frac{2x_3}{h}\right)^2\right] Q_\alpha \tag{3.55}$$

for $\gamma \neq \alpha$. The generalized tractions at a boundary point can be defined as

$$p_\alpha = M_{\alpha\beta} n_\beta, \quad p_3 = Q_\alpha n_\alpha, \quad t_\alpha = N_{\alpha\beta} n_\beta \tag{3.56}$$

where n_β are the components of the outward normal vector to the shell boundary.

3.5.2 Stress Resultant-Strain Relationships

Considering the hypothesis Q_α / R_α is negligible in the force equilibrium equations, the relationships between stress resultants and strain components are given by the following expressions (see [1]):

$$M_{\alpha\beta} = D \frac{1-\nu}{2} \left(2\chi_{\alpha\beta} + \frac{2\nu}{1-\nu} \chi_{\gamma\gamma} \delta_{\alpha\beta}\right)$$

$$Q_\alpha = C \gamma_{\alpha 3}$$

$$N_{\alpha\beta} = B \frac{1-\nu}{2} \left(\varepsilon_{\alpha\beta} + \varepsilon_{\beta\alpha} + \frac{2\nu}{1-\nu} \varepsilon_{\gamma\gamma} \delta_{\alpha\beta}\right) \tag{3.57}$$

where $2\chi_{\alpha\beta} = w_\alpha + w_{3,\alpha}$. Finally, by replacing equations (3.48)–(3.50) in Eq. (3.57), the above relationships are expressed in terms of the generalized displacement components as

$$M_{\alpha\beta} = D\frac{1-\nu}{2}\left(w_{\alpha,\beta} + w_{\beta,\alpha} + \frac{2\nu}{1-\nu}w_{\gamma,\gamma}\delta_{\alpha\beta}\right)$$

$$Q_\alpha = C(w_\alpha + w_{3,\alpha})$$

$$N_{\alpha\beta} = B\frac{1-\nu}{2}\left(u_{\alpha,\beta} + u_{\beta,\alpha} + \frac{2\nu}{1-\nu}u_{\gamma,\gamma}\delta_{\alpha\beta}\right) + B[(1-\nu)\kappa_{\alpha\beta} + \nu\delta_{\alpha\beta}\kappa_{\phi\phi}]w_3. \tag{3.58}$$

3.5.3 Equilibrium Equations

Consider the differential plate element shown in Fig. 3.5, which is acted upon by the domain forces q_i on its midsurface. The equilibrium forces and moments along the axes x_1, x_2, and x_3 are given by the following equations:

$$\begin{aligned}
\frac{\partial N_{11}}{\partial x_1} + \frac{\partial N_{12}}{\partial x_2} + \frac{Q_1}{R_1} + q_1 &= I_0\frac{\partial^2 u_1}{\partial t^2} + I_1\frac{\partial^2 w_1}{\partial t^2}\\
\frac{\partial N_{12}}{\partial x_1} + \frac{\partial N_{22}}{\partial x_2} + \frac{Q_2}{R_2} + q_2 &= I_0\frac{\partial^2 u_1}{\partial t^2} + I_1\frac{\partial^2 w_2}{\partial t^2}\\
\frac{\partial M_{11}}{\partial x_1} + \frac{\partial M_{21}}{\partial x_2} - Q_1 &= I_2\frac{\partial^2 w_1}{\partial t^2} + I_1\frac{\partial^2 u_1}{\partial t^2}\\
\frac{\partial M_{12}}{\partial x_1} + \frac{\partial M_{22}}{\partial x_2} - Q_2 &= I_2\frac{\partial^2 w_2}{\partial t^2} + I_1\frac{\partial^2 u_2}{\partial t^2}\\
\frac{\partial Q_1}{\partial x_1} + \frac{\partial Q_2}{\partial x_2} - \left(\frac{N_{11}}{R_1} + \frac{N_{22}}{R_2}\right) + q_3 &= I_0\frac{\partial^2 w_3}{\partial t^2}
\end{aligned} \tag{3.59}$$

where q_α are in-plane (membrane) forces per unit area and q_3 are pressure forces.

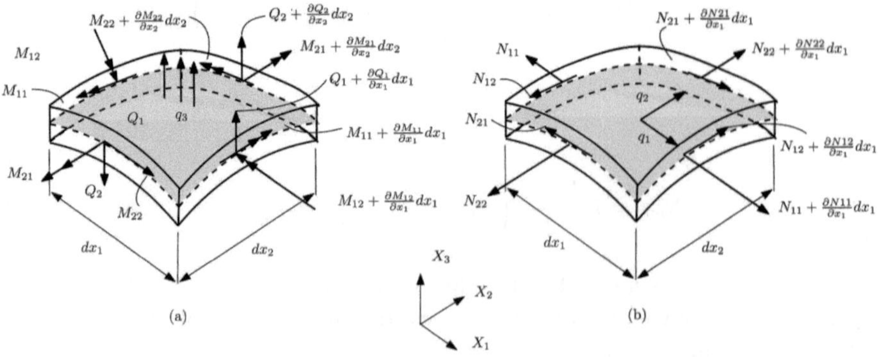

Fig. 3.5 Thick shell differential element

3.5.4 Equilibrium Equations in Terms of Displacements

Substituting the Eq. (3.58) into the Eq. (3.59) gives the equilibrium equations in terms of the displacement components:

$$B\nabla^2 u_1 + \frac{B}{2}(1+\nu)\frac{\partial}{\partial x_2}\left(-\frac{\partial u_1}{\partial x_2} + \frac{\partial u_2}{\partial x_1}\right) + q_1 + B(\kappa_{11} + \nu\kappa_{22})\frac{\partial w_3}{\partial x_1}$$
$$= I_0\frac{\partial^2 u_1}{\partial t^2} + I_1\frac{\partial^2 w_1}{\partial t^2}$$

$$\frac{B}{2}(1+\nu)\frac{\partial}{\partial x_1}\left(\frac{\partial u_1}{\partial x_2} - \frac{\partial u_2}{\partial x_1}\right) + B\nabla^2 u_2 + q_2 + B(\kappa_{11} + \nu\kappa_{22})\frac{\partial w_3}{\partial x_2}$$
$$= I_0\frac{\partial^2 u_1}{\partial t^2} + I_1\frac{\partial^2 w_2}{\partial t^2}$$

$$D\nabla^2 w_1 + \frac{D}{2}(1+\nu)\frac{\partial}{\partial x_2}\left(-\frac{\partial w_1}{\partial x_2} + \frac{\partial w_2}{\partial x_1}\right) - Cw_1 - C\frac{\partial w_3}{\partial x_1}$$
$$= I_2\frac{\partial^2 w_1}{\partial t^2} + I_1\frac{\partial^2 u_1}{\partial t^2}$$

$$\frac{D}{2}(1+\nu)\frac{\partial}{\partial x_1}\left(\frac{\partial w_1}{\partial x_2} - \frac{\partial w_2}{\partial x_1}\right) + D\nabla^2 w_2 - Cw_2 - C\frac{\partial w_3}{\partial x_2}$$
$$= I_2\frac{\partial^2 w_2}{\partial t^2} + I_1\frac{\partial^2 u_2}{\partial t^2}$$

$$C\nabla^2 w_3 + C\frac{\partial w_1}{\partial x_1} + C\frac{\partial w_2}{\partial x_2} + q_3 - B(\kappa_{11} + \nu\kappa_{22})\frac{\partial u_1}{\partial x_1} -$$
$$B(\kappa_{11} + \nu\kappa_{22})\frac{\partial u_2}{\partial x_2} - B(\kappa_{11}^2 + \kappa_{22}^2 + 2\nu\kappa_{11}\kappa_{22})w_3 = I_0\frac{\partial^2 w_3}{\partial t^2} \tag{3.60}$$

The essential and natural boundary conditions are given by

$$\begin{gathered} u_n = u_n^*, \quad u_s = u_s^*, \quad w_3 = w_3^*, \quad w_n = w_n^*, \quad w_s = w_s^*, \quad \text{in } \Gamma_w \\ N_{nn} = N_{nn}^*, \quad N_{ns} = N_{ns}^*, \quad Q_n = Q_n^*, \quad M_{nn} = M_{nn}^*, \quad M_{ns} = M_{ns}^*, \quad \text{in } \Gamma_t \end{gathered} \tag{3.61}$$

and the initial conditions at $t = 0$:

$$u_n = u_n^0, \quad u_s = u_s^0, \quad w_3 = w_3^0, \quad w_n = w_n^0, \quad w_s = w_s^0, \tag{3.62}$$
$$\dot{u}_n = u_n^0, \quad \dot{u}_s = u_s^0, \quad \dot{w}_3 = w_3^0, \quad \dot{w}_n = w_n^0, \quad \dot{w}_s = w_s^0 \text{ in } \Omega. \tag{3.63}$$

3.5.5 Non-linear Equations

For the analysis of large deflections, nonlinear contributions from out-of-plane displacement are included in the strain-displacement relationships. Therefore, the in-plane normal strain tensor can be written as [10]

$$\varepsilon_{\alpha\beta} = \frac{1}{2}\left(u_{\alpha,\beta} + u_{\beta,\alpha}\right) + \frac{w_3}{R_\alpha}\delta_{\alpha\beta} + \frac{1}{2}w_{3,\alpha}w_{3,\beta} = \varepsilon_{\alpha\beta}^{(l)} + \varepsilon_{\alpha\beta}^{(n)}$$

The in-plane and transverse shear deformations and the curvature relations with the derivatives of the generalized displacements are the same as those established above, for $M_{\alpha\beta}$ and for Q_α. Considering Eq. (3.5.5), the in-plane internal resultant force $N_{\alpha\beta}$, given by Eq. (3.59), is given by

$$N_{\alpha\beta} = B\frac{1-v}{2}\left(u_{\alpha,\beta} + u_{\beta,\alpha} + \frac{2v}{1-v}u_{\gamma,\gamma}\delta_{\alpha\beta}\right) + B\left[(1-v)k_{\alpha\beta} + v\delta_{\alpha\beta}k_{\phi\phi}\right]w_3$$
$$+ B\frac{1-v}{2}\left(w_{3,\alpha}w_{3,\beta} + \frac{v}{1-v}w_{3,\gamma}w_{3,\gamma}\delta_{\alpha\beta}\right) = N_{\alpha\beta}^{(i)} + N_{\alpha\beta}^{(ii)} + N_{\alpha\beta}^{(n)}.$$

For ease of representation, the term $N_{\alpha\beta}$ is split into $N_{\alpha\beta}^{(i)}$, which is due to in-plane displacements, $N_{\alpha\beta}^{(ii)}$, which is due to layer curvature and out-of-plane displacement, and $N_{\alpha\beta}^{(n)}$, which is due to the nonlinear contribution of out-of-plane displacement.

Therefore, the equilibrium equations in terms of displacements are obtained as follows:

$$L_{ik}^b w_k + f_i^b = 0$$

and

$$L_{\alpha\beta}^m u_\beta + f_\alpha^m = 0$$

where L_{ik}^b is the Navier differential operator for shear-deformable plate bending problems:

$$L_{\alpha\beta}^b = \frac{D(1-v)}{2}\left[\left(\nabla^2 - \lambda^2\right)\delta_{\alpha\beta} + \frac{(1+v)}{(1-v)}\frac{\partial}{\partial x_\alpha}\frac{\partial}{\partial x_\beta}\right] \tag{3.64}$$

$$L_{\alpha 3}^b = -\frac{(1-v)D}{2}\lambda^2\frac{\partial}{\partial x_\alpha} \tag{3.65}$$

$$L_{3\alpha}^b = -L_{\alpha 3}^b \tag{3.66}$$

and

$$L^b_{33} = \frac{(1-v)D}{2}\lambda^2\nabla^2$$

With $f^b_\alpha = 0$ and $f^b_3 = q^*_3$, while $L^m_{\alpha\beta}$ is the Navier differential operator for two-dimensional plane stress problems:

$$L^m_{\alpha\beta} = \frac{B(1-v)}{2}\left[\nabla^2\delta_{\alpha\beta} + \frac{(1+v)}{(1-v)}\frac{\partial}{\partial x_\alpha}\frac{\partial}{\partial x_\beta}\right]$$

with $f^m_\alpha = q^*_\alpha$.

The terms q^*_i are equivalent body forces

$$q^*_\alpha = q_\alpha + N^{(ii)}_{\alpha\beta,\beta} + N^{(n)}_{\alpha\beta,\beta}$$

and

$$q^*_3 = q_3 - k_{\alpha\beta}N_{\alpha\beta} + \left(N_{\alpha\beta}w_{3,\beta}\right)_{,\alpha}$$

These composed of the body forces, the contribution of the curvature terms, and the contribution of the nonlinear terms and are given as follows:

$$q^*_\alpha = q_\alpha + B\left[(1-v)k_{\alpha\beta} + v\delta_{\alpha\beta}k_{\gamma\gamma}\right]w_{3,\beta} + \tag{3.67}$$

$$B\frac{1-v}{2}\left(w_{3,\alpha\beta}w_{3,\beta} + w_{3,\alpha}w_{3,\beta\beta} + \frac{2v}{1-v}w_{3,\gamma}w_{3,\gamma\beta}\delta_{\alpha\beta}\right) \tag{3.68}$$

and

$$q^*_3 = q_3 - k_{\alpha\beta}\left\{B\frac{1-v}{2}\left(u_{\alpha,\beta} + u_{\beta,\alpha} + \frac{2v}{1-v}u_{\gamma,\gamma}\delta_{\alpha\beta}\right) + \right.$$
$$\left. B\left[(1-v)k_{\alpha\beta} + v\delta_{\alpha\beta}k_{\phi\phi}\right]w_3 + B\frac{1-v}{2}\left(w_{3,\alpha}w_{3,\beta} + \frac{v}{1-v}w_{3,\gamma}w_{3,\gamma}\delta_{\alpha\beta}\right)\right\} \tag{3.69}$$

$$+\left[w_{3,\beta}\left\{B\frac{1-v}{2}\left(u_{\alpha,\beta} + u_{\beta,\alpha} + \frac{2v}{1-v}u_{\gamma,\gamma}\delta_{\alpha\beta}\right) + \right.\right.$$
$$\left.\left. B\left[(1-v)k_{\alpha\beta} + v\delta_{\alpha\beta}k_{\gamma\gamma}\right]w_3 + B\frac{1-v}{2}\left(w_{3,\alpha}w_{3,\beta} + \frac{v}{1-v}w_{3,\gamma}w_{3,\gamma}\delta_{\alpha\beta}\right)\right\}\right]_{,\alpha}. \tag{3.70}$$

3.6 Thick Laminated Plates

It is assumed that there are no relative displacements between the laminate layers and that the kinematic hypotheses of the FSDT are fulfilled, so the equations presented in Sect. 3.5.1 remain valid here. It is also assumed that each layer is orthotropic concerning the material axes and that it obeys Hooke's Law. As in isotropic plates, here the shear stress components σ_{13} and σ_{23} are not negligible. In this way, the following stress and strain relationships are obtained in addition to Eq. (3.20) for an orthotropic laminate in the material axes system $x_1' - x_2'$ [2] (see Fig. 3.3).

$$\begin{Bmatrix} \sigma_{44} \\ \sigma_{55} \end{Bmatrix}^k = \begin{bmatrix} \bar{Q}_{44} & \bar{Q}_{45} \\ \bar{Q}_{45} & \bar{Q}_{55} \end{bmatrix}^k \begin{Bmatrix} \varepsilon_4 \\ \varepsilon_5 \end{Bmatrix}^k \tag{3.71}$$

where the components $\bar{Q}_{44}^k$ and $\bar{Q}_{55}^k$ are given by

$$\begin{aligned} \bar{Q}_{44} &= Q_{44}\cos^2\theta + Q_{55}\sin^2\theta \\ \bar{Q}_{45} &= (Q_{55} - Q_{44})\cos\theta\sin\theta \\ \bar{Q}_{55} &= Q_{55}\cos^2\theta + Q_{44}\sin^2\theta \end{aligned} \tag{3.72}$$

Thus, the equations for membrane forces, moments, and shear forces are given by

$$\begin{Bmatrix} N_{11} \\ N_{22} \\ N_{12} \end{Bmatrix} = \sum_{k=1}^{N} \int_{x_3^k}^{x_3^{k+1}} \begin{Bmatrix} \sigma_{11} \\ \sigma_{22} \\ \sigma_{12} \end{Bmatrix}^k dx_3 = \sum_{k=1}^{N} \int_{x_3^k}^{x_3^{k+1}} \begin{bmatrix} \bar{Q}_{11} & \bar{Q}_{12} & \bar{Q}_{16} \\ \bar{Q}_{21} & \bar{Q}_{22} & \bar{Q}_{26} \\ \bar{Q}_{16} & \bar{Q}_{26} & \bar{Q}_{66} \end{bmatrix}^k \times$$
$$\begin{Bmatrix} u_{1,1} + w_{1,1} \\ u_{2,2} + w_{2,1} \\ u_{2,1} + u_{1,2} + w_{1,2} + w_{2,1} \end{Bmatrix} dx_3 =$$

$$\begin{Bmatrix} N_{11} \\ N_{22} \\ N_{12} \end{Bmatrix} = \begin{bmatrix} A_{11} & A_{12} & A_{16} \\ A_{21} & A_{22} & A_{26} \\ A_{16} & A_{26} & A_{66} \end{bmatrix} \begin{Bmatrix} u_{1,1} \\ u_{2,2} \\ u_{2,1} + u_{1,2} \end{Bmatrix} + \begin{bmatrix} B_{11} & B_{12} & B_{16} \\ B_{21} & B_{22} & B_{26} \\ B_{16} & B_{26} & B_{66} \end{bmatrix} \begin{Bmatrix} w_{1,1} \\ w_{2,1} \\ w_{1,2} + w_{2,1} \end{Bmatrix} \tag{3.73}$$

$$\begin{Bmatrix} M_{11} \\ M_{22} \\ M_{12} \end{Bmatrix} = \sum_{k=1}^{N} \int_{x_3^k}^{x_3^{k+1}} \begin{Bmatrix} \sigma_{11} \\ \sigma_{22} \\ \sigma_{12} \end{Bmatrix}^k z dx_3 = \sum_{k=1}^{N} \int_{x_3^k}^{x_3^{k+1}} \begin{bmatrix} \bar{Q}_{11} & \bar{Q}_{12} & \bar{Q}_{16} \\ \bar{Q}_{21} & \bar{Q}_{22} & \bar{Q}_{26} \\ \bar{Q}_{16} & \bar{Q}_{26} & \bar{Q}_{66} \end{bmatrix}^k \times$$
$$\begin{Bmatrix} u_{1,1} + w_{1,1} \\ u_{2,2} + w_{2,1} \\ u_{2,1} + u_{1,2} + w_{1,2} + w_{2,1} \end{Bmatrix} z dx_3 =$$

$$\begin{Bmatrix} M_{11} \\ M_{22} \\ M_{12} \end{Bmatrix} = \begin{bmatrix} B_{11} & B_{12} & B_{16} \\ B_{21} & B_{22} & B_{26} \\ B_{16} & B_{26} & B_{66} \end{bmatrix} \begin{Bmatrix} u_{1,1} \\ u_{2,2} \\ u_{2,1} + u_{1,2} \end{Bmatrix} + \begin{bmatrix} D_{11} & D_{12} & D_{16} \\ D_{21} & D_{22} & D_{26} \\ D_{16} & D_{26} & D_{66} \end{bmatrix} \begin{Bmatrix} w_{1,1} \\ w_{2,1} \\ w_{1,2} + w_{2,1} \end{Bmatrix} \tag{3.74}$$

$$\begin{Bmatrix} Q_2 \\ Q_1 \end{Bmatrix} = K \sum_{k=1}^{N} \int_{x_3^k}^{x_3^{k+1}} \begin{Bmatrix} \sigma_{23} \\ \sigma_{13} \end{Bmatrix} dx_3 = K \begin{bmatrix} A_{44} & A_{45} \\ A_{45} & A_{55} \end{bmatrix} \begin{Bmatrix} w_{3,2} + w_2 \\ w_{3,1} + w_1 \end{Bmatrix} \tag{3.75}$$

where the terms A_{44}, A_{45}, and A_{55} are given by

$$(A_{44}, A_{45}, A_{55}) = \sum_{k=1}^{N} (\bar{Q}_{44}^k, \bar{Q}_{45}^k, \bar{Q}_{55}^k)(x_{3(k+1)} - x_{3(k)}) \tag{3.76}$$

The equilibrium equations for thick plates are identical to those presented for isotropic plates in Sect. 3.5.3. By substituting the kinematic relationships given by Eqs. (3.48)–(3.51) into Eqs. (3.74) and (3.75) and then substituting into Eq. (3.29), the equations for a laminated thick plate are given by

$$\begin{aligned} &A_{11}\frac{\partial^2 u_1}{\partial x_1^2} + A_{12}\frac{\partial^2 u_2}{\partial x_2 \partial x_1} + A_{16}\left(\frac{\partial^2 u_1}{\partial x_2 \partial x_1} + \frac{\partial^2 u_2}{\partial x_1^2}\right) + B_{11}\frac{\partial^2 w_1}{\partial x_1^2} + B_{12}\frac{\partial^2 w_2}{\partial x_2 \partial x_1} \\ &+B_{16}\left(\frac{\partial^2 w_1}{\partial x_2 \partial x_1} + \frac{\partial^2 w_2}{\partial x_1^2}\right) + A_{16}\frac{\partial^2 u_1}{\partial x_2 \partial x_1} + A_{26}\frac{\partial^2 u_2}{\partial x_2^2} + A_{66}\left(\frac{\partial^2 u_1}{\partial x_2^2} + \frac{\partial^2 u_2}{\partial x_1 \partial x_2}\right) \\ &+B_{16}\frac{\partial^2 w_1}{\partial x_1 \partial x_2} + B_{26}\frac{\partial^2 w_2}{\partial x_2^2} + B_{66}\left(\frac{\partial^2 w_2}{\partial x_2^2} + \frac{\partial^2 w_1}{\partial x_2 \partial x_1}\right) = I_0\frac{\partial^2 u_1}{\partial t^2} + I_1\frac{\partial^2 w_1}{\partial t^2} \end{aligned} \tag{3.77}$$

$$\begin{aligned} &A_{16}\frac{\partial^2 u_1}{\partial x_1^2} + A_{26}\frac{\partial^2 v_2}{\partial x_2 \partial x_1} + A_{66}\left(\frac{\partial^2 u_1}{\partial x_2 \partial x_1} + \frac{\partial^2 u_2}{\partial x_1^2}\right) + B_{16}\frac{\partial^2 w_1}{\partial x_1^2} + B_{26}\frac{\partial^2 w_2}{\partial x_2 \partial x_1} \\ &+B_{66}\left(\frac{\partial^2 w_2}{\partial x_2^2} + \frac{\partial^2 w_1}{\partial x_2 \partial x_1}\right) + A_{12}\frac{\partial^2 u_1}{\partial x_2 \partial x_1} + A_{22}\frac{\partial^2 u_2}{\partial x_2^2} + A_{26}\left(\frac{\partial^2 u_2}{\partial x_2 \partial x_1} + \frac{\partial^2 u_1}{\partial x_2^2}\right) \\ &+B_{12}\frac{\partial^2 w_1}{\partial x_1 \partial x_2} + B_{22}\frac{\partial^2 w_2}{\partial x_2^2} + B_{26}\left(\frac{\partial^2 w_1}{\partial x_2^2} + \frac{\partial^2 w_2}{\partial x_2 \partial x_1}\right) = I_0\frac{\partial^2 u_2}{\partial t^2} + I_1\frac{\partial^2 w_2}{\partial t^2} \end{aligned} \tag{3.78}$$

$$
\begin{aligned}
&B_{11}\frac{\partial^2 u_1}{\partial x_1^2} + B_{12}\frac{\partial^2 u_2}{\partial x_2 \partial x_1} + B_{16}\left(\frac{\partial^2 u_1}{\partial x_2 \partial x_1} + \frac{\partial^2 u_2}{\partial x_1^2}\right) + D_{11}\frac{\partial^2 w_1}{\partial x_1^2} + D_{12}\frac{\partial^2 w_2}{\partial x_2 \partial x_1} \\
&+D_{16}\left(\frac{\partial^2 u_1}{\partial x_2 \partial x_1} + \frac{\partial^2 w_2}{\partial x_1^2}\right) + B_{16}\frac{\partial^2 u_1}{\partial x_2 \partial x_1} + B_{26}\frac{\partial^2 u_2}{\partial x_2^2} + B_{66}\left(\frac{\partial^2 u_1}{\partial x_2^2} + \frac{\partial^2 u_2}{\partial x_1 \partial x_2}\right) \\
&+D_{16}\frac{\partial^2 w_1}{\partial x_1 \partial x_2} + D_{26}\frac{\partial^2 w_2}{\partial x_2^2} + D_{66}\left(\frac{\partial^2 w_1}{\partial x_2^2} + \frac{\partial^2 w_2}{\partial x_2 \partial x_1}\right) \\
&-KA_{55}\left(\frac{\partial w_3}{\partial x_1} + w_1\right) - KA_{45}\left(\frac{\partial w_3}{\partial x_2} + w_2\right) \\
&= I_2\frac{\partial^2 w_1}{\partial t^2} + I_1\frac{\partial^2 u_1}{\partial t^2}
\end{aligned}
\tag{3.79}
$$

$$
\begin{aligned}
&B_{16}\frac{\partial^2 u_1}{\partial x_1^2} + B_{26}\frac{\partial^2 u_2}{\partial x_2 \partial x_1} + B_{66}\left(\frac{\partial^2 u_1}{\partial x_2 \partial x_1} + \frac{\partial^2 u_2}{\partial x_1^2}\right) + D_{16}\frac{\partial^2 w_1}{\partial x_1^2} + D_{26}\frac{\partial^2 w_2}{\partial x_2 \partial x_1} \\
&+D_{66}\left(\frac{\partial^2 w_1}{\partial x_2 \partial x_1} + \frac{\partial^2 w_2}{\partial x_1^2}\right) + B_{12}\frac{\partial^2 u_1}{\partial x_2 \partial x_1} + B_{22}\frac{\partial^2 u_2}{\partial x_2^2} + B_{26}\left(\frac{\partial^2 u_1}{\partial x_2^2} + \frac{\partial^2 u_2}{\partial x_1 \partial x_2}\right) \\
&+D_{12}\frac{\partial^2 w_1}{\partial x_1 \partial x_2} + D_{22}\frac{\partial^2 w_2}{\partial x_2^2} + D_{26}\left(\frac{\partial^2 w_1}{\partial x_2^2} + \frac{\partial^2 w_2}{\partial x_2 \partial x_1}\right) \\
&-KA_{45}\left(\frac{\partial w_3}{\partial x_1} + w_1\right) - KA_{44}\left(\frac{\partial w_3}{\partial x_2} + w_2\right) \\
&= I_2\frac{\partial^2 w_2}{\partial t^2} + I_1\frac{\partial^2 u_2}{\partial t^2}
\end{aligned}
\tag{3.80}
$$

$$
\begin{aligned}
&KA_{55}\left(\frac{\partial^2 w_3}{\partial x_1^2} + \frac{\partial^2 w_1}{\partial x_1}\right) + KA_{45}\left(\frac{\partial^2 w_3}{\partial x_2 \partial x_1} + \frac{\partial w_2}{\partial x_1}\right) \\
&+KA_{45}\left(\frac{\partial^2 w_3}{\partial x_2 \partial x_1} + \frac{\partial w_1}{\partial x_2}\right) + KA_{44}\left(\frac{\partial^2 w_3}{\partial x_2^2} + \frac{\partial w_2}{\partial x_2}\right) \\
&= I_0\frac{\partial^2 w_3}{\partial t^2}
\end{aligned}
\tag{3.81}
$$

These equations' boundary conditions and initial conditions are given in (3.63).

For special symmetric laminates ($\theta = 0°$ and $\theta = 90°$), $A_{\alpha 6} = 0$, $D_{\alpha 6} = 0$, and $B_{ij} = 0$, so the previous system of equations reduces to

$$
A_{11}\frac{\partial^2 u_1}{\partial x_1^2} + A_{12}\frac{\partial^2 u_2}{\partial x_2 \partial x_1} + A_{66}\left(\frac{\partial^2 u_1}{\partial x_2^2} + \frac{\partial^2 u_2}{\partial x_1 \partial x_2}\right) = I_0\frac{\partial^2 u_1}{\partial t^2} + I_1\frac{\partial^2 w_1}{\partial t^2} \tag{3.82}
$$

$$A_{66}\left(\frac{\partial^2 u_1}{\partial x_2 \partial x_1}+\frac{\partial^2 u_2}{\partial x_1^2}\right)+A_{12}\frac{\partial^2 u_1}{\partial x_2 \partial x_1}+A_{22}\frac{\partial^2 u_2}{\partial x_2^2}=I_0\frac{\partial^2 u_2}{\partial t^2}+I_1\frac{\partial^2 w_2}{\partial t^2} \quad (3.83)$$

$$D_{11}\frac{\partial^2 w_1}{\partial x_1^2}+D_{12}\frac{\partial^2 w_2}{\partial x_2 \partial x_1}+D_{66}\left(\frac{\partial^2 w_1}{\partial x_2^2}+\frac{\partial^2 w_2}{\partial x_2 \partial x_1}\right)-KA_{55}\left(\frac{\partial w_3}{\partial x_1}+w_1\right)$$
$$-KA_{45}\left(\frac{\partial w_3}{\partial x_2}+w_2\right)=I_2\frac{\partial^2 w_1}{\partial t^2}+I_1\frac{\partial^2 u_1}{\partial t^2} \quad (3.84)$$

$$D_{66}\left(\frac{\partial^2 w_1}{\partial x_2 \partial x_1}+\frac{\partial^2 w_2}{\partial x_1^2}\right)+D_{12}\frac{\partial^2 w_1}{\partial x_1 \partial x_2}+D_{22}\frac{\partial^2 w_2}{\partial x_2^2}-KA_{45}\left(\frac{\partial w_3}{\partial x_1}+w_1\right)$$
$$-KA_{44}\left(\frac{\partial w_3}{\partial x_2}+w_2\right)=I_2\frac{\partial^2 w_2}{\partial t^2}+I_1\frac{\partial^2 u_2}{\partial t^2} \quad (3.85)$$

$$KA_{55}\left(\frac{\partial^2 w_3}{\partial x_1^2}+\frac{\partial^2 w_1}{\partial x_1}\right)+KA_{45}\left(\frac{\partial^2 w_3}{\partial x_2 \partial x_1}+\frac{\partial w_2}{\partial x_1}\right)+KA_{45}\left(\frac{\partial^2 w_3}{\partial x_2 \partial x_1}+\frac{\partial w_1}{\partial x_2}\right)$$
$$+KA_{44}\left(\frac{\partial^2 w_3}{\partial x_2^2}+\frac{\partial w_2}{\partial x_2}\right)=I_0\frac{\partial^2 w_3}{\partial t^2}. \quad (3.86)$$

3.6.1 Equilibrium Equations in Terms of Displacements for Thick Laminated Shell

Finally, in a similar way as presented for isotropic thick shell, the equilibrium equations for laminated thick shells, using the considerations given above, are written as (see [2]) see Fig. 3.6.

$$A_{11}\left(\frac{\partial^2 u_1}{\partial x_1^2}+\frac{1}{R_1}\frac{\partial w_3}{\partial x_1}\right)+A_{12}\left(\frac{\partial^2 u_2}{\partial x_1 \partial x_2}+\frac{1}{R_2}\frac{\partial w_3}{\partial x_1}\right)+A_{16}\left(\frac{\partial^2 u_2}{\partial x_1^2}+\frac{\partial^2 u_1}{\partial x_1 \partial x_2}\right)$$
$$+B_{11}\frac{\partial^2 w_1}{\partial x_1^2}+B_{12}\frac{\partial^2 w_2}{\partial x_1 \partial x_2}+B_{16}\left[\frac{\partial^2 w_2}{\partial x_1^2}+\frac{\partial^2 w_1}{\partial x_1 \partial x_2}-C_0\left(\frac{\partial^2 u_2}{\partial x_1^2}-\frac{\partial^2 u_1}{\partial x_1 \partial x_2}\right)\right]$$
$$+A_{16}\left(\frac{\partial^2 u_1}{\partial x_1 \partial x_2}+\frac{1}{R_1}\frac{\partial w_3}{\partial x_2}\right)+A_{26}\left(\frac{\partial^2 u_2}{\partial x_2^2}+\frac{1}{R_2}\frac{\partial w_3}{\partial x_2}\right)+A_{66}\left(\frac{\partial^2 u_2}{\partial x_1 \partial x_2}+\frac{\partial^2 u_1}{\partial x_2^2}\right)$$
$$+B_{16}\frac{\partial^2 w_2}{\partial x_1 \partial x_2}+B_{26}\frac{\partial^2 w_2}{\partial x_2^2}+B_{66}\left[\frac{\partial^2 w_2}{\partial x_1 \partial x_2}+\frac{\partial^2 w_1}{\partial x_2^2}-C_0\left(\frac{\partial^2 u_2}{\partial x_1 \partial x_2}-\frac{\partial^2 u_2}{\partial x_2^2}\right)\right]$$

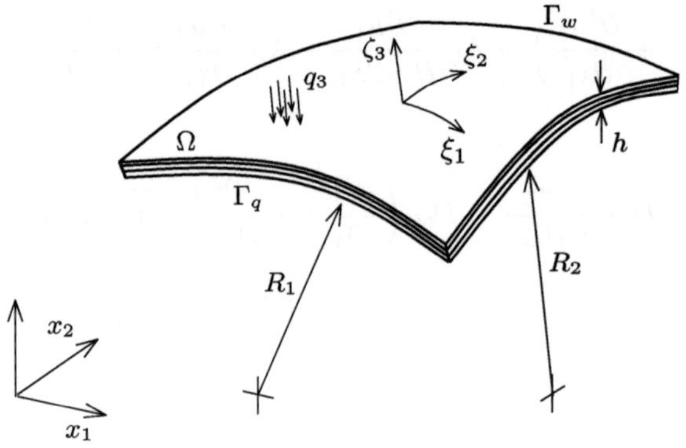

Fig. 3.6 Laminated thick shell geometry

$$+\frac{1}{R_1}\left[K_sA_{44}\left(w_1+\frac{\partial w_3}{\partial x_1}-\frac{u_1}{R_1}\right)+K_sA_{45}\left(w_2+\frac{\partial w_3}{\partial x_2}-\frac{u_2}{R_2}\right)\right]$$
$$=I_0\frac{\partial^2 u_1}{\partial t^2}+I_1\frac{\partial^2 w_1}{\partial t^2} \tag{3.87}$$

$$A_{12}\left(\frac{\partial^2 u_1}{\partial x_1\partial x_2}+\frac{1}{R_1}\frac{\partial w_0}{\partial x_2}\right)+A_{22}\left(\frac{\partial^2 u_2}{\partial x_2^2}+\frac{1}{R_2}\frac{\partial w_3}{\partial x_2}\right)+A_{26}\left(\frac{\partial^2 u_2}{\partial x_1\partial x_2}+\frac{\partial^2 u_1}{\partial x_2^2}\right)$$
$$+B_{12}\frac{\partial^2 w_1}{\partial x_1\partial x_2}+B_{22}\frac{\partial^2 w_2}{\partial x_2^2}+B_{26}\left[\frac{\partial^2 w_2}{\partial x_1\partial x_2}+\frac{\partial^2 w_1}{\partial x_2^2}-C_0\left(\frac{\partial^2 u_2}{\partial x_1\partial x_2}-\frac{\partial^2 u_1}{\partial x_2^2}\right)\right]$$
$$+A_{16}\left(\frac{\partial^2 u_1}{\partial x_1^2}+\frac{1}{R_1}\frac{\partial w_3}{\partial x_1}\right)+A_{26}\left(\frac{\partial^2 u_2}{\partial x_1\partial x_2}+\frac{1}{R_2}\frac{\partial w_3}{\partial x_1}\right)+A_{66}\left(\frac{\partial^2 u_2}{\partial x_1^2}+\frac{\partial^2 u_1}{\partial x_1\partial x_2}\right)$$
$$+B_{16}\frac{\partial^2 w_1}{\partial x_1^2}+B_{26}\frac{\partial^2 w_2}{\partial x_1\partial x_2}+B_{66}\left[\frac{\partial^2 w_2}{\partial x_1^2}+\frac{\partial^2 w_1}{\partial x_1\partial x_2}-C_0\left(\frac{\partial^2 u_2}{\partial x_1}-\frac{\partial^2 u_2}{\partial x_1\partial x_2}\right)\right]$$
$$+\frac{1}{R_2}\left[K_sA_{45}\left(w_1+\frac{\partial w_3}{\partial x_1}-\frac{u_1}{R_1}\right)+K_sA_{55}\left(w_2+\frac{\partial w_3}{\partial x_2}-\frac{u_2}{R_2}\right)\right]$$
$$=I_0\frac{\partial^2 u_2}{\partial t^2}+I_1\frac{\partial^2 w_2}{\partial t^2} \tag{3.88}$$

$$B_{11}\left(\frac{\partial^2 u_1}{\partial x_1^2}+\frac{1}{R_1}\frac{\partial w_3}{\partial x_1}\right)+B_{12}\left(\frac{\partial^2 u_2}{\partial x_1\partial x_2}+\frac{1}{R_2}\frac{\partial w_3}{\partial x_1}\right)+B_{16}\left(\frac{\partial^2 u_2}{\partial x_1^2}+\frac{\partial^2 u_1}{\partial x_1\partial x_2}\right)$$
$$+D_{11}\frac{\partial^2 w_1}{\partial x_1^2}+D_{12}\frac{\partial^2 w_2}{\partial x_1\partial x_2}+D_{16}\left[\frac{\partial^2 w_2}{\partial x_1^2}+\frac{\partial^2 w_1}{\partial x_1\partial x_2}-C_0\left(\frac{\partial^2 u_2}{\partial x_1^2}-\frac{\partial^2 u_1}{\partial x_1\partial x_2}\right)\right]$$
$$+B_{16}\left(\frac{\partial^2 u_1}{\partial x_1\partial x_2}+\frac{1}{R_1}\frac{\partial w_3}{\partial x_2}\right)+B_{26}\left(\frac{\partial^2 u_2}{\partial x_2^2}+\frac{1}{R_2}\frac{\partial w_3}{\partial x_2}\right)+B_{66}\left(\frac{\partial^2 u_2}{\partial x_1\partial x_2}+\frac{\partial^2 u_1}{\partial x_2^2}\right)$$

$$+D_{16}\frac{\partial^2 w_1}{\partial x_1 \partial x_2}+D_{26}\frac{\partial^2 w_2}{\partial x_2^2}+D_{66}\left[\frac{\partial^2 w_2}{\partial x_1 \partial x_2}+\frac{\partial^2 w_1}{\partial x_2^2}-C_0\left(\frac{\partial^2 u_2}{\partial x_1 \partial x_2}-\frac{\partial^2 u_1}{\partial x_2^2}\right)\right]$$
$$+K_s A_{45}\left(w_2+\frac{\partial w_3}{\partial x_2}-\frac{u_2}{R_2}\right)+K_s A_{55}\left(w_1+\frac{\partial w_3}{\partial x_1}-\frac{u_1}{R_2}\right)=I_2\frac{\partial^2 w_1}{\partial t^2}+I_1\frac{\partial^2 u_1}{\partial t^2} \tag{3.89}$$

$$B_{16}\left(\frac{\partial^2 u_1}{\partial x_1^2}+\frac{1}{R_1}\frac{\partial w_3}{\partial x_1}\right)+B_{26}\left(\frac{\partial^2 u_2}{\partial x_1 \partial x_2}+\frac{1}{R_2}\frac{\partial w_3}{\partial x_1}\right)+B_{66}\left(\frac{\partial^2 u_2}{\partial x_1^2}+\frac{\partial^2 u_1}{\partial x_1 \partial x_2}\right)$$
$$+D_{16}\frac{\partial^2 w_1}{\partial x_1^2}+D_{26}\frac{\partial^2 w_2}{\partial x_1 \partial x_2}+D_{66}\left[\frac{\partial^2 w_2}{\partial x_1^2}+\frac{\partial^2 w_1}{\partial x_1 \partial x_2}-C_0\left(\frac{\partial^2 u_2}{\partial x_1^2}-\frac{\partial^2 u_1}{\partial x_1 \partial x_2}\right)\right]$$
$$+B_{12}\left(\frac{\partial^2 u_1}{\partial x_1 \partial x_2}+\frac{1}{R_1}\frac{\partial w_3}{\partial x_2}\right)+B_{22}\left(\frac{\partial^2 u_2}{\partial x_2^2}+\frac{1}{R_2}\frac{\partial w_3}{\partial x_2}\right)+B_{66}\left(\frac{\partial^2 u_2}{\partial x_1 \partial x_2}+\frac{\partial^2 u_1}{\partial x_2^2}\right)$$
$$+D_{12}\frac{\partial^2 w_1}{\partial x_1 \partial x_2}+D_{22}\frac{\partial^2 w_2}{\partial x_2^2}+D_{26}\left[\frac{\partial^2 w_2}{\partial x_1 \partial x_2}+\frac{\partial^2 w_1}{\partial x_2^2}-C_0\left(\frac{\partial^2 u_2}{\partial x_1 \partial x_2}-\frac{\partial^2 u_1}{\partial x_2^2}\right)\right]$$
$$+K_s A_{44}\left(w_2+\frac{\partial w_3}{\partial x_2}-\frac{u_2}{R_2}\right)+K_s A_{45}\left(w_1+\frac{\partial w_3}{\partial x_1}-\frac{u_1}{R_2}\right)=I_2\frac{\partial^2 w_2}{\partial t^2}+I_1\frac{\partial^2 u_2}{\partial t^2} \tag{3.90}$$

$$K_s A_{45}\left(\frac{\partial^2 w_3}{\partial x_1 \partial x_2}+\frac{\partial w_2}{\partial x_1}-\frac{1}{R_1}\frac{\partial u_2}{\partial x_1}\right)+K_s A_{55}\left(\frac{\partial^2 w_3}{\partial x_1^2}+\frac{\partial w_1}{\partial x_1}-\frac{1}{R_1}\frac{\partial u_1}{\partial x_1}\right)$$
$$+K_s A_{44}\left(\frac{\partial^2 w_3}{\partial x_2^2}+\frac{\partial w_2}{\partial x_2}-\frac{1}{R_2}\frac{\partial u_2}{\partial x_1}\right)+KsA_{45}\left(\frac{\partial^2 w_3}{\partial x_1 \partial x_2}+\frac{\partial w_1}{\partial x_2}-\frac{1}{R_1}\frac{\partial u_2}{\partial x_2}\right)$$
$$-\frac{1}{R_1}\left\{A_{11}\left(\frac{\partial u_1}{\partial x_1}+\frac{w_3}{R_1}\right)+A_{12}\left(\frac{\partial u_2}{\partial x_2}+\frac{w_3}{R_2}\right)+A_{16}\left(\frac{\partial u_2}{\partial x_1}+\frac{\partial u_1}{\partial x_2}\right)\right.$$
$$\left.+B_{11}\frac{\partial w_1}{\partial x_1}+B_{12}\frac{\partial w_2}{\partial x_2}+B_{16}\left[\frac{\partial w_2}{\partial x_1}+\frac{\partial w_1}{\partial x_2}-C_0\left(\frac{\partial u_2}{\partial x_1}+\frac{\partial u_1}{\partial x_2}\right)\right]\right\}$$
$$-\frac{1}{R_2}\left\{A_{12}\left(\frac{\partial u_1}{\partial x_1}+\frac{w_3}{R_1}\right)+A_{22}\left(\frac{\partial u_2}{\partial x_2}+\frac{w_3}{R_2}\right)+A_{26}\left(\frac{\partial u_2}{\partial x_1}+\frac{\partial u_1}{\partial x_2}\right)\right.$$
$$\left.+B_{12}\frac{\partial w_1}{\partial x_1}+B_{22}\frac{\partial w_2}{\partial x_2}+B_{26}\left[\frac{\partial w_2}{\partial x_1}+\frac{\partial w_1}{\partial x_2}-C_0\left(\frac{\partial u_2}{\partial x_1}+\frac{\partial u_1}{\partial x_2}\right)\right]\right\}$$
$$+q_3=I_1\frac{\partial^2 w_3}{\partial t^2} \tag{3.91}$$

These equations' boundary conditions and initial conditions are given in (3.63).

3.7 Summary

In this chapter, the classical theory of plates and shells is presented, the hypotheses that govern it are presented, and the governing equations are developed from them for both isotropic and laminated materials. Likewise, the governing equations for problems involving large plate deflections have been developed following the Reissner-Midlin theory for isotropic materials and its extension to the case of thick shells, which will allow the analysis of large deflections and post-buckling.

The governing equations have been developed in their most general form, considering dynamic response. The equations developed here are the basis for the integral formulations of the boundary element method for isotropic and laminated plates and shells, which will be developed in subsequent chapters.

References

1. Dirgantara T (2002) Boundary element analysis of cracks in shear deformable plates and shells. WIT Press, Southampton
2. Reddy JN (2004) Mechanics of laminated composite plates and shells: theory and analysis, 2nd edn. CRC Press, New York
3. Reissner E (1947) On bending of elastic plates. Q Appl Math 5:55–68
4. Reissner E (1955) On transverse vibration of thin, shallow elastic shells. Q Appl Math 13(2):169–176
5. Steigmann DJ, Birsan N, Shirani M (2024) Lecture notes on the theory of plates and shells: classical and modern developments (Solid mechanics and its applications), vol 274. Springer
6. Timoshenko SP, Woinowsky-Krieger S (1964) The theory of plates and shells. Springer
7. Wempner G, Talaslidis D (2019) Mechanics of solids and shells: theories and approximations (Mechanical and aerospace engineering series), 1st ed. Springer
8. Wen PH, Aliabadi MH (2000) Application of dual reciprocity method to plates and shells. Eng Anal Bound Elem 24:583–590
9. Wen PH, Aliabadi MH, Young A (2005) Large deflection analysis of Reissner plate by boundary element method. Comput Struct 83:870–879
10. Wen PH, Aliabadi MH, Young A (2006) Post buckling analysis of Reissner plates by the boundary element method. J Strain Anal Eng Des 41:239–252

Chapter 4
Boundary Element Analysis of Isotropic Thin Plates

Abstract This chapter presents the boundary element formulation of thin plates. The development of the boundary integral formulation, part of the governing equations given in the last chapter, is also presented. This integral form presents the discretization of the integrals using the boundary element method and the application of the Dual Reciprocity Boundary Element Method to transform the domain integrals into boundary integrals. At the end of the chapter, application examples for static problems are presented.

4.1 Introduction

The application of the Boundary Element Method for the analysis of thin plates dates back to the works of [4, 6, 9, 12, 16] presents BEM formulations for thin plates rested in elastic foundations. Transient dynamic analysis is addressed by [3, 7]. Modal and frequency analysis is presented in [13]. Formulations based on the Dual Reciprocity Boundary Element Method were proposed by [10, 11, 15].

This chapter presents the boundary element formulation of thin plates. The development of the boundary integral formulation, part of the governing equations given in the last chapter, is also presented. This integral form presents the discretization of the integrals using the boundary element method and the application of the Dual Reciprocity Boundary Element Method to transform the domain integrals into boundary integrals. At the end of the chapter, application examples for static problems are presented.

4.2 Equilibrum Equations

Consider an isotropic plate of thickness h with its midplane Ω located at $x_1 - x_2$, as shown in Fig. 4.1. The x_3 axis is taken positive downward from the midplane. The upper and lower surfaces of the plate are located at $z_3 = -h/2$ and $z_3 = h/2$, respectively. The plate boundary is defined by $\Gamma = \Gamma_u + \Gamma_t$, where Γ_u are subjected

J. Useche Vivero, *The Boundary Element Method for Engineering Analysis of Plates and Shells*, Solid Mechanics and Its Applications 281,
https://doi.org/10.1007/978-3-031-93050-8_4

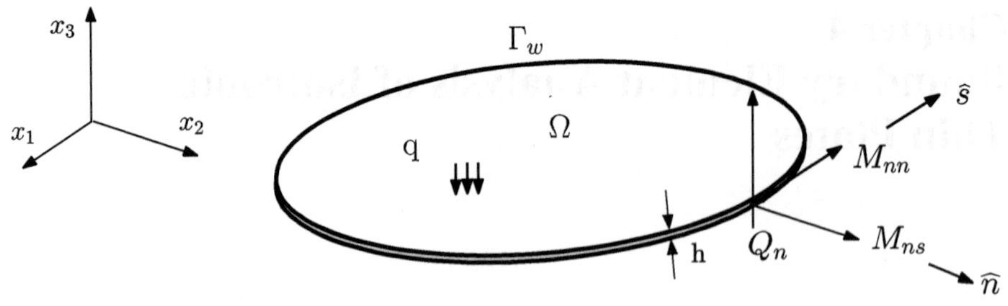

Fig. 4.1 Thin plate geometry

to defined displacements and Γ_t to imposed shear forces and bending moments. The plate complies with the hypotheses posed in the Kirchhoff plate theory, as presented in Chap. 3.

As presented in the previous chapter, the equilibrium equations for deflection are given by [14, 17]:

$$D\frac{\partial^4 w_3}{\partial x_1^4} + 2D\frac{\partial^4 w_3}{\partial x_1^2 \partial x_2^2} + D\frac{\partial^4 w_3}{\partial x_2^4} - q(x_1, x_2) = I_0\frac{\partial^2 w_3^0}{\partial t^2} + I_2 \tag{4.1}$$

where $D = 2h^3E/(3(1 - \upsilon^2))$ is the bending stiffness factor.

4.3 Boundary Element Integral Equations

The boundary integral equations are obtained directly from the Maxwell-Betti reciprocity theorem. For this, let us consider two distinct deformed states of a plate, defined by the variables (w, θ_i, M_i, Q, and q) and (W^*, θ_i^*, M_i^*, Q^*, and q^*). The Maxwell-Betti reciprocity theorem is written as [12, 18]:

$$\int_\Gamma (Q^*w + M_i^*\theta_i)d\Gamma + \int_\Omega (q^*w)d\Omega = \int_\Gamma (QW^* + M_i\theta_i^*)d\Gamma + \int_\Omega (qW^*)d\Omega \tag{4.2}$$

Based on the tangential standard components of rotation θ_n and momentum M_n, θ_s and M_s, respectively, in the previous equation, we can write the term $M_i\theta_i^*$ in the first integral on the right-hand side as

$$\int_\Gamma M_i \theta_i^* d\Gamma = \int_\Gamma (M^n n_i - M^s s_i)(\theta^{*n} n_i - \theta^{*s} s_i) d\Gamma$$
$$= \int_\Gamma \left(M^n \frac{\partial W^*}{\partial n} + M^s \frac{\partial W^*}{\partial s} \right) d\Gamma \tag{4.3}$$

The second term of the integral on the right-hand side can be rewritten as

$$\int_\Gamma M^s \frac{\partial W^*}{\partial s} d\Gamma = \int_\Gamma M^s W^* d\Gamma - \int_\Gamma W^* \frac{\partial M^s}{\partial s} d\Gamma \tag{4.4}$$

For closed boundarys with continuity C^1, the first integral is zero. However, for boundarys with a corner at point C, this integral must be evaluated as

$$\int_\Gamma M^s W^* d\Gamma = \left(M^s \big|_c^- - M^s \big|_c^+ \right) W^* = \sum_{i=1}^{N_c} R_{c_i} w_{c_i}^* \tag{4.5}$$

where N_c is the total number of corners in the boundary. Thus, the integral (4.4) can be written as

$$\int_\Gamma M_i \theta_i^* d\Gamma = \int_\Gamma \left(M^n \frac{\partial W^*}{\partial n} - \frac{\partial M^s}{\partial s} W^* \right) d\Gamma + \sum_{i=1}^{N_c} R_{c_i} w_{c_i}^* \tag{4.6}$$

and similarly:

$$\int_\Gamma M_i^* \theta_i d\Gamma = \int_\Gamma \left(M^{n*} \frac{\partial W^*}{\partial n} - \frac{\partial M^{s*}}{\partial s} w \right) d\Gamma + \sum_{i=1}^{N_c} R_{c_i}^* w_{c_i} \tag{4.7}$$

Finally, replacing these equations and the relationships between V and Q we obtain:

$$\int_\Gamma \left(V^* w + M^* \frac{\partial W^*}{\partial n} \right) d\Gamma + \int_\Omega (q^* w) d\Omega + \sum_{i=1}^{N_c} R_{c_i} w_{c_i}^* =$$
$$\int_\Gamma \left(V W^* + M \frac{\partial W^*}{\partial n} \right) d\Gamma + \int_\Omega (q W^*) d\Omega + \sum_{i=1}^{N_c} R_{c_i}^* w_{c_i} \tag{4.8}$$

Taking the quantities $()^*$ as fundamental solutions of Eq. (4.8) (with $q^* = \delta(\mathbf{x}', \mathbf{x})$), we obtain the boundary integral equation:

$$cw(\mathbf{x}') + \int_{\Gamma} \left(V^*(\mathbf{x}', \mathbf{x})w + M^{n*}(\mathbf{x}', \mathbf{x})\frac{\partial W^*}{\partial n} \right) d\Gamma + \sum_{i=1}^{N_c} R_{c_i} w^*_{c_i} =$$

$$\int_{\Gamma} \left(V W^*(\mathbf{x}', \mathbf{x}) + M^{n*}(\mathbf{x}', \mathbf{x})\frac{\partial W^*}{\partial n} \right) d\Gamma + \int_{\Omega} (q(\mathbf{x})W^*(\mathbf{x}', \mathbf{x}))d\Omega + \sum_{i=1}^{N_c} R^*_{c_i} w_{c_i} \tag{4.9}$$

As seen in Chap. 2, the constant c is introduced to consider that the Dirac delta function can be applied in the domain, on the boundary, or outside the domain. If the Dirac delta function is applied at a point where the boundary is smooth, then $c = 1/2$.

The variables in Eq. (4.9) are displacements $w(\mathbf{x})$, rotations $\partial w(\mathbf{x})/\partial n$, moments $m_n(\mathbf{x})$, and shear forces $V_n(\mathbf{x})$. Some of these variables are known for a given boundary condition. However, to have an equal number of equations and unknown variables, it is necessary to write an integral equation corresponding to the derivative of the displacement $w(\mathbf{x}')$. The directions of the axes of this coordinate system coincide with the normals and tangents to the boundary at the placement point. Thus, we have

$$\frac{1}{2}\frac{\partial w(\mathbf{x}')}{\partial n_1} + \int_{\Gamma} \left[\frac{\partial V^*}{\partial n_1}(\mathbf{x}', \mathbf{x})w(\mathbf{x}) - \frac{\partial m^*_n}{\partial n_1}(\mathbf{x}', \mathbf{x})\frac{\partial w}{\partial n}(\mathbf{x}) \right] d\Gamma +$$

$$\sum_{i=1}^{N_c} \frac{\partial R^*_{c_i}}{\partial n_1}(\mathbf{x}', \mathbf{x})w_{c_i}(\mathbf{x}) = \int_{\Gamma} \left\{ V_n(\mathbf{x})\frac{\partial W^*}{\partial n_1}(\mathbf{x}', \mathbf{x}) - m_n(\mathbf{x})\frac{\partial}{\partial n_1}\left[\frac{\partial W^*}{\partial n}(\mathbf{x}', \mathbf{x}) \right] \right\} d\Gamma +$$

$$\sum_{i=1}^{N_c} R_{c_i}(\mathbf{x})\frac{\partial W^*_{c_i}}{\partial n_1}(\mathbf{x}', \mathbf{x}) + \int_{\Omega} q(\mathbf{x})\frac{\partial W^*}{\partial n_1}(\mathbf{x}', \mathbf{X})d\Omega. \tag{4.10}$$

4.3.1 Fundamental Solutions

The fundamental solution for displacement $W^*(\mathbf{x})$ at any field point $\mathbf{x}$ on an infinite plate generated by a unit load applied at the source point $\mathbf{x}'$ is given by [1, 18]:

$$W^*(\mathbf{x}', \mathbf{x}) = \frac{1}{8\pi D} r^2 \ln(r) \tag{4.11}$$

The fundamental solutions for rotation, moments, and shear forces are obtained:

$$\theta^*_i(\mathbf{x}', \mathbf{x}) = \frac{x_i \ln(r)}{4\pi D} \tag{4.12}$$

Therefore, the normal derivative of the displacement is

$$\frac{\partial W^*}{\partial n}(\mathbf{x}', \mathbf{x}) = -\theta_i^*(\mathbf{x}', \mathbf{x}) n_i(\mathbf{x}) = -\frac{x_i \ln(r)}{4\pi D} n_i \tag{4.13}$$

Using these results, the fundamental solution for bending moment is given by

$$m_{ij}^*(\mathbf{x}', \mathbf{x}) = \frac{1}{4\pi}\left\{\frac{(1-\nu)x_i x_j}{r^2} + \delta_{ij}[(1+\nu)\ln(r) + \nu]\right\} \tag{4.14}$$

$$M^{n*}(\mathbf{x}', \mathbf{x}) = -\frac{1}{4\pi}\left\{\frac{(1-\nu)x_i x_j}{r^2} + \delta_{ij}[(1+\nu)\ln(r) + \nu]\right\} n_i n_j \tag{4.15}$$

$$M^{s*}(\mathbf{x}', \mathbf{x}) = \frac{1}{4\pi}\left\{\frac{(1-\nu)x_i x_j}{r^2} + \delta_{ij}[(1+\nu)\ln(r) + \nu]\right\} s_i n_j \tag{4.16}$$

Finally, the fundamental solutions q^* and V^* are given by the expressions:

$$q^*(\mathbf{x}', \mathbf{x}) = -\frac{x_i}{2\pi r^2} \tag{4.17}$$

and therefore:

$$V^*(\mathbf{x}', \mathbf{x}) = -\frac{x_i}{2\pi r^2} n_i - \frac{\partial M^{s*}}{\partial s}(\mathbf{x}). \tag{4.18}$$

4.4 The Cell Integration Method

The treatment of the domain integral is present in Eqs. (4.9) and (4.10). In this method, the domain is divided into cells (Fig. 4.2), such that the domain integrals can be represented discretely as [2, 5, 18]:

$$\int_\Omega W_{ij}^*(\mathbf{x}', \mathbf{X}) q_j(\mathbf{X}) d\Omega = \sum_{m=1}^{M_c} \int_{\Omega_m} W_{ij}^*(\mathbf{x}', \mathbf{X}) q_j(\mathbf{X}) d\Omega_m \tag{4.19}$$

and

$$\int_\Omega \frac{\partial W_{ij}^*(\xi_k, \eta_l)}{\partial n_1} q_j(\mathbf{X}) d\Omega = \sum_{m=1}^{M_c} \int_{\Omega_m} \frac{\partial W_{ij}^*(\xi_k, \eta_l)}{\partial n_1} q_j(\mathbf{X}) d\Omega_m \tag{4.20}$$

where M_c is the number of cells used, the integral can be evaluated numerically using Gaussian integration with $nk \times nl$ integration points:

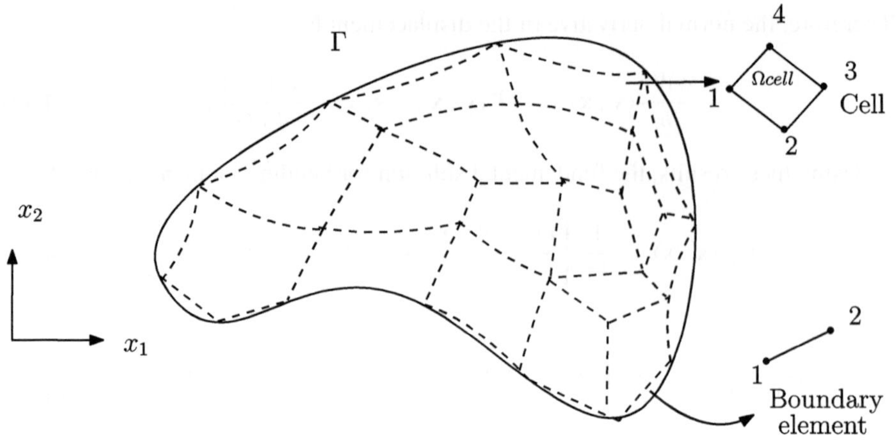

Fig. 4.2 Cell integration method

$$\int_{\Omega} q(\mathbf{x}) W^*(\mathbf{x}', \mathbf{X}) d\Omega = \sum_{m=1}^{M_c} \left[\sum_{k=1}^{nk} \sum_{l=1}^{nl} W_{ij}^*(\xi_k, \eta_l) q_j(\xi_k, \eta_l) J_{cell} w_{kl} \right] \tag{4.21}$$

and

$$\int_{\Omega} q(\mathbf{x}) \frac{\partial W^*}{\partial n_1}(\mathbf{x}', \mathbf{X}) d\Omega = \sum_{m=1}^{M_c} \left[\sum_{k=1}^{nk} \sum_{l=1}^{nl} \frac{\partial W_{ij}^*(\xi_k, \eta_l)}{\partial n_1} q_j(\xi_k, \eta_l) J_{cell} w_{kl} \right] \tag{4.22}$$

where J_{cell} is the Jacobian of the transformation. In general, cells with constant integration produce good results in the treatment of domain integrals in plate and shell equations. Therefore, the boundary integral equation for plates using the cell method is written as

$$\begin{aligned} c w(\mathbf{x}') + \int_{\Gamma} \left(V^*(\mathbf{x}', \mathbf{x}) w + M^{n*}(\mathbf{x}', \mathbf{x}) \frac{\partial W^*}{\partial n} \right) d\Gamma + \sum_{i=1}^{N_c} R_{c_i} W_{c_i}^* = \\ \int_{\Gamma} \left(V W^*(\mathbf{x}', \mathbf{x}) + M^{n*}(\mathbf{x}', \mathbf{x}) \frac{\partial W^*}{\partial n} \right) d\Gamma \\ + \sum_{m=1}^{M_c} \left[\sum_{k=1}^{nk} \sum_{l=1}^{nl} W_{ij}^*(\xi_k, \eta_l) q_j(\xi_k, \eta_l) J_{cell} w_{kl} \right] + \sum_{i=1}^{N_c} R_{c_i}^* w_{c_i} \end{aligned} \tag{4.23}$$

and

$$\frac{1}{2}\frac{\partial w(\mathbf{x}')}{\partial n_1}+\int_\Gamma\left[\frac{\partial V^*}{\partial n_1}(\mathbf{x}',\mathbf{x})w(\mathbf{x})-\frac{\partial m_n^*}{\partial n_1}(\mathbf{x}',\mathbf{x})\frac{\partial w}{\partial n}(\mathbf{x})\right]d\Gamma+$$

$$\sum_{i=1}^{N_c}\frac{\partial R_{c_i}^*}{\partial n_1}(\mathbf{x}',\mathbf{x})w_{c_i}(\mathbf{x})=\int_\Gamma\left\{V_n(\mathbf{x})\frac{\partial W^*}{\partial n_1}(\mathbf{x}',\mathbf{x})-m_n(\mathbf{x})\frac{\partial}{\partial n_1}\left[\frac{\partial W^*}{\partial n}(\mathbf{x}',\mathbf{x})\right]\right\}d\Gamma$$

$$+\sum_{i=1}^{N_c}R_{c_i}(\mathbf{x})\frac{\partial W_{c_i}^*}{\partial n_1}(\mathbf{x}',\mathbf{x})+\sum_{m=1}^{M_c}\left[\sum_{k=1}^{nk}\sum_{l=1}^{nl}W_{ij}^*(\xi_k,\eta_l)q_j(\xi_k,\eta_l)\,J_{cell}w_{kl}\right] \tag{4.24}$$

Domain integrals with weak singularity are usually treated by using quadrilateral elements with constant integration and their partition into cells (Fig. 4.2). Consider triangular cells with constant integration as shown in Figs. 4.2. The integral in one of the triangles can be transformed into the integral over a square as

$$\int_\Omega W_{ij}^*(\mathbf{x}',\mathbf{X})q_j(\mathbf{X})\approx\sum_{m=1}^{M_c}\left[\sum_{m=1}^{4}\int_{-1}^{1}\int_{-1}^{1}W_{ij}^*(\xi,\eta)q_j(\xi,\eta)J(\xi,\eta)J(\chi_m,\zeta_m)^c d\chi_m d\zeta_m\right] \tag{4.25}$$

and

$$\int_\Omega\frac{\partial W_{ij}^*}{\partial n_1}(\mathbf{x}',\mathbf{X})q_j(\mathbf{X})$$

$$\approx\sum_{m=1}^{M_c}\left[\sum_{m=1}^{4}\int_{-1}^{1}\int_{-1}^{1}\frac{\partial W_{ij}^*}{\partial n_1}(\xi,\eta)q_j(\xi,\eta)J(\xi,\eta)J(\chi_m,\zeta_m)^c d\chi_m d\zeta_m\right] \tag{4.26}$$

Table 4.1 presents the transformation functions ξ, η, and the Jacobian J^c as a function of the local coordinates (χ^c, ζ^c) of each cell.

Table 4.1 Triangle-square transformation

Triángulo	ξ	η	J^c
I	$\xi=\frac{1}{2}(1+\chi_1)$	$\eta=\frac{1}{2}(1+\chi_1)\zeta_1$	$J^c(\chi_1)=\frac{1}{4}(1+\chi_1)$
II	$\xi=\frac{1}{2}(1+\zeta_2)\chi_2$	$\eta=\frac{1}{2}(1+\zeta_2)$	$J^c(\zeta_2)=\frac{1}{4}(1+\zeta_2)$
III	$\xi=\frac{1}{2}(\chi_3-1)$	$\eta=\frac{1}{2}(1-\chi_3)\zeta_3$	$J^c(\chi_3)=\frac{1}{4}(1-\chi_3)$
IV	$\xi=\frac{1}{2}(1-\zeta_4)\chi_4$	$\eta=\frac{1}{2}(\zeta_4-1)$	$J^c(\zeta_4)=\frac{1}{4}(1-\zeta_4)$

4.5 Numerical Examples

Rectangular Plate Under Distributed Load

Consider a thin isotropic plate with a thickness of $h/a = 0.01$, clamped on three of its edges, as shown in Fig. 4.3. The dimensions are $a/b = 2$. The elastic properties of the plate are $E = 2.1 \times 109$ and $\upsilon = 0.3$. The plate is subjected to a uniformly distributed pressure $q_3 = 0.1$.

In the analysis, a model with 20 discontinuous quadratic elements is used for the interpolation of displacements and tractions along the boundary. In order to evaluate the domain integral related to the distributed pressure term, 30 bi-linear four-node cells are used (see Fig. 4.3). The same figure also presents the results obtained for the computed deflection.

Finally, Fig. 4.4 presents the deflection of a circular plate with radius $R = 1.0$ (Fig. 4.4). The plate is simply supported along its edge and subjected to a uniformly distributed load $q_3 = 1.0$ over the entire domain. The plate thickness is $h = 0.01$. The elastic properties of the plate are $E = 2.1 \times 109$ and $\upsilon = 0.3$.

4.6 Summary

The application of the Boundary Element Method for the analysis of thin plates was presented. This formulation is based on the governing equation of the plate, which includes inertial effects. Domain integrals were treated using the cell method. The

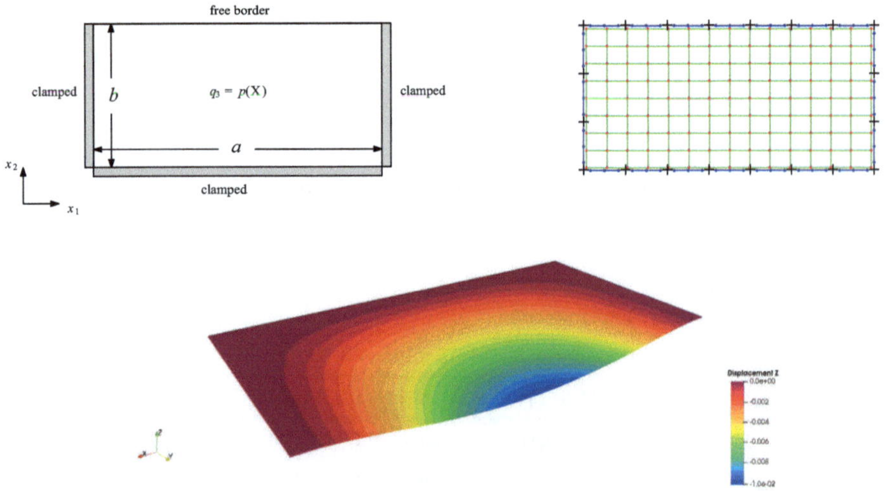

Fig. 4.3 Rectangular plate under distributed load (Left). Right: Boundary element model showing cells (green ones) and internal collocation points (in red). Below: Transversal displacement obtained from the BEM model

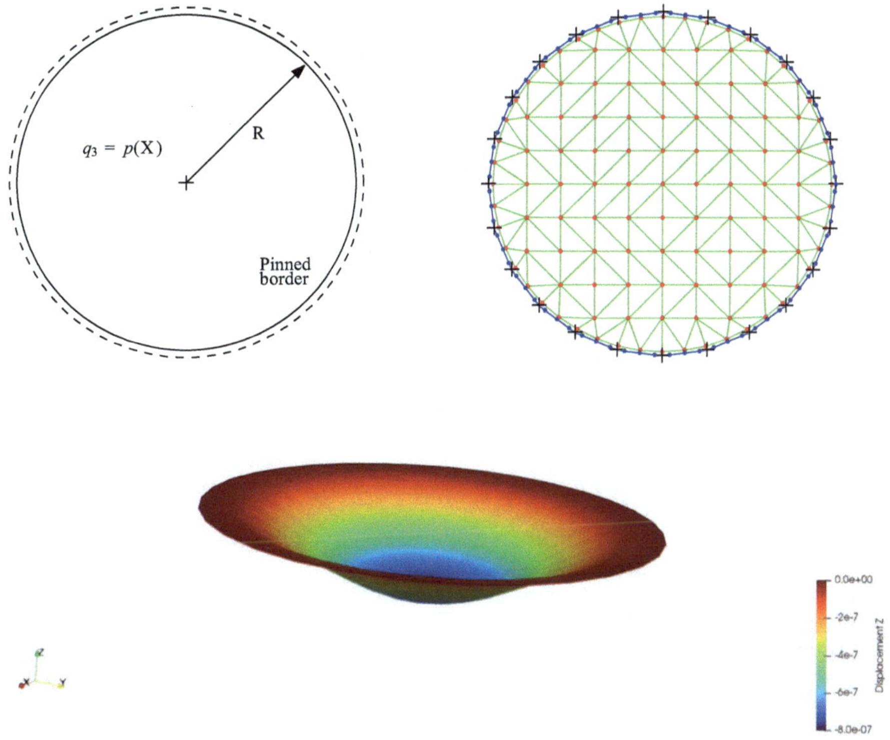

Fig. 4.4 Rectangular plate under distributed load (Left). Right: Boundary element model showing cells (green ones) and internal collocation points (in red). Below: Transversal displacement obtained from the BEM model

analysis of static problems is presented through examples. The reader is invited to explore the use of the Radial Integration Method for handling the domain integral associated with the q_3 term (for this, see the work of [8]). Likewise, the dynamic analysis of isotropic Kirchhoff plates using the BEM can be studied in [7].

Although thin-plate theory has many engineering applications, its main limitation lies in its failure to consider the effects of transverse shear deformation on plate deflection and the development of shear stresses associated with these deformations. These deformations and stresses play a determining role in the behavior of plates with discontinuities, such as holes, perforations, and cracks, sites where Kirchhoff's hypotheses lose validity. For this purpose, the Reissner-Mindlin plate theory, which considers deformation and shear stresses in the plate, will be discussed in the next chapter.

References

1. Banerjee PK (1994) The boundary element methods in engineering, 2nd edn. McGraw-Hill, Berkshire, England
2. Banerjee PK (1993) The finite element methods in engineering, 2nd edn. McGraw-Hill, London
3. Beskos E (1990) Dynamic analysis of plates and shallow shells by the D/BEM. Stud Appl Mecha 24:177–196
4. Bezine G (1988) A new boundary element method for bending of plates on elastic foundations. Int J Solids Struct 24(6):557–565
5. Brebbia CA, Wrobel LC (1996) Boundary element techniques in engineering. Springer
6. Bézine G (1978) Boundary integral formulation for plate flexure with arbitrary boundary conditions. Mech Res Commun 5(4):197–206
7. Dominguez J (1993) Boundary elements in dynamics. Computational mechanics, New York
8. Gao X (2002) The radial integration method for evaluation of domain integrals with boundary only discretization. Eng Anal Bound Elem 26:905–916
9. Costa Jr JA, Brebbia C (1984) Plate bending problems using BEM: in boundary elements, vol VI, 6th edn. Springer
10. Katsikadelis JT, Armenakas AE (1988) Plates bending analysis by the dual reciprocity boundary elements. Eng Anal 5(1):36–40
11. Partridge PW, Brebbia CA, Wrobel LC (1992) The dual reciprocity boundary element method. Computational Mechanics Publications, Southampton
12. París F, de León S (1987) Boundary element method applied to the analysis of thin plates. Comput Struct 25(2):225–233
13. Providakis CP, Beskos DE (1989) Free and forced vibration of plates by boundary elements. Comput Method Appl Mech Eng 74:231–250
14. Reddy JN (2004) Mechanics of laminated composite plates and shells: theory and analysis, 2nd edn. CRC Press, New York
15. Silva A, Venturini WS (1988) Dual reciprocity process applied to solve bending plates on elastic foundations: boundary elements X, vol 10. Springer
16. Stern M (1979) A general boundary integral formulation for the numerical solution of plate bending problems. Int J Solids Struct 15(10):769–782
17. Timoshenko SP, Woinowsky-Krieger S (1964) The theory of plates and shells. Springer
18. Wrobel LC, Aliabadi MH (2002) The boundary element method. Volume 2: Applications in solid and structures. Wiley, New York

Chapter 5
Boundary Element Analysis of Isotropic Thick Plates

Abstract This chapter presents the direct boundary element formulation of isotropic Reissner plates. The Reissner plate theory and governing equations for shear deformable plates are presented. Based on these equations, the boundary integral formulation is developed. Fundamental solutions for displacements and tractions for infinite elastic plane under transversal loads and in-plane bending moments are showed. Boundary element method is applied to discretizate the integral boundary equation. Procedures for treatment of the singular integrals are showed. Finally, numerical examples are showed and results are discussed.

5.1 Introduction

The Boundary Element formulation for Reissner-Mindlin plates was initially proposed by [10], where he developed the fundamental solutions to the governing equation. Wen et al. [12–18] extend this formulation for the static and dynamic analysis of plates and plate-on-foundation problems by Pasternak as presented by [6]. Later, Useche and Albuquerque [9] applies Dual Reciprocity Boundary Element Method for the transient, harmonic, and modal analysis of thick plates.

This chapter presents the direct boundary element formulation of isotropic Reissner plates. The Reissner plate theory and governing equations for shear deformable plates are presented. Based on these equations, the boundary integral formulation is developed. Fundamental solutions, developed by [10], for displacements and tractions for infinite elastic plane under transversal loads and in-plane bending moments are showed. Boundary element method is applied to discretizate the integral boundary equation. Procedures for treatment of the singular integrals are showed. Finally, numerical examples are showed and results are discussed. Preliminary conclusions are established. This chapter is mainly based on the work of [2, 9, 11, 15, 16].

J. Useche Vivero, *The Boundary Element Method for Engineering Analysis of Plates and Shells*, Solid Mechanics and Its Applications 281,
https://doi.org/10.1007/978-3-031-93050-8_5

5.2 Basic Hypotesis

Consider an isotropic plate of thickness h with its midplane Ω located at $x_1 - x_2$, as shown in Fig. 5.1. The x_3-axis is taken positive downward from the midplane. The top and bottom surfaces of the plate are located at $z_3 = -h/2$ and $z_3 = h/2$, respectively. The plate boundary is defined by $\Gamma = \Gamma_u + \Gamma_t$, where Γ_u are subjected to defined displacements and Γ_t imposed shear forces and bending moments. The plate follows the plate hypotheses given in the Reissner-Mindlin theory.

The equilibrium equations in the Reissner-Mindlin thick plate theory, as presented in Chap. 3, are as follows [7, 9, 10, 15]:

$$D\nabla^2 w_1 + \frac{D}{2}(1+\nu)\frac{\partial}{\partial x_2}\left(-\frac{\partial w_1}{\partial x_2} + \frac{\partial w_2}{\partial x_1}\right) - Cw_1 - C\frac{\partial w_3}{\partial x_1} = I_2\frac{\partial^2 w_1}{\partial t^2}$$

$$\frac{D}{2}(1+\nu)\frac{\partial}{\partial x_1}\left(\frac{\partial w_1}{\partial x_2} - \frac{\partial w_2}{\partial x_1}\right) + D\nabla^2 w_2 - Cw_2 - C\frac{\partial w_3}{\partial x_2} = I_2\frac{\partial^2 w_2}{\partial t^2}$$

$$C\nabla^2 w_3 + C\frac{\partial w_1}{\partial x_1} + C\frac{\partial w_2}{\partial x_2} + q_3 = I_0\frac{\partial^2 w_3}{\partial t^2} \quad (5.1)$$

where $C = D(1-\upsilon)/2\lambda^2$. In the first two equations the terms I_1 and I_2 have been omitted as they are considered small compared to the rotational accelerations. The essential and natural boundary conditions are given by

$$\begin{aligned} &u_n = u_n^*, \quad u_s = u_s^*, \quad w_3 = w_3^*, \quad w_n = w_n^*, \quad w_s = w_s^*, \quad \text{in } \Gamma_w \\ &N_{nn} = N_{nn}^*, \quad N_{ns} = N_{ns}^*, \quad Q_n = Q_n^*, \quad M_{nn} = M_{nn}^*, \quad M_{ns} = M_{ns}^*, \quad \text{in } \Gamma_t \end{aligned} \quad (5.2)$$

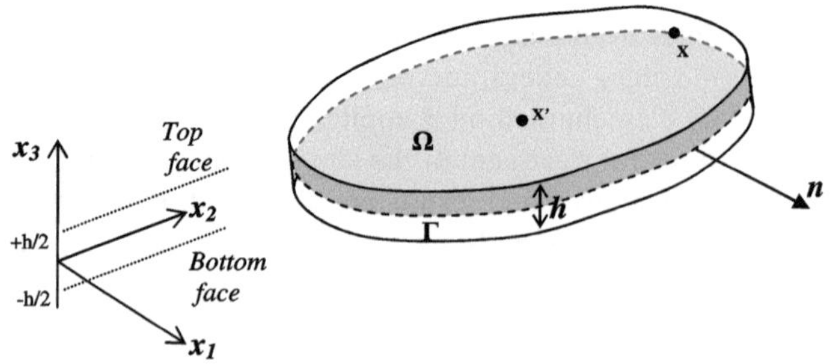

Fig. 5.1 Thick plate geometry

and the initial conditions at $t = 0$:

$$u_n = u_n^0, \quad u_s = u_s^0, \quad w_3 = w_3^0, \quad w_n = w_n^0, \quad w_s = w_s^0, \tag{5.3}$$

$$\dot{u}_n = u_n^0, \quad \dot{u}_s = u_s^0, \quad \dot{w}_3 = w_3^0, \quad \dot{w}_n = w_n^0, \quad \dot{w}_s = w_s^0 \text{ in } \Omega. \tag{5.4}$$

where superindex "0" intends for known values at $t = 0$

5.3 Boundary Integral Formulation

The integral equation can be derived by considering the integral representation of the governing equations through the following integral identity [3, 15]:

$$\int_\Omega \left[\left(M_{\alpha\beta,\beta} - Q_\alpha \right) W_\alpha^* + \left(Q_{\alpha,\alpha} + q \right) W_3^* \right] d\Omega = 0 \tag{5.5}$$

where W_i^* $(i = \alpha, 3)$ are weighting functions. Applying Green's theorem to this equation gives

$$\int_\Gamma p_j W_j^* d\Gamma + \int_\Omega q W_3^* d\Omega + \int_\Omega Q_\alpha \left(W_\alpha^* + W_{3,\alpha}^* \right) d\Omega + \int_\Omega M_{\alpha\beta} W_{\alpha,\beta}^* d\Omega = 0 \tag{5.6}$$

where $j = \alpha, 3$. Replacing the stress resultants ($M_{\alpha\beta}$ and Q_α) by the generalized displacements and their derivatives and applying Green's theorem again for the integral $M_{\alpha\beta}$, we obtain

$$\begin{aligned} &\int_\Gamma p_j W_j^* d\Gamma + \int_\Omega Q_\alpha \left(w_\alpha + w_{3,\alpha} \right) d\Omega \\ &- \int_\Gamma W_{\alpha,\beta}^* \left\{ \frac{D(1-\nu)}{2} \left(w_\alpha n_\beta + w_\beta n_\alpha + \frac{2\nu}{(1-\nu)} w_\gamma n_\gamma \delta_{\alpha\beta} \right) \right\} d\Gamma \\ - \int_\Gamma \frac{D(1-\nu)}{2} &\left(w_\alpha W_{\alpha,\beta\beta}^* + w_\beta W_{\alpha,\beta\alpha}^* + \frac{2\nu}{(1-\nu)} w_\gamma W_{\alpha,\beta\gamma}^* \delta_{\alpha\beta} \right) d\Gamma + \int_\Omega q_3 W_3^* d\Omega \end{aligned} \tag{5.7}$$

The second integral on the left side of the above equation can be solved using Green's theorem as follows:

$$\int_\Omega Q_\alpha \left(w_\alpha + w_{3,\alpha} \right) d\Omega = \int_\Gamma p_3 w_3 d\Gamma + \int_\Omega \left(Q_\alpha^* w_\alpha - Q_{\alpha,\alpha}^* w_3 \right) d\Omega \tag{5.8}$$

Substituting Eq. (5.7) into Eq. (5.8) and grouping them, we obtain

$$\int_{\Gamma} \left(W_j^* p_j - P_j^* w_j\right) d\Gamma + \int_{\Omega} W_3^* q_3 d\Omega + \int_{\Omega} \left[\left(M_{\alpha\beta,\beta}^* - Q_\alpha^*\right) w_\alpha + Q_{\alpha,\alpha}^* w_3\right] d\Omega + \int_{\Omega} W_k^* \left(\mathbf{x}', \mathbf{x}\right) \Lambda_{ij} \ddot{w}_k d\Omega \tag{5.9}$$

where inertial term: $\Lambda_{\alpha\beta} = 1/12\rho h^3 \delta_{\alpha\beta}$ and $\Lambda_{33} = \rho h$ was included as body force. This equation represents the Betti reciprocity theorem for Reissner plates. It should be noted that the weighting functions can be chosen to represent an arbitrary state. This state is defined for generalized concentrated loads: two bending moments ($\alpha = 1, 2$) and a concentrated shear force ($i = 3$) at an arbitrary point $\mathbf{x}' \in \Omega$. Then, Eq. (5.9) can be rewritten after introducing the load direction i as follows:

$$\int_{\Gamma} W_{ij}^* \left(\mathbf{x}', \mathbf{x}\right) p_j \left(\mathbf{x}\right) d\Gamma - \int_{\Gamma} P_{ij}^* \left(\mathbf{x}', \mathbf{x}\right) w_j \left(\mathbf{x}\right) d\Gamma + \int_{\Omega} W_{i3}^* \left(\mathbf{x}', \mathbf{x}\right) q_3 d\Omega$$
$$+ \int_{\Omega} \left[\left(M_{i\alpha\beta,\beta}^* \left(\mathbf{x}', \mathbf{x}\right) - Q_{i\alpha}^* \left(\mathbf{x}', \mathbf{x}\right)\right)\right] w_\alpha \left(\mathbf{x}\right) d\Omega + \int_{\Omega} Q_{i\alpha,\alpha}^* \left(\mathbf{x}', \mathbf{x}\right) w_3 \left(\mathbf{x}\right) d\Omega +$$
$$\int_{\Omega} W_{ij}^* \left(\mathbf{x}', \mathbf{x}\right) \Lambda_{ij} \ddot{w}_k d\Omega \left(\mathbf{x}\right) d\Omega \tag{5.10}$$

Choosing the weighting function as the fundamental solution to the equations:

$$\begin{aligned} M_{i\alpha\beta,\beta}^* \left(\mathbf{x}', \mathbf{x}\right) - Q_{i\alpha}^* \left(\mathbf{x}', \mathbf{x}\right) &= -\delta \left(\mathbf{x}', \mathbf{x}\right) \delta_{i\alpha} \\ Q_{i\alpha,\alpha}^* \left(\mathbf{x}', \mathbf{x}\right) &= -\delta \left(\mathbf{x}', \mathbf{x}\right) \delta_{i3} \end{aligned} \tag{5.11}$$

And making use of the Dirac delta property:

$$\int_{\Omega} \delta \left(\mathbf{x}', \mathbf{x}\right) w_i \left(\mathbf{x}\right) d\Omega = w_i \left(\mathbf{x}'\right) \tag{5.12}$$

Then Eq. (5.10) can be written for an internal source point $\mathbf{x}'$ as (Fig. 5.2):

$$c_{ij} w_j(\mathbf{x}') + \int_{\Gamma} P_{ij}^* \left(\mathbf{x}', \mathbf{x}\right) w_j \left(\mathbf{x}\right) d\Gamma = \int_{\Gamma} W_{ij}^* \left(\mathbf{x}', \mathbf{x}\right) p_j \left(\mathbf{x}\right) d\Gamma$$
$$+ \int_{\Omega} W_{i3}^* \left(\mathbf{x}', \mathbf{x}\right) q_3 \left(\mathbf{x}\right) d\Omega + \int_{\Omega} W_{ij}^* \left(\mathbf{x}', \mathbf{x}\right) \Lambda_{ij} \ddot{w}_k d\Omega \left(\mathbf{x}\right) d\Omega \tag{5.13}$$

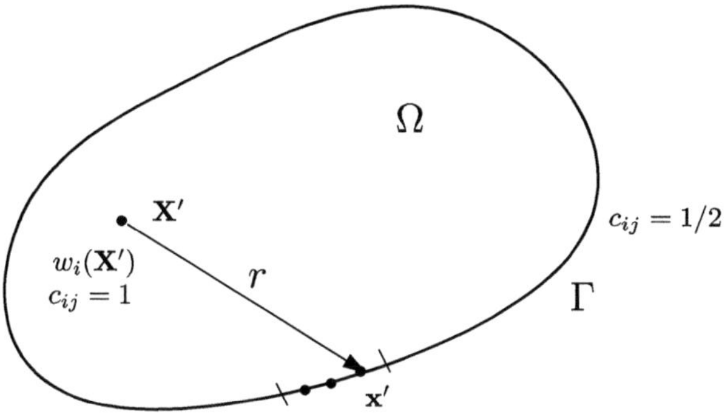

Fig. 5.2 Internal colocation point

where $W_{ij}^*\left(\mathbf{x}', \mathbf{x}\right)$ and $P_{ij}^*\left(\mathbf{x}', \mathbf{x}\right)$ are the fundamental solutions for the displacements and tractions, respectively, which represent the displacement and traction at point $\mathbf{x}$ in the j direction due to the unit load applied at $\mathbf{x}'$ in the i, direction, respectively. Weeën gives the expressions for these kernels [10] as follows:

$$W_{\alpha\beta}^* = \frac{1}{8\pi D\left(1-v\right)}\{[8B\left(z\right) - \left(1-v\right)\left(2\ln\left(z\right)-1\right)]\,\delta_{\alpha\beta} - [8A\left(z\right) - 2\left(1-v\right)]\,r_{,\alpha}r_{,\beta}\} \tag{5.14}$$

$$W_{\alpha 3}^* = -W_{3\alpha}^* = \frac{1}{8\pi D}\left(2\ln\left(z\right)-1\right)rr_{,\alpha} \tag{5.15}$$

$$W_{\alpha\beta}^* = \frac{1}{8\pi D\left(1-v\right)\lambda^2}\left[\left(1-v\right)z^2\left(\ln\left(z\right)-1\right) - 8\ln\left(z\right)\right] \tag{5.16}$$

and

$$P_{\alpha\beta}^* = -\frac{1}{4\pi r}\left[\left(4A\left(z\right)+2zK_1\left(z\right)+1-v\right)\left(\delta_{\alpha\gamma}r_{,n}+r_{,\alpha}n_{\gamma}\right) + \left(4A\left(z\right)+1+v\right)r_{,\gamma}n_{\alpha} - 2\left(8A\left(z\right)+2zK_1\left(z\right)+1-v\right)r_{,\alpha}r_{,\gamma}r_{,n}\right] \tag{5.17}$$

$$P_{\gamma 3}^* = \frac{\lambda^2}{2\pi}\left[B\left(z\right)n_{\gamma} - A\left(z\right)r_{,\gamma}r_{,n}\right] \tag{5.18}$$

$$P_{3\alpha}^* = \frac{-\left(1-v\right)}{8\pi}\left[\left(2\frac{\left(1+v\right)}{\left(1-v\right)}\ln\left(z\right)-1\right)n_{\alpha} + 2r_{,\alpha}r_{,n}\right] \tag{5.19}$$

$$P_{33}^{*} = \frac{-1}{2\pi r} r_{,n} \tag{5.20}$$

where

$$A(z) = K_0(z) + \frac{2}{z}\left[K_1(z) - \frac{1}{z}\right] \tag{5.21}$$

$$B(z) = K_0(z) + \frac{1}{z}\left[K_1(z) - \frac{1}{z}\right] \tag{5.22}$$

where $K_0(z)$ and $K_1(z)$ are modified Bessel functions, $z = \lambda r$, r is the absolute distance between the source and field points; $r_{,\alpha} = r_\alpha / r$ with $r_\alpha = x_\alpha(\mathbf{x}) - x_\alpha(\mathbf{x}')$ and $r_{,n} = r_{,\alpha} n_\alpha$. $A(z)$ s a smooth function, while $B(z)$ is a weakly singular $O(ln(r))$ Therefore, W_{ij}^{*} is a weak singularity and P_{ij}^{*} has a strong singularity (Cauchy principal value) $O(1/r)$.

5.4 Boundary Element Discretization

The boundary is discretized into N_e elements, on which the unknowns are approximated by a second-degree interpolation polynomial using discontinuous quadratic elements. After discretization, Eq. (5.23) can be rewritten as

$$\begin{aligned} c_{ij}\left(\mathbf{x}'\right) w_i\left(\mathbf{x}'\right) + \sum_{j=1}^{N_e}\sum_{m=1}^{3} w_j^m \int_{-1}^{+1} P_{ij}^{*}\left(\mathbf{x}', \mathbf{x}(\xi)\right) \Phi^m(\xi) J_j(\xi)\, d\xi \\ = \sum_{j=1}^{N_e}\sum_{m=1}^{3} p_j^m \int_{-1}^{+1} W_{ij}^{*}\left(\mathbf{x}', \mathbf{x}(\xi)\right) \Phi^m(\xi) J_j(\xi)\, d\xi \\ + q \sum_{j=1}^{N_e} \int_{-1}^{+1} V_{i,\alpha}^{*}\left(\mathbf{x}', \mathbf{x}(\xi)\right) n_\alpha(\xi) J_j(\xi)\, d\xi \end{aligned} \tag{5.23}$$

where J is the Jacobian of the transformation and Φ is the shape function of the element, as presented in Chap. 1 (for explanation about treatment of domain integral with uniform constant load (q) see Sect. 5.7.1). For a general quadratic element, as shown in Fig. 5.3, we have

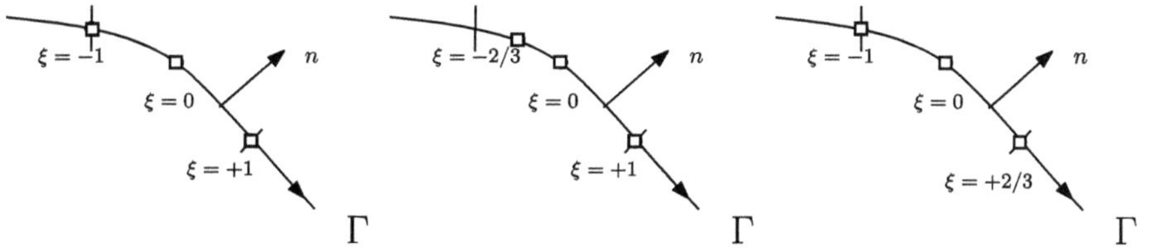

Fig. 5.3 General quadratic element

$$J\left(\xi\right) = \sqrt{\frac{\partial x_\theta\left(\xi\right)}{\partial\xi}\frac{\partial x_\theta\left(\xi\right)}{\partial\xi}} \tag{5.24}$$

$$n_\alpha\left(\xi\right) = \frac{1}{J\left(\xi\right)}\frac{\partial x_\beta\left(\xi\right)}{\partial\xi}\epsilon_{\alpha\beta 3} \tag{5.25}$$

where $\epsilon_{\alpha\beta 3}$ is the permutation symbol. After performing the placement process, Eq. (5.23) can be written as follows:

$$\mathbf{H}\ddot{\mathbf{w}} + \mathbf{H}\mathbf{w} = \mathbf{G}\mathbf{p} + \mathbf{Q}. \tag{5.26}$$

5.5 Internal Stress Resultants

The stress resultants at the domain point $\mathbf{x}'$ can be evaluated from Eq. (5.23), using the resulting stress-displacement relationships:

$$M_{\alpha\beta}\left(\mathbf{x}'\right) = \int_\Gamma W^*_{\alpha\beta k}\left(\mathbf{x}',\mathbf{x}\right)p_k d\Gamma - \int_\Gamma P^*_{\alpha\beta k}\left(\mathbf{x}',\mathbf{x}\right)w_k d\Gamma + \int_\Omega W^*_{\alpha\beta 3}\left(\mathbf{x}',\mathbf{x}\right)q_3 d\Omega \tag{5.27}$$

$$Q_\alpha = \int_\Gamma W^*_{3\beta k}\left(\mathbf{x}',\mathbf{x}\right)p_k d\Gamma - \int_\Gamma P^*_{3\beta k}\left(\mathbf{x}',\mathbf{x}\right)w_k d\Gamma + \int_\Omega W^*_{3\beta 3}\left(\mathbf{x}',\mathbf{x}\right)q_3 d\Omega \tag{5.28}$$

The fundamental solutions $W^*_{i\beta k}$, $P^*_{i\beta k}$ and $Q^*_{i\beta}$, are linear combinations of the first derivatives of W^*_{ij}, P^*_{ij}, and $V^*_{i,\beta}$, and are given by (see [2, 10]):

$$W^*_{\alpha\beta\gamma} = \frac{1}{4\pi r}\left[(4A(z) + 2zK_1(z) + 1 - v)\left(\delta_{\beta\gamma}r_{,\alpha} + \delta_{\alpha\gamma}r_{,\beta}\right)\right.$$
$$\left.-2(8A(z) + 2zK_1(z) + 1 - v)r_{,\alpha}r_{,\beta}r_{,\gamma} + (4A(z) + 2zK_1(z) + 1 - v)\delta_{\alpha\beta}r_{,\gamma}\right] \tag{5.29}$$

$$W^*_{\alpha\beta 3} = \frac{-(1-v)}{8\pi}\left[\left(2\frac{(1+v)}{(1-v)}\ln(z) - 1\right)\delta_{\alpha\beta} + 2r_{,\alpha}r_{,\beta}\right]$$
$$W^*_{3\beta 3} = \frac{1}{2\pi r}r_{,\beta} \tag{5.30}$$

$$P^*_{\alpha\beta\gamma} = \frac{D(1-v)}{4\pi r^2}\left\{(4A(z) + 2zK_1(z) + 1 - v)\left(\delta_{\gamma\alpha}n_\beta + \delta_{\gamma\beta}n_\alpha\right)\right.$$
$$+(4A(z) + 1 + 3v)\delta_{\alpha\beta}n_\gamma - \left(16A(z) + 6zK_1(z) + z^2K_0(z) + 2 - 2v\right) \tag{5.31}$$

$$P^*_{\alpha\beta 3} = \frac{D(1-v)\lambda^2}{4\pi r}\left[(2A(z) + zK_1(z))\left(r_{,\beta}n_\alpha + r_{,\alpha}n_\beta\right)\right.$$
$$-2(4A(z) + zK_1(z))r_{,\alpha}r_{,\beta}r_{,n} + 2A(z)\delta_{\alpha\beta}r_{,n} \tag{5.32}$$

$$P^*_{3\beta 3} = \frac{D(1-v)\lambda^2}{4\pi r^2}\left[\left(z^2B(z) + 1\right)n_\beta - \left(z^2A(z) + 2\right)r_{,\beta}r_{,n}\right] \tag{5.33}$$

$$Q^*_{\alpha\beta} = \frac{-r}{64\pi}\left\{(4\ln(z) - 3)\left[(1-v)\left(r_{,\beta}n_\alpha + r_{,\alpha}n_\beta\right) + (1+3v)\delta_{\alpha\beta}r_{,n}\right]\right.$$
$$\left.+4\left[(1-v)r_{,\alpha}r_{,\beta} + v\delta_{\alpha\beta}\right]r_{,n}\right\} \tag{5.34}$$

$$Q^*_{3\beta} = \frac{1}{8\pi}\left[(2\ln(z) - 1)n_\beta + 2r_{,\beta}r_{,n}\right]. \tag{5.35}$$

5.6 Treatment of the Singularities

The influence matrix **G** and the load vector **Q** contain weak singularity terms, which can be canceled using a nonlinear coordinate transformation [8]. Furthermore, element subdivisions are used for improved numerical accuracy. The influence matrix **H**, on the other hand, contains strong singularity terms, which can be evaluated indirectly by considering basic rigid-body displacements [2, 10, 15]:

$$\begin{aligned} w_1 &= C, \quad w_2 = 0, \quad w_3 = -Cr_1 \\ w_1 &= 0, \quad w_2 = C, \quad w_3 = -Cr_1 \\ w_1 &= 0, \quad w_2 = 0, \quad w_3 = C \end{aligned} \tag{5.36}$$

where C is an arbitrary constant, this gives

$$\begin{aligned} c_{i\beta}\left(\mathbf{x}'\right) &= -\int_{\Gamma} \left[P_{i\beta}^{*}\left(\mathbf{x}', \mathbf{x}\right) + r_{\beta} P_{i3}\left(\mathbf{x}', \mathbf{x}\right)\right] d\Gamma \\ c_{i3}\left(\mathbf{x}'\right) &= -\int_{\Gamma} P_{i3}\left(\mathbf{x}', \mathbf{x}\right) d\Gamma. \end{aligned} \tag{5.37}$$

5.6.1 *Telles Transformation and Element Sub-division*

The Telles transformation efficiently handles quasi-singular integrals in thick plate (and shell) equations [8]. It is used in conjunction with the element subdivision technique. This method employs a nonlinear transformation that automatically moves the Gaussian integration points to the minimum integration distance found (between the source point and the integration point), causing the Jacobian to smooth out the singularity.

Consider the integral:

$$I = \int_{-1}^{+1} f(\xi) d\xi \tag{5.38}$$

where the function $f(\xi)$ is singular at the point $\bar{\xi}$. Let us consider the third-degree transformation polynomial of the form:

$$\xi = a\gamma^3 + b\gamma^2 + c\gamma + d \tag{5.39}$$

With the following conditions:

$$\frac{d\xi}{d\gamma}\Big|_{\bar{\xi}} = 0, \quad \frac{d^2\xi}{d\gamma^2}\Big|_{\bar{\xi}} = 0, \quad \xi(1) = 1, \quad \xi(-1) = -1 \tag{5.40}$$

The first condition implies that the Jacobian of the transformation has a minimum at $\bar{\xi}$. Under these conditions, the coefficients of the transformation polynomial are given by

$$a = 1/Q, \quad b = -3\bar{\gamma}/Q, \quad c = 3\bar{\gamma}^2/Q, d = -b, \quad Q = 1 + 3\bar{\gamma}^2 \tag{5.41}$$

where $\bar{\gamma}$ is given by

$$\bar{\gamma} = \left\{\bar{\xi}\xi^* + |\xi^*|\right\}^{1/3} + \left\{\bar{\xi}\xi^* - |\xi^*|\right\}^{1/3} + \bar{\xi} \tag{5.42}$$

Where, $\xi^* = \bar{\xi}^2 - 1$. Following table presents the implementation of this technique.

Kernels calculation Matlab function

```
function [x,Jt] = telles(gamm,eet)

%telles: CALCULA LA TRANFORMACION NO LINEAL DE TELLES.
%DESARROLLADO POR JAIRO F. USECHE, M.SC., UNICAMP 2005

 %----- CONSTANTES PARA TRANSFORMACION DE TELLES

 eest = eet^2 - 1;
 term1 = eet*eest + abs(eest);
 if term1 < 0
     term1 = (-term1)^(1/3);
     term1 = -term1;
 else
     term1 = term1^(1/3);
 end

 term2 = eet*eest - abs(eest);
 if term2 < 0
     term2 = (-term2)^(1/3);
     term2 = -term2;
 else
     term2 = term2^(1/3);
 end
 GAMM = term1 + term2 + eet;

 Q = 1 + 3*GAMM^2;
 A = 1/Q;
 B = -3*GAMM/Q;
 C = 3*GAMM^2/Q;
 D = -B;

 %----- TRANFORMACION DE COORDENADA Y JACOBIANO
 x = A*gamm^3 + B*gamm^2 + C*gamm + D;
 Jt = 3*A*gamm^2 + 2*B*gamm + C;
return
```

The Telles method is generally used in conjunction with the element subdivision method. In this method, the element to be integrated is partitioned into n subelements depending on the minimum distance from the field point to the component. In this way, this scheme concentrates the number of Gauss points near the quasi-singular point by subdividing the element based on a previously determined minimum distance (usually 1/4 of the length of the component containing the quasi-singularity). The combination of Telles integration and the element subdivision scheme increases the efficiency in the calculation of quasi-singularities that appear in the boundary integrals for displacements and traction for thick plates and shells and should, therefore, be used.

Kernels calculation Matlab function

```
function [He,Cije,Ge,Qe] = IntSingHGQ(nod,Xe,Ye,Xi,NDIV,BOOL,flag)

x1 = Xe(1); x2 = Xe(2); x3 = Xe(3);
y1 = Ye(1); y2 = Ye(2); y3 = Ye(3);
xi = Xi(1); yi = Xi(2);
xo1 = x1; xo3 = x3; yo1 = y1; yo3 = y3;
[lixo,Ind] = find(BOOL~=0);
%----- puntos de gauss
[eg,Wg] = PtosGauss12;
Je = jacobiano(x1,x3,y1,y3);
Le = 2*Je; DEL = Le/(2*NDIV);
INCx = DEL*(x3-x1)/Le; INCy = DEL*(y3-y1)/Le;

He=0; Cije=0; Ge=0; Qe=0;
for k=1:NDIV
    %coordenadas del sub-elemento
    x2 = x1 + INCx; y2 = y1 + INCy;
    x3 = x2 + INCx; y3 = y2 + INCy;
    Jac = jacobiano(x1,x3,y1,y3);
    if abs(x1-x3)~=0
        eet = (x1+x3-2*Xi(1))/(x1-x3);
    else
        eet = (y1+y3-2*Xi(2))/(y1-y3);
    end
%---- INTEGRACION GAUSSIANA & TELLES
    for ii = 1 : 12
        [et,Jt] = telles(eg(ii),eet);
        [R,RX,RY,n] = RNormPos(et,x1,x3,y1,y3,xi,yi,Jac);
        %---- transf. de et para evaluacion de fforma disc.
        if abs(x1-x3) ~=0
            x = 0.5*x1*(1-et) + 0.5*x3*(1+et);
            ee = (xo1 + xo3 - 2*x)/(xo1 - xo3);
        else
            y = 0.5*y1*(1-et) + 0.5*y3*(1+et);
            ee = (yo1 + yo3 - 2*y)/(yo1 - yo3);
        end
        [FHe,FGe,FQe,FCij] = Kerndis(R,RX,RY,n,ee);
        He = FHe*Jt*Wg(ii)*Jac + He;
        Cije = FCij*Jt*Wg(ii)*Jac + Cije;
        Ge = FGe*Jt*Wg(ii)*Jac + Ge;
        Qe = FQe*Jt*Wg(ii)*Jac + Qe;
    end
    x1 = x3; y1 = y3;
end
return
```

5.7 Domain Integral Treatment

5.7.1 Uniform Load

In the case of a uniform charge, the domain integral can be transferred to the edge integral, applying the divergence theorem, to obtain [2, 6]:

$$\int_{\Omega} W_{ij}^{*}\left(\mathbf{x}',\mathbf{X}\right)q_j(\mathbf{X})d\Omega = q_j\int_{\Gamma} V_{i,j}^{*}\left(\mathbf{x}',\mathbf{x}\right)n_{\alpha}(\mathbf{x})d\Gamma \tag{5.43}$$

where V_i^* are the particular solutions of the equation $V_{i,\theta\theta}^* = W_{i3}^*$. According with Mindlin, the term: $v/((1-v)\lambda^2)$ has negligible contribution to the results. For the sake of simplicity, this term will be ignored in this work. The expressions for $V_{i,\beta}^*$ are given as follows:

$$V_{\alpha,\beta}^* = \frac{r^2}{128\pi D}\left[(4\ln(z)-5)\,\delta_{\alpha\beta} + 2\,(4\ln(z)-3)\,r_{,\alpha}r_{,\beta}\right] \tag{5.44}$$

$$V_{3,\beta}^* = \frac{-rr_{,\beta}}{128\pi D\,(1-v)\,\lambda^2}\left[32\,(2\ln(z)-1) - z^2\,(1-v)\,(4\ln(z)-5)\right] \tag{5.45}$$

If the body forces are not uniform, integration techniques such as those described in the following sections should be used.

5.7.2 The Dual Reciprocity Boundary Element Method

The dual reciprocity method has been successfully used in linear thick plate bending problems [13–15], as well as in nonlinear problems in plates and thick shells [1, 19] to transform domain integrals to boundary integrals without requiring the domain to be discretized.

This method is based on the representation of the evaluated functions, q_i in the domain as a linear combination of radial basis functions written in general form (see Chap. 1):

$$q_i(\mathbf{x}) \approx \sum_{m=1}^{N+L} \alpha_i^k f_m(r) \tag{5.46}$$

where α_i^k are unknown coefficients, f^k are radial approximation functions; N is the total number of placement points on the boundary, and L is the number of placement points in the domain (see Fig. 5.4). The approximation functions used in plates and shells are of the type $c + r$, r^3 and $r^2\ln(r)$, the first one with $c = 1$ being the

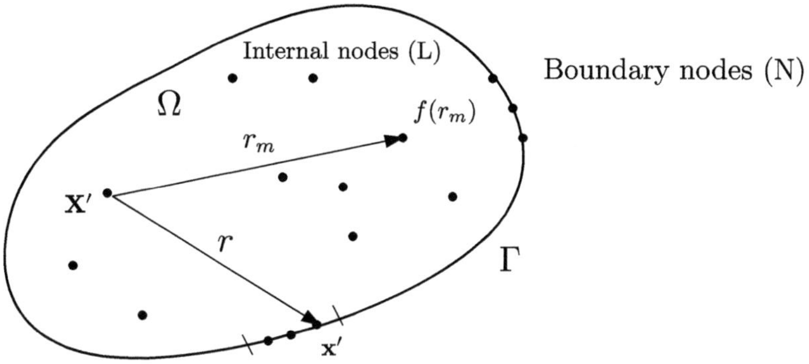

Fig. 5.4 Dual reciprocity boundary element method

most used. In these $r^2 = (x_1 - x_{k1})^2 + (x_2 - x_{k2})^2$, where $\mathbf{x}_k = (x_{k1}, x_{k2})$ are the coordinates where the value of q_i is known.

Thus, the domain integral in Eq. (5.23) can be rewritten as

$$\int_{\Omega} W_{ij}^* q_j d\Omega = \sum_{m=1}^{N+L} \alpha_i^m \int_{\Omega} W_{ij}^* f_m(r) \tag{5.47}$$

Considering the particular fundamental solution of Eq. (5.1) for $q_i = f_m$:

$$L_{\alpha k}^{adj} \hat{w}_{mj} = f_m(r), \quad L_{3k}^{adj} \hat{w}_{mj} = f_m(r) \tag{5.48}$$

Therefore, the integral equation in this case is written as

$$c_{ij} \hat{w}_{mj} \left(\mathbf{x}'\right) + \int_{\Gamma} P_{ij}^* \left(\mathbf{x}', \mathbf{x}\right) \hat{w}_{mj} \left(\mathbf{x}\right) d\Gamma = \int_{\Gamma} W_{ij}^* \left(\mathbf{x}', \mathbf{x}\right) \hat{p}_{mj} \left(\mathbf{x}\right) d\Gamma + \int_{\Omega} W_{ij}^* (\mathbf{x}', \mathbf{x}) f_m d\Omega \tag{5.49}$$

With which we obtain

$$\int_{\Omega} W_{ij}^* (\mathbf{x}', \mathbf{x}) f_m d\Omega = c_{ij} \hat{w}_{mj} \left(\mathbf{x}'\right) + \int_{\Gamma} P_{ij}^* \left(\mathbf{x}', \mathbf{x}\right) \hat{w}_{mj} \left(\mathbf{x}\right) d\Gamma - \int_{\Gamma} W_{ij}^* \left(\mathbf{x}', \mathbf{x}\right) \hat{p}_{mj} \left(\mathbf{x}\right) d\Gamma \tag{5.50}$$

Finally, multiplying both sides by α_i^m and adding gives the transformation of the domain integral into boundary integrals:

$$\int_{\Omega} W_{ij}^{*} q_j d\Omega = \sum_{m=1}^{N+L} \alpha_i^m \int_{\Omega} W_{ij}^{*}(\mathbf{x}', \mathbf{x}) f_m d\Omega$$

$$= \sum_{m=1}^{N+L} \alpha_i^m \left[c_{ij} \hat{w}_{mj} \left(\mathbf{x}'\right) + \int_{\Gamma} P_{ij}^{*} \left(\mathbf{x}', \mathbf{x}\right) \hat{w}_{mj} \left(\mathbf{x}\right) d\Gamma \right.$$

$$\left. - \int_{\Gamma} W_{ij}^{*} \left(\mathbf{x}', \mathbf{x}\right) \hat{p}_{mj} \left(\mathbf{x}\right) d\Gamma \right] \tag{5.51}$$

Similarly, the domain integral related to the inertia term $I_0 \ddot{w}_3$, present in Eq. (5.23) can be transformed by DRBEM as (see [9]):

$$I_0 \int_{\Omega} W_{ij}^{*} \ddot{w}_k d\Omega = \sum_{m=1}^{N+L} \beta_i^m \int_{\Omega} W_{ij}^{*}(\mathbf{x}', \mathbf{x}) f_m d\Omega$$

$$= \sum_{m=1}^{N+L} \beta_i^m \left[c_{ij} \hat{w}_{mj} \left(\mathbf{x}'\right) + \int_{\Gamma} P_{ij}^{*} \left(\mathbf{x}', \mathbf{x}\right) \hat{w}_{mj} \left(\mathbf{x}\right) d\Gamma - \right.$$

$$\left. \int_{\Gamma} W_{ij}^{*} \left(\mathbf{x}', \mathbf{x}\right) \hat{p}_{mj} \left(\mathbf{x}\right) d\Gamma \right] \tag{5.52}$$

Integral related to $I_2 \ddot{w}_\alpha$ are treated in similar way. The following interpolation polynomial has been used for $\ddot{w}_k$:

$$w_k \approx \sum_{m=1}^{N+L} \beta_i^m f_m(r) \tag{5.53}$$

In the previous equations, the coefficients α_i^m and β_i^m are obtained by evaluating the accelerations w_k at $N + L$ placement points, which gives a system of equations with α_i^m and β_i^m as unknowns. Thus,

$$\mathbf{q}_j = \mathbf{Q}\boldsymbol{\alpha}, \rightarrow \boldsymbol{\alpha} = \mathbf{Q}^{-1}\mathbf{q}_j \tag{5.54}$$

$$\ddot{\mathbf{w}}_\beta = \mathbf{F}\boldsymbol{\beta}, \rightarrow \boldsymbol{\beta} = \mathbf{F}^{-1}\ddot{\mathbf{w}}_\beta \tag{5.55}$$

Therefore, the boundary integral equation for Reissner plates using the dual reciprocity method is written as

$$c_{ij}\left(\mathbf{x}'\right) w_i\left(\mathbf{x}'\right) + \sum_{j=1}^{N_e}\sum_{m=1}^{3} w_j^m \int_{-1}^{+1} P_{ij}^*\left(\mathbf{x}', \mathbf{x}\left(\xi\right)\right) \Phi^m\left(\xi\right) J_j\left(\xi\right) d\xi =$$

$$\sum_{j=1}^{N_e}\sum_{m=1}^{3} p_j^m \int_{-1}^{+1} W_{ij}^*\left(\mathbf{x}', \mathbf{x}\left(\xi\right)\right) \Phi^m\left(\xi\right) J_j\left(\xi\right) d\xi$$

$$+\sum_{m=1}^{N+L} \alpha_i^m \left[c_{ij}\hat{w}_{mj}\left(\mathbf{x}'\right) + \int_{\Gamma} P_{ij}^*\left(\mathbf{x}', \mathbf{x}\right) \hat{w}_{mj}\left(\mathbf{x}\right) d\Gamma - \int_{\Gamma} W_{ij}^*\left(\mathbf{x}', \mathbf{x}\right) \hat{p}_{mj}\left(\mathbf{x}\right) d\Gamma \right] +$$

$$\sum_{m=1}^{N+L} \beta_i^m \left[c_{ij}\hat{w}_{mj}\left(\mathbf{x}'\right) + \int_{\Gamma} P_{ij}^*\left(\mathbf{x}', \mathbf{x}\right) \hat{w}_{mj}\left(\mathbf{x}\right) d\Gamma - \int_{\Gamma} W_{ij}^*\left(\mathbf{x}', \mathbf{x}\right) \hat{p}_{mj}\left(\mathbf{x}\right) d\Gamma \right] \tag{5.56}$$

After performing the collocation process, above equations can be written as follows:

$$\mathbf{M}\ddot{\mathbf{w}} + \mathbf{H}\mathbf{w} = \mathbf{G}\mathbf{p} + \mathbf{Q} \tag{5.57}$$

The particular solutions $\hat{w}_{mj}$ and $\hat{p}_{mj}$ for rotations w_α considering the function $f(r) = r - \lambda^2 r^3/9$ are given by

For rotation w_1:

$$\begin{aligned} \hat{w}_{11} = a_0 r^3 + a_1 r(4x_1^2 + x_2^2) + a2(6x_1^2 + x_2^2) \\ \hat{w}_{12} = b_0 x_1 x_2 r + b_1 x_1 x_2 r^3 \\ \hat{w}_{13} = -c_0 x_1 r^3 \left(5\frac{\lambda^2 r^2}{7}\right) \end{aligned} \tag{5.58}$$

where $q_0 = -2/9D(1-\nu), a_1 = (1+\nu)/45D(1-\nu), a_2 = \lambda^2/1575D, b_0 = (1+\nu)/15D(1-\nu), b_1 = \lambda^2/315D, c_0 = -1/255D$.

The particular solution $\hat{p}_{mj}$ is given by

$$\hat{p}_{1\alpha} = M_{1\alpha\beta} n_\beta, \quad \hat{p}_{13} = Q_{1\beta} n_\beta \tag{5.59}$$

where

$$\begin{aligned}
\hat{M}_{111} &= D\left(\hat{w}_{11,1} + \nu\hat{w}_{12,2}\right) \\
\hat{M}_{122} &= D\left(\nu\hat{w}_{11,1} + \hat{w}_{12,2}\right) \\
\hat{M}_{112} &= \frac{D(1-\nu)}{2}\left(\hat{w}_{11,2} + \hat{w}_{12,1}\right) \\
\hat{Q}_{11} &= \frac{D(1-\nu)\lambda^2}{2}\left(\hat{w}_{11} + \hat{w}_{13,1}\right) \\
\hat{Q}_{12} &= \frac{D(1-\nu)\lambda^2}{2}\left(\hat{w}_{12} + \hat{w}_{13,2}\right)
\end{aligned} \tag{5.60}$$

and

$$\begin{aligned}
\hat{w}_{11,1} &= 3a_0 r x_1 + a_1(4x_1^2 + x_2^2)\frac{x_1}{r} + 8a_1 x_1 r + 3a_2 x_1 r(6x_1^2 + x_2^2) + 12a_2 r^3 x_1 \\
\hat{w}_{11,2} &= 3a_0 r x_2 + a_1(4x_1^2 + x_2^2)\frac{x_2}{r} + 2a_1 x_2 r + 3a_2 x_2 r(6x_1^2 + x_2^2) + 2a_2 r^3 x_2 \\
\hat{w}_{12,1} &= b_0 r x_2 + b_0 x_1^2 \frac{x_2}{r} + b_1(r^3 x_2 + 3b_1 r x_1^2 x_2) \\
\hat{w}_{12,2} &= b_0\left(r x_1 + x_2^2 \frac{x_1}{r}\right) + b_1(r^3 x_1 + 3r x_1 x_2^2) \\
\hat{w}_{13,1} &= -(5r^3 + 15r x_1^2 - \lambda^2(r^5 + 5x_1^2 r^3))\frac{c_0}{7} \\
\hat{w}_{13,2} &= -5r x_1 x_2\left(3 - \frac{\lambda^2 r^2 c_0}{7}\right)
\end{aligned} \tag{5.61}$$

For rotation w_2:

$$\begin{aligned}
\hat{w}_{21} &= b_0 r x_1 x_2 + b_1 r^3 x_1 x_2 \\
\hat{w}_{22} &= a_0 r^3 + a_1 r\left(4x_2^2 + x_1^2\right) + a_2 r^3\left(6x_2^2 + x_1^2\right) \\
\hat{w}_{23} &= -c_0 x_2 r^3\left(5 - \frac{\lambda^2 r^2}{7}\right)
\end{aligned} \tag{5.62}$$

The particular solution $\hat{p}_{mj}$ is given by

$$\hat{p}_{2\alpha} = M_{2\alpha\beta} n_\beta, \quad \hat{p}_{23} = Q_{2\beta} n_\beta \tag{5.63}$$

where

$$
\begin{aligned}
\hat{M}_{211} &= D\left(\hat{w}_{21,1} + \nu\hat{w}_{22,2}\right) \\
\hat{M}_{222} &= D\left(\nu\hat{w}_{21,1} + \hat{w}_{22,2}\right) \\
\hat{M}_{212} &= \frac{D(1-\nu)}{2}\left(\hat{w}_{21,2} + \hat{w}_{22,1}\right) \\
\hat{Q}_{21} &= \frac{D(1-\nu)\lambda^2}{2}\left(\hat{w}_{21} + \hat{w}_{23,1}\right) \\
\hat{Q}_{22} &= \frac{D(1-\nu)\lambda^2}{2}\left(\hat{w}_{22} + \hat{w}_{23,2}\right)
\end{aligned}
\tag{5.64}
$$

And,

$$
\begin{aligned}
\hat{w}_{21,1} &= b_0 r x_2 + b_0\frac{x_1x_2}{r} + b_1x_2r^3 + 3b_1x_1^2x_2r \\
\hat{w}_{21,2} &= b_0 r x_1 + b_0\frac{x_1x_2^2}{r} + b_1x_1r^3 + 3b_1x_1x_2^2r \\
\hat{w}_{22,1} &= 3a_0rx_1 + a_1\frac{x_1}{r}(4x_2^2 + x_1^2) + 2a_1rx_1 + 3a_2rx_1(6x_2^2 + x^2) + 2a_2r^3x_1 \\
\hat{w}_{22,2} &= 3a_0rx_2 + a_1\frac{x_2}{r}(4x_2^2 + x_1^2) + 8a_1rx_2 + 3a_2rx_2(6x_2^2 + x^2) + 12a_2r^3x_2 \\
\hat{w}_{23,1} &= -15c_0rx_1x_2 + c_0 5/7x_2^2\lambda^2r^3 \\
\hat{w}_{23,2} &= -c_0r^3(5 - \lambda^2r^2/7) - 15c_0x_2^2r + 5/7c_0x_2^2\lambda^2r^3
\end{aligned}
\tag{5.65}
$$

and $f(r)c + r$, for deflection, w_3, are given by

$$
\begin{aligned}
\hat{w}_{31} &= -\left(\frac{cr^2}{16} + \frac{r^3}{45}\right)\frac{x_1}{D} \\
\hat{w}_{32} &= -\left(\frac{cr^2}{16} + \frac{r^3}{45}\right)\frac{x_2}{D} \\
\hat{w}_{33} &= -\left(\frac{cr^2}{2} + \frac{2r^3}{9}\right)\frac{1}{(1-\nu)\lambda^2D} + \left(\frac{cr^4}{64} + \frac{r^5}{225}\right)\frac{1}{D}
\end{aligned}
\tag{5.66}
$$

The particular solution $\hat{p}_{mj}$ is given by

$$
\hat{p}_{3\alpha} = M_{3\alpha\beta}n_\beta, \quad \hat{p}_{33} = Q_{3\beta}n_\beta \tag{5.67}
$$

where

$$
\begin{aligned}
\hat{M}_{311} &= D\left(\hat{w}_{31,1} + \nu\hat{w}_{32,2}\right)\\
\hat{M}_{322} &= D\left(\nu\hat{w}_{31,1} + \hat{w}_{32,2}\right)\\
\hat{M}_{312} &= \frac{D(1-\nu)}{2}\left(\hat{w}_{31,2} + \hat{w}_{32,1}\right)\\
\hat{Q}_{31} &= \frac{D(1-\nu)\lambda^2}{2}\left(\hat{w}_{31} + \hat{w}_{33,1}\right)\\
\hat{Q}_{32} &= \frac{D(1-\nu)\lambda^2}{2}\left(\hat{w}_{32} + \hat{w}_{33,2}\right)
\end{aligned}
\tag{5.68}
$$

And,

$$
\begin{aligned}
\hat{M}_{311} &= -\left(\frac{1}{8} + \frac{r}{15}\right)\left(x_1^2 + \nu x_2^2\right) - (1+\nu)\left(\frac{r^2}{16} + \frac{r^3}{45}\right)\\
\hat{M}_{322} &= -\left(\frac{1}{8} + \frac{r}{15}\right)\left(x_2^2 + \nu x_1^2\right) - (1+\nu)\left(\frac{r^2}{16} + \frac{r^3}{45}\right)\\
\hat{M}_{312} &= -(1-\nu)\left(\frac{1}{8} + \frac{r}{15}\right)x_1x_2\\
\hat{Q}_{31} &= -\left(1 + \frac{2r}{3}\right)\frac{x_1}{2}\\
\hat{Q}_{32} &= -\left(1 + \frac{2r}{3}\right)\frac{x_2}{2}
\end{aligned}
\tag{5.69}
$$

Other integration techniques in the boundary element method for plane elasticity have been explored by [4, 5], which are based on the use of radial basis functions. This method has proven to be a viable alternative to DRBEM as it does not require fundamental solutions. However, its use in thick plate and shell problems has not yet been explored.

5.7.3 *Transient Analysis*

To obtain the time response using Eq. (5.57), the Houbolt integration scheme is used in this work. In this method, the acceleration is approximate in the form (see [9]):

$$
\ddot{\mathbf{d}}^{\tau+\Delta\tau} = \frac{1}{\Delta\tau^2}(2\mathbf{d}^{\tau+\Delta\tau} - 5\mathbf{d}^{\tau} + 4\mathbf{d}^{\tau-\Delta\tau} - \mathbf{d}^{\tau-2\Delta\tau}) \tag{5.70}
$$

where $\Delta\tau$ is the time-step. Now, Eq. (5.57) can be written at time $\tau + \Delta\tau$ as

$$
\mathbf{M}\ddot{\mathbf{d}}_{\tau+\Delta\tau} + \mathbf{H}\mathbf{d}_{\tau+\Delta\tau} = \mathbf{G}\mathbf{f}_{\tau+\Delta\tau} + \mathbf{b}_{\tau+\Delta\tau} \tag{5.71}
$$

Replacing Eq. (5.70) into the last equation finally we have

$$
\begin{aligned}
(2\mathbf{M} + \Delta\tau^2\mathbf{H})\mathbf{d}^{\tau+\Delta\tau} - \frac{1}{\Delta\tau^2}(\mathbf{G}\mathbf{f}^{\tau+\Delta\tau} + \mathbf{b}^{\tau+\Delta\tau}) = \\
\mathbf{M}(5\mathbf{d}^{\tau} - 4\mathbf{d}^{\tau-\Delta\tau} - \mathbf{d}^{\tau-2\Delta\tau}).
\end{aligned}
\tag{5.72}
$$

5.7.4 Modal and Harmonic Analysis

In order to obtain the harmonic response using Eq. (5.57), the system was initially partitioned according to the type of applied boundary condition, and statically condensed in such way that the final equation system could be solved for unknown displacements only. Partitioning the boundary $\Gamma = \Gamma_1 \bigcup \Gamma_2$, and considering $\mathbf{u} = \mathbf{w} = 0$ and $\ddot{\mathbf{u}} = \ddot{\mathbf{w}} = 0$ in Γ_1 and $\mathbf{p} = 0$ in Γ_2, Eq. (5.57) can be written as

$$
\hat{\mathbf{M}}\ddot{\mathbf{d}}_2 + \hat{\mathbf{H}}\mathbf{d}_2 = \mathbf{b} \tag{5.73}
$$

where,

$$
\begin{aligned}
\hat{\mathbf{M}} &= \mathbf{M}_{22} - \mathbf{G}_{12}\mathbf{G}_{11}^{-1}\mathbf{M}_{12} \\
\hat{\mathbf{H}} &= \mathbf{H}_{22} - \mathbf{G}_{12}\mathbf{G}_{11}^{-1}\mathbf{H}_{12}
\end{aligned}
\tag{5.74}
$$

Assuming that displacements and distributed forces are harmonic functions of time, τ, they can be expanded in the form:

$$
\mathbf{d}(\tau) = \mathbf{\Phi} e^{-i\omega\tau}, \; \ddot{\mathbf{d}}(\tau) = -\omega^2 \mathbf{\Phi} e^{-i\omega\tau}, \; \mathbf{b}(\tau) = \mathbf{B} e^{-i\omega\tau} \tag{5.75}
$$

with ω being frequency of excitation, $\mathbf{\Phi}$ and $\mathbf{B}$ are the amplitude of displacements and distributed forces, respectively. Replacing these expressions into Eq. (5.73) we obtain:

$$
(\hat{\mathbf{H}} - \omega^2\hat{\mathbf{M}})\mathbf{\Phi} = \mathbf{F} \tag{5.76}
$$

This equation is solved by incrementing ω in small steps from an initial value. At each step, the above equation is solved and the plate displacements and rotations for each frequency are obtained.

On the other hand, for modal analysis, the above equation is rewritten as

$$
(\hat{\mathbf{H}} - \omega^2\hat{\mathbf{M}})\mathbf{\Phi} = \mathbf{0} \tag{5.77}
$$

This generalized eigenvalue problem can be reduced to a standard one by inversion of matrix $\hat{\mathbf{H}}$, obtaining the system:

$$
\mathbf{A}\mathbf{\Phi} = \bar{\omega}\mathbf{\Phi} \tag{5.78}
$$

where $\mathbf{A} = \hat{\mathbf{H}}^{-1}\hat{\mathbf{M}}$ and $\bar{\omega} = 1/\omega^2$.

5.8 Numerical Examples

Simply Supported Thin Square Plate

A clamped thick square plate of 4 m side simply supported on all sides is considered (see Fig. 5.5). The BEM analysis is performed using the Reissner plate theory. The Young modulus was 0.1 MPa and the Poisson's ratio is 0.33. Thickness of the plate 1 m. A uniform pressure of 102 Pa is applied over the plate domain. Boundary plate was meshing using discontinuous quadratic boundary element. Was found that 16 element boundary mesh generate an error of 0.02% in the internal forces when compared with analytic solution using the Kirchhoff plate theory. Figure 5.5 shows the transversal displacement component w_3 distribution. Table 5.1 presents the internal bending moments and shear forces at points A, B and C. Good coorelations when comparing with analytic solution is obtained.

Circular Plate Under Transient Pressure Load

In this example, the dynamic response of a clamped circular plate under transient pressure load $P(t)$ given by

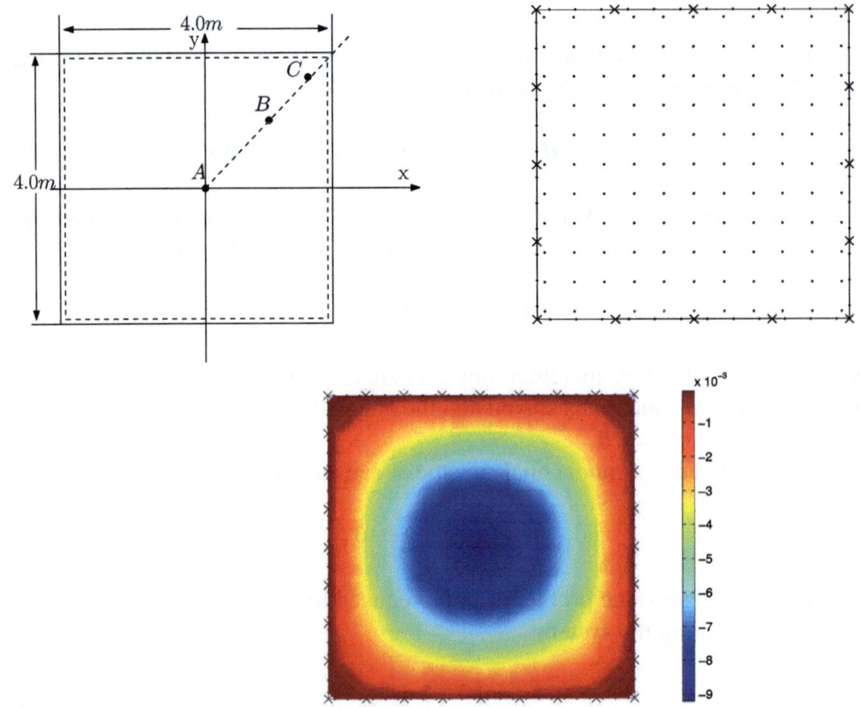

Fig. 5.5 Simply supported thin square plate. Right: Boundary element mesh for simply supported thin plate. Above: Transversal displacements for simply supported thin plate

$$P(t) = \begin{cases} 40t & , \quad 0 \le t \le 0.025 \\ -40(t - 0.075) - 1 & , \quad 0.025 < t \le 0.075 \\ 40(t - 0.1) & , \quad 0.075 < t \le 0.1 \end{cases} \tag{5.79}$$

The plate have a relation of $h/r = 0.01$ (where r is the radius of the plate and h is the thickness), elastic constants $E = 210 \times 10^3$ MPa, $\nu = 0.33$, and mass density $\rho = 7853$ Kg/m^3. For this problem, the boundary conditions are $w_3 = 0$ and $w_n = w_s = 0$ for normal and tangential rotations. A BEM model considering 32 boundary elements and 421, 571, and 691 DRM, as presented in Fig. 5.6, was used for the analysis. Deflection at the center of the plate was used as a reference parameter.

A convergence analysis for the central deflection versus time step was carried out using a BEM model with 145 DRM internal collocation points. A convergence solution for the central deflection of the plate is obtained for a time step of 1.5×10^{-3} s. Figure 5.6 shows the history of central deflection for time interval from 0 s to 2.0 s and time step $\Delta t = 1.5 \times 10^{-3}$ s for different BEM meshes, together with a finite element solution. The finite element solution was obtained using the Finite Element

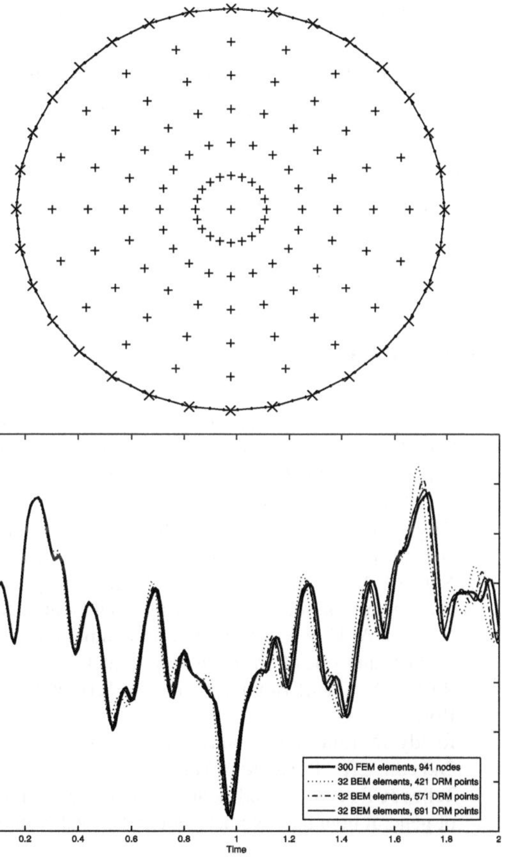

Fig. 5.6 Boundary element model for circular plate. Below: Time history of central transversal displacement for circular plate under dynamic pressure load as function of DRM collocation points

Table 5.1 Bending moments and shear forces at internal points for thin square plate

Point	$M_{11}(tf.m/m)$		$M_{12}(tf.m/m)$		$M_{22}(tf.m/m)$		$Q_1(tf/m)$		$Q_2(tf/m)$	
	BEM	Exact	BEM	Exact	BEM	Exact	BEM	Exact	BEM	Exact
Point A	−0.4904	−0.4905	0.0000	0.0000	−0.4905	−0.4904	0.0000	0.0000	0.0000	0.0000
Point B	−0.3766	0.3770	0.0814	0.0815	−0.3766	−0.3770	0.2175	0.2177	0.2175	0.2177
Point C	−0.1143	−0.1142	0.2521	0.2522	−0.1142	−0.1143	0.2682	0.2683	0.2682	0.2683

Method. Again, DRM provides results with very good agreement when compared with the finite element solution.

5.9 Summary

In this chapter, the Boundary Element Method has been presented and applied to analyze isotropic deformable plates in shear. Three boundary integral formulations have been established involving rotations, deflections, and generalized forces (moments and shear). In addition, boundary integral equations for internal resultant troops have also been established. Fundamental solutions for displacements and tractions, given by [10] are presented. The implementation of the cell method and the DRBEM for the treatment of domain integrals are given. The MATLAB(R) implementation of the DRBEM is shown. The Dual Reciprocity Method treats the domain integral related to inertial terms, considering linear and angular acceleration. Numerical examples are presented, and good correlations of results are obtained.

References

1. Baiz PM, Aliabadi MH (2007) Buckling analysis of shear deformable shallow shells by the boundary element method. Eng Anal Bound Elem 31(4):361–372
2. Dirgantara T (2002) Boundary element analysis of cracks in shear deformable plates and shells. WIT Press, Southampton
3. Dirgantara T, Aliabadi MH (2001) Dual boundary element formulation for fracture mechanics analysis of shear deformable shells. Int J Solids Struct 38(44–45):7769–7800
4. Narvaez A, Useche J (2020) The kriging integration method applied to the boundary element analysis of Poisson problems. Eng Anal Bound Elem 121:1–20
5. Narvaez A, Useche J (2022) A new radial basis integration method applied to the boundary element analysis of 2d scalar wave equations. Eng Anal Bound Elem 136:77–92
6. Rashed Y (2000) Boundary element formulation for thick plates. WIT Press, Southampton, Boston
7. Reddy JN (2004) Mechanics of laminated composite plates and shells: theory and analysis, 2nd edn. CRC Press, New York
8. Telles JCF (1987) A self-adaptive coordinate transformation for efficient numerical evaluation of general boundary element integrals. Int J Numer Methods Eng 24:959–973

9. Useche J, Albuquerque EL (2012) Dynamic analysis of shear deformable plates using the dual reciprocity method. Eng Anal Bound Elem 36:627–632
10. Vander Weeën F (1982) Application of the direct boundary integral equation method to Reissner's plate model. Int J Numer Methods Eng 67:1–10
11. Wen PH, Aliabadi MH (2000) Application of dual reciprocity method to plates and shells. Eng Anal Bound Elem 24:583–590
12. Wen PH, Aliabadi MH (2006) Boundary element frequency domain formulation for dynamic analysis of Mindlin plates. Int J Numer Methods Eng 67(11):1617–1640
13. Wen PH, Aliabadi MH, Rooke DP (1998) A new method for transformation of domain integrals to the boundary integrals in boundary element method. Commun Numer Methods Eng 14(11):1055–1065
14. Wen PH, Aliabadi MH, Young A (1999) Transformation of domain integrals to boundary integrals in Bem analysis of shear deformable plate bending problems. Comput Mech 24:304–309
15. Wen PH, Aliabadi MH, Young A (2000) Application of dual reciprocity method to plates and shells. Eng Anal Bound Elem 24(7–8):583–590
16. Wen PH, Aliabadi MH, Young A (2000) The boundary element method for dynamic plate bending problems. Int J Solid Struct 37(37):5177–5188
17. Wen PH, Aliabadi MH, Young A (2000) Boundary element method for dynamic plate bending problems. Int J Solid Struct 37(37):5177–5188
18. Wen PH, Adetoro M, Xu Y (2008) The fundamental solution of Mindlin plates with damping in the Laplace domain and its applications. Eng Anal Bound Elem 32(10):870–882
19. Wen PH, Aliabadi MH, Young A (2005) Large deflection analysis of Reissner plate by boundary element method. Comput Struct 83:870–879

Chapter 6
Boundary Element Analysis of Thick Isotropic Shells

Abstract This chapter presents the application of the Boundary Element Method to the solution of shell governing equations given by first-order shear theory for isotropic materials. The BEM formulation employed is based on the coupling of the Reissner plate boundary integral formulation and the plane elasticity formulation, which are coupled by domain integrals related to curvatures and displacements. The formulation uses fundamental thick-plate and plane stress solutions, allowing for a more straightforward application of the boundary element method. The Dual Reciprocity Boundary Element is used to transform domain-to-boundary integrals.

6.1 Introduction

The first application of the boundary element method to shell analysis dates back to the work of [6, 10, 11]. However, these works were limited to thin plates without shear deformation. It was only until the work of [5] where a direct BEM formulation for shallow shells involving shear deformation based on fundamental shell solutions is presented. However, these formulations involve specialized and complex mathematical scaffolding. A less complex BEM formulation for static and dynamic plate and shella analysis is presented in the works of [7, 20]. These formulations are based on the coupling of the Reissner plate boundary integral formulation and the plane elasticity formulation, which are coupled by domain integrals related to curvatures and displacements. The formulation uses fundamental thick-plate and plane stress solutions, allowing for a more straightforward application of the boundary element method. This approach has been used in the static analysis of shallow shells with quadratic mid surface by [3, 15–18]. Useche and Albuquerque [12], Useche and Harnish [13], Useche and Medina [14] follow this methodology and apply the Dual Reciprocity Boundary Element to the dynamic analysis of thick shallow shells.

This chapter presents the application of the Boundary Element Method to the solution of shell governing equations given by first-order shear theory for isotropic materials. The BEM formulation employed is based on the coupling of the Reissner plate boundary integral formulation and the plane elasticity formulation, which are

J. Useche Vivero, *The Boundary Element Method for Engineering Analysis of Plates and Shells*, Solid Mechanics and Its Applications 281,
https://doi.org/10.1007/978-3-031-93050-8_6

coupled by domain integrals related to curvatures and displacements. Fundamental solutions for the Reissner plate and isotropic plane stress are employed. The DRBEM is used to transform domain-to-boundary integrals.

6.2 Basic Hypothesis

Consider an isotropic shell of uniform thickness h and mass density ρ, occupying the area Ω, in the x_1x_2 plane, bounded by the boundary $\Gamma = \Gamma_w \bigcup \Gamma_q$ with $\Gamma = \Gamma_w \bigcap \Gamma_q \equiv 0$, as presented in Fig. 6.1. The shell follows the assumptions given by the isotropic thick shell theory presented in Chap. 3.

The dynamic flexural response for the surface plate was modeled by coupling classical Reissner plate theory and two-dimensional plane stress elasticity as presented in [17]. In this book, shells with midsurfaces defined by the function are considered:

$$x_3 = -\frac{1}{2}\left(\kappa_{11}x_1^2 + \kappa_{22}x_2^2\right) \tag{6.1}$$

where $\kappa_{11} = 1/R_1$ and $\kappa_{22} = 1/R_2$ are the principal curvatures of the surface in the directions of the x_1 and x_2 axes, respectively.

The equilibrium equations in terms of the displacement components, as presented in Chap. 3, are given by [8, 12]:

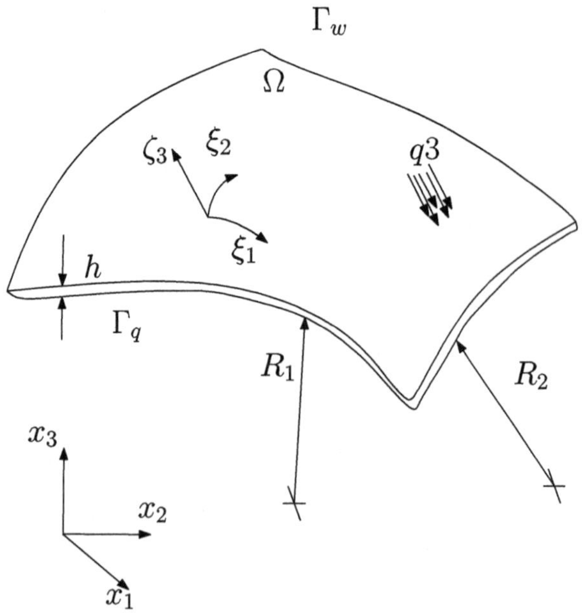

Fig. 6.1 Shallow shell geometry

$$B\nabla^2 u_1 + \frac{B}{2}(1+\nu)\frac{\partial}{\partial x_2}\left(-\frac{\partial u_1}{\partial x_2} + \frac{\partial u_2}{\partial x_1}\right) + q_1^* = I_0\frac{\partial^2 u_1}{\partial t^2} + I_1\frac{\partial^2 w_1}{\partial t^2} \tag{6.2}$$

$$\frac{B}{2}(1+\nu)\frac{\partial}{\partial x_1}\left(\frac{\partial u_1}{\partial x_2} - \frac{\partial u_2}{\partial x_1}\right) + B\nabla^2 u_2 + q_2^* = I_0\frac{\partial^2 u_1}{\partial t^2} + I_1\frac{\partial^2 w_2}{\partial t^2} \tag{6.3}$$

$$D\nabla^2 w_1 + \frac{D}{2}(1+\nu)\frac{\partial}{\partial x_2}\left(-\frac{\partial w_1}{\partial x_2} + \frac{\partial w_2}{\partial x_1}\right) - Cw_1 - C\frac{\partial w_3}{\partial x_1} = I_2\frac{\partial^2 w_1}{\partial t^2} + I_1\frac{\partial^2 u_1}{\partial t^2}$$

$$\frac{D}{2}(1+\nu)\frac{\partial}{\partial x_1}\left(\frac{\partial w_1}{\partial x_2} - \frac{\partial w_2}{\partial x_1}\right) + D\nabla^2 w_2 - Cw_2 - C\frac{\partial w_3}{\partial x_2} = I_2\frac{\partial^2 w_2}{\partial t^2} + I_1\frac{\partial^2 u_2}{\partial t^2}$$

$$C\nabla^2 w_3 + C\frac{\partial w_1}{\partial x_1} + C\frac{\partial w_2}{\partial x_2} + q_3^* = I_0\frac{\partial^2 w_3}{\partial t^2} \tag{6.4}$$

where

$$q_1^* = q_1 + B(\kappa_{11} + \nu\kappa_{22})\frac{\partial w_3}{\partial x_1}, \quad q_2^* = q_2 + B(\nu\kappa_{11} + \kappa_{22})\frac{\partial w_3}{\partial x_2} \tag{6.5}$$

$$q_3^* = q_3 - B(\kappa_{11} + \nu\kappa_{22})\frac{\partial u_1}{\partial x_1} - B(\kappa_{11} + \nu\kappa_{22})\frac{\partial u_2}{\partial x_2} - B(\kappa_{11}^2 + \kappa_{22}^2 + 2\nu\kappa_{11}\kappa_{22})w_3 \tag{6.6}$$

The first two equations in (6.4) correspond to the plane stress equations, and the remaining ones represent the plate bending equations. As can be seen, the curvature terms are contained in the terms q_1^*, q_2^*, and q_3^*, which correspond to body forces. The essential and natural boundary conditions are given by

$$\begin{aligned} u_n = u_n^*, \quad u_s = u_s^*, \quad w_3 = w_3^*, \quad w_n = w_n^*, \quad w_s = w_s^*, \quad \text{in } \Gamma_w \\ N_{nn} = N_{nn}^*, \quad N_{ns} = N_{ns}^*, \quad Q_n = Q_n^*, \quad M_{nn} = M_{nn}^*, \quad M_{ns} = M_{ns}^*, \quad \text{in } \Gamma_t \end{aligned} \tag{6.7}$$

and the initial conditions at $t = 0$:

$$u_n = u_n^0, \quad u_s = u_s^0, \quad w_3 = w_3^0, \quad w_n = w_n^0, \quad w_s = w_s^0, \tag{6.8}$$

$$\dot{u}_n = u_n^0, \quad \dot{u}_s = u_s^0, \quad \dot{w}_3 = w_3^0, \quad \dot{w}_n = w_n^0, \quad \dot{w}_s = w_s^0 \text{ in } \Omega. \tag{6.9}$$

6.2.1 Bending Integral Equations

The integral equation can be derived by considering the integral representation of the governing equations (6.4) through the following identities [2]:

$$\int_{\Omega} \left[\left(M_{\alpha\beta,\beta} - Q_{\alpha}\right) W_{\alpha}^{*} + \left(Q_{\alpha,\alpha} - \kappa_{\alpha\beta} N_{\alpha\beta} + q\right) W_{3}^{*}\right] d\Omega + \int_{\Omega} W_{j}^{*} \Lambda_{jk}^{b} \ddot{w}_{k} d\Omega + \int_{\Omega} W_{\alpha}^{*} \Lambda_{\alpha\beta}^{bm} \ddot{u}_{\beta} d\Omega = 0 \quad (6.10)$$

and

$$\int_{\Omega} \left(N_{\alpha\beta,\beta} + q_{\alpha}\right) U_{\alpha}^{*} d\Omega + \int_{\Omega} U_{\theta}^{*} \Lambda_{\theta\alpha}^{m} \ddot{u}_{\alpha} d\Omega + \int_{\Omega} U_{\theta}^{*} \Lambda_{\theta\alpha}^{bm} \ddot{w}_{\alpha} d\Omega = 0$$

where U^{*} and W^{*} are the weighting functions and Ω is the projected domain of the shell onto the $x_1 - x_2$ plane with boundary Γ. Tensors Λ_{ij}^{b}, Λ_{ij}^{bm} and Λ_{ij}^{m} are defined as: $\Lambda_{\alpha\beta}^{b} = I_2 \delta_{\alpha\beta}$ and $\Lambda_{33}^{b} = I_0$; $\Lambda_{\alpha\beta}^{bm} = I_1 \delta_{\alpha\beta}$; $\Lambda_{\alpha\beta}^{m} = I_0 \delta_{\alpha\beta}$; $\delta_{\alpha\beta}$ is the Kronecker's delta and I_i is the mass inertias.

Applying Green's identity, Eq. (6.10) can be expressed as

$$\int_{\Omega} M_{\alpha\beta} n_{\beta} W_{\alpha}^{*} d\Gamma - \int_{\Omega} M_{\alpha\beta} W_{\alpha,\beta}^{*} d\Omega - \int_{\Omega} Q_{\alpha} W_{\alpha}^{*} d\Omega + \int_{\Omega} Q_{\alpha} n_{\alpha} W_{3}^{*} d\Gamma - \int_{\Omega} Q_{\alpha} n_{\alpha} W_{3,\alpha}^{*} d\Omega - \int_{\Omega} \kappa_{\alpha\beta} N_{\alpha\beta} W_{3}^{*} d\Omega + \int_{\Omega} q_{3} W_{3}^{*} d\Omega + \int_{\Omega} W_{j}^{*} \Lambda_{jk}^{b} \ddot{w}_{k} d\Omega + \int_{\Omega} W_{\alpha}^{*} \Lambda_{\alpha\beta}^{bm} \ddot{u}_{\beta} d\Omega = 0 \quad (6.11)$$

Substituting the relations for p_j above equation can be rewritten as

$$\int_{\Omega} p_{j} W_{j}^{*} d\Gamma - \int_{\Omega} M_{\alpha\beta} W_{\alpha,\beta}^{*} d\Omega - \int_{\Omega} Q_{\alpha} \left(W_{\alpha} + W_{3,\alpha}^{*}\right) d\Omega - \int_{\Omega} \kappa_{\alpha\beta} N_{\alpha\beta} W_{3}^{*} d\Omega + \int_{\Omega} q_{3} W_{3}^{*} d\Omega + \int_{\Omega} W_{j}^{*} \Lambda_{jk}^{b} \ddot{w}_{k} d\Omega + \int_{\Omega} W_{\alpha}^{*} \Lambda_{\alpha\beta}^{bm} \ddot{u}_{\beta} d\Omega = 0 \quad (6.12)$$

where $j = 1, 2, 3$. Replacing the moment and shear force resultants $M_{\alpha\beta}$ and Q_{α} in terms of the generalized displacements and their derivatives, the above equation becomes

$$\int_{\Omega} p_j W_j^* d\Gamma - \int_{\Gamma} W_{\alpha,\beta}^* \left\{ \frac{D(1-\nu)}{2} \left(w_{\alpha,\beta} + w_{\beta,\alpha} + \frac{2v}{(1-\nu)} \delta_{\alpha\beta} w_{\gamma,\gamma} \right) \right\} d\Gamma$$
$$- \int_{\Omega} C(w_\alpha + w_{3,\alpha})(W_\alpha^* + W_{3,\alpha}^*) d\Omega - \int_{\Omega} \kappa_{\alpha\beta} N_{\alpha\beta} W_3^* d\Omega + \int_{\Omega} q_3 W_3^* d\Omega +$$
$$\int_{\Omega} W_j^* \Lambda_{jk}^b \ddot{w}_k d\Omega + \int_{\Omega} W_\alpha^* \Lambda_{\alpha\beta}^{bm} \ddot{u}_\beta d\Omega = 0 \tag{6.13}$$

Applying Green's identity to $M_{\alpha\beta}$ and rearranging terms, we obtain

$$\int_{\Omega} p_j W_j^* d\Gamma - \int_{\Gamma} W_{\alpha,\beta}^* \left\{ \frac{D(1-\nu)}{2} \left(w_{\alpha,\beta} + w_{\beta,\alpha} + \frac{2v}{(1-\nu)} \delta_{\alpha\beta} w_{\gamma,\gamma} \right) \right\} d\Gamma$$
$$+ \int_{\Omega} \frac{D(1-\nu)}{2} \left(w_\alpha W_{\alpha,\beta\beta}^* + w_\beta W_{\alpha,\beta\alpha}^* + \frac{2v}{(1-\nu)} w_\gamma W_{\alpha,\beta\gamma}^* \delta_{\alpha\beta} \right) d\Omega$$
$$- \int_{\Omega} Q_\alpha (w_\alpha + w_{3,\alpha}) d\Omega - \int_{\Omega} \kappa_{\alpha\beta} N_{\alpha\beta} W_3^* d\Omega + \int_{\Omega} q_3 W_3^* d\Omega$$
$$+ \int_{\Omega} W_j^* \Lambda_{jk}^b \ddot{w}_k d\Omega + \int_{\Omega} W_\alpha^* \Lambda_{\alpha\beta}^{bm} \ddot{u}_\beta d\Omega = 0 \tag{6.14}$$

Using Green's identity and manipulating terms, this equation can finally be rewritten as

$$\int_{\Gamma} \left(W_j^* p_j - P_j^* w_j \right) d\Gamma - \int_{\Gamma} \kappa_{\alpha\beta} B \frac{1-\nu}{2} \left(u_\alpha n_\beta + u_\beta n_\alpha + \frac{2\nu}{1-\nu} u_\gamma n_\gamma \delta_{\alpha\beta} \right) W_3^* d\Gamma$$
$$+ \int_{\Omega} \kappa_{\alpha\beta} B \frac{1-\nu}{2} \left(u_\alpha W_{3,\beta}^* + u_\beta W_{3,\alpha}^* + \frac{2\nu}{1-\nu} u_\gamma u_\gamma W_{3,\gamma}^* \delta_{\alpha\beta} \right) d\Omega$$
$$- \int_{\Omega} \kappa_{\alpha\beta} B[(1-\nu)\kappa_{\alpha\beta} + \nu \delta_{\alpha\beta} \kappa_{\gamma\gamma}] w_3 W_3^* d\Omega$$
$$+ \int_{\Omega} [(M_{\alpha\beta,\beta} - Q_\alpha^*) w_\alpha + Q_{\alpha,\alpha}^* w_3] d\Omega +$$
$$\int_{\Omega} W_3^* q_3 d\Omega + \int_{\Omega} W_j^* \Lambda_{jk}^b \ddot{w}_k d\Omega + \int_{\Omega} W_\alpha^* \Lambda_{\alpha\beta}^{bm} \ddot{u}_\beta d\Omega = 0 \tag{6.15}$$

It should be noted that the functions W_j^* and P_j^* can be chosen arbitrarily. If the state is defined for generalized concentrated loads: two bending moments ($i = \alpha = 1, 2$) and a concentrated shear force ($i = 3$) at an arbitrary source point $\mathbf{X} \in \Omega$ as was done in the thick plate case, then the above equation can be rewritten after introducing the load direction i as follows:

$$\begin{aligned}
&\int_\Gamma W_{ij}^*(\mathbf{X}',\mathbf{x})\,p_j(\mathbf{x})\,d\Gamma - \int_\Gamma P_{ij}^*(\mathbf{X}',\mathbf{x})\,w_j(\mathbf{x})\,d\Gamma + \int_\Omega W_{i3}^*(\mathbf{X}',\mathbf{X})\,q_3 d\Omega \\
&- \int_\Gamma \kappa_{\alpha\beta} B\left(\frac{\nu}{1-\nu}\right)\left[u_\alpha(\mathbf{x})n_\beta + u_\beta(\mathbf{x})n_\alpha + \frac{\nu}{1-\nu}u_\gamma(\mathbf{X})n_\gamma\delta_{\alpha\beta}\right] W_{i3}^*(\mathbf{X}',\mathbf{x})\,d\Gamma \\
&- \int_\Omega \kappa_{\alpha\beta} B\left(\frac{\nu}{1-\nu}\right)\Big[u_\alpha(\mathbf{X})\,W_{i3,\beta}^*(\mathbf{X}',\mathbf{X}) + u_\beta(\mathbf{X})\,W_{i3,\alpha}^*(\mathbf{X}',\mathbf{X}) + \\
&\qquad \frac{\nu}{1-\nu}u_\gamma(\mathbf{X})n_\gamma W_{i3,\gamma}^*(\mathbf{X}',\mathbf{x})\,\delta_{\alpha\beta}\Big]\,d\Omega \\
&- \int_\Gamma \kappa_{\alpha\beta} B\left[(1-\nu)\kappa_{\alpha\beta} + \nu\delta_{\alpha\beta}\kappa_{\gamma\gamma} w_3(\mathbf{X})\,W_{i3}^*(\mathbf{X}',\mathbf{X})\right] d\Omega \\
&+ \int_\Omega \left[M_{i\alpha\beta,\beta}^*(\mathbf{X}',\mathbf{X}) - Q_{i\alpha}^*(\mathbf{X}',\mathbf{X})\right] w_\alpha(\mathbf{X})\,d\Omega + \int_\Omega Q_{i\alpha,\alpha}^*(\mathbf{X}',\mathbf{X})\,w_3(\mathbf{X})\,d\Omega \\
&+ \int_\Omega W_j^* \Lambda_{jk}^b \ddot{w}_k d\Omega + \int_\Omega W_\alpha^* \Lambda_{\alpha\beta}^{bm} \ddot{u}_\beta d\Omega = 0
\end{aligned} \tag{6.16}$$

where $\mathbf{x}'$ represents the placement point on the boundary. Again, choosing the weighting function as the fundamental solution to the governing equations and making use of the Dirac delta property, then Eq. (6.16) can be written for any point in the domain or on the boundary as

$$\begin{aligned}
&c_{ij}(\mathbf{x}')\,w_j(\mathbf{x}') + \fint_\Gamma P_{ij}^*(\mathbf{x}',\mathbf{x})\,w_j(\mathbf{x})\,d\Gamma = \int_\Gamma W_{ij}^*(\mathbf{x}',\mathbf{x})\,p_j(\mathbf{x})\,d\Gamma \\
&- \int_\Gamma \kappa_{\alpha\beta} B\left(\frac{1-\nu}{2}\right)\left[u_\alpha(\mathbf{x})n_\beta + u_\beta(\mathbf{x})n_\alpha + \frac{2\nu}{1-\nu}u_\gamma(\mathbf{x})n_\gamma\delta_{\alpha\beta}\right] W_{i3}^*(\mathbf{x}',\mathbf{x})\,d\Gamma \\
&+ \int_\Omega \kappa_{\alpha\beta} B\left(\frac{1-\nu}{2}\right)\Big[u_\alpha(\mathbf{X})\,W_{i3,\beta}^*(\mathbf{x}',\mathbf{X}) + u_\beta(\mathbf{X})\,W_{i3,\alpha}^*(\mathbf{x}',\mathbf{X}) + \\
&\qquad \frac{2\nu}{1-\nu}u_\gamma(\mathbf{X})n_\gamma W_{i3,\gamma}^*(\mathbf{x}',\mathbf{X})\,\delta_{\alpha\beta}\Big]\,d\Omega \\
&- \int_\Gamma \kappa_{\alpha\beta} B[(1-\nu)\kappa_{\alpha\beta} + \nu\delta_{\alpha\beta}\kappa_{\gamma\gamma}]w_3(\mathbf{X})\,W_{i3}^*(\mathbf{x}',\mathbf{X})\,d\Omega
\end{aligned}$$

$$
+\int_{\Omega} W_{i3}^{*}\left(\mathbf{x}', \mathbf{X}\right) q_3\left(\mathbf{X}\right) d\Omega +
$$

$$
+\int_{\Omega} W_{ij}(\mathbf{x}', \mathbf{x}) \Lambda_{ij}^{b} \ddot{w}_j(\mathbf{X}) d\Omega + \int_{\Omega} W_{i\alpha}(\mathbf{x}', \mathbf{X}) \Lambda_{i\alpha}^{bm} \ddot{u}_\alpha(\mathbf{X}) d\Omega \quad (6.17)
$$

where $⨍$ represents the Cauchy principal integral, it represents the boundary integral for the bending equations. The fundamental solutions W_{ij}^{*} and P_{ij}^{*} are the fundamental thick-plate solutions introduced in the previous chapter. Tensors Λ_{ij}^{b}, Λ_{ij}^{bm} and Λ_{ij}^{m} are defined as (see [12, 13]): $\Lambda_{\alpha\beta}^{b} = I_2\delta_{\alpha\beta}$ and $\Lambda_{33}^{b} = I_0$; $\Lambda_{\alpha\beta}^{bm} = I_1\delta_{\alpha\beta}$. In this formulation, the terms related to the shell curvature appear in domain integrals and can be treated as *equivalent body system*.

Similarly, for in-plane elasticity equations, applying Green's theorem, it can be written as [1]:

$$
\int_{\Gamma} N_{\alpha\beta}^{i} n_\beta U_\alpha^{*} d\Gamma - \int_{\Omega} N_{\alpha\beta}^{i} U_{\alpha,\beta}^{*} d\Omega + \int_{\Omega} N_{\alpha\beta,\beta}^{ii} U_\alpha^{*} d\Omega +
$$

$$
\int_{\Omega} q_\alpha U_\alpha^{*} d\Omega + \int_{\Omega} U_\theta^{*} \Lambda_{\theta\alpha}^{m} \ddot{u}_\alpha d\Omega + \int_{\Omega} U_\theta^{*} \Lambda_{\theta\alpha}^{bm} \ddot{w}_\alpha d\Omega = 0 \quad (6.18)
$$

where

$$
N_{\alpha\beta}^{i} = B\frac{1-\nu}{2}\left(u_{\alpha,\beta} + u_{\beta,\alpha} + \frac{2\nu}{1-\nu} u_{\gamma\gamma}\delta_{\alpha\beta}\right) \quad (6.19)
$$

$$
N_{\alpha\beta}^{ii} = B[(1-\nu)\kappa_{\alpha\beta} + \nu\alpha_{\alpha\beta}\kappa_{\gamma\gamma}] \quad (6.20)
$$

Taking into account the relations $t_\alpha^{i} = N_{\alpha\beta}^{i} n_\beta$ and $t_\alpha^{ii} = N_{\alpha\beta}^{ii} n_\beta$, replacing in Eq. (6.18) and applying Green's theorem to the second and third domain integrals, we finally obtain the integral equation for plane stress:

$$
c_{\theta\alpha}\left(\mathbf{x}'\right) u_\theta\left(\mathbf{x}'\right) + ⨍_{\Gamma} T_{\theta\alpha}^{*}\left(\mathbf{x}', \mathbf{x}\right) u_\alpha\left(\mathbf{x}\right) d\Gamma
$$

$$
+\int_{\Omega} U_{\theta\alpha,\beta}^{*}\left(\mathbf{x}', \mathbf{x}\right) B\left[(1-\nu)\kappa_{\alpha\beta} + \nu\delta_{\alpha\beta}\kappa_{\gamma\gamma}\right] w_3\left(\mathbf{X}\right) d\Omega
$$

$$
= \int_{\Gamma} U_{\theta\alpha}^{*}\left(\mathbf{x}', \mathbf{x}\right) t_\alpha\left(\mathbf{x}\right) d\Gamma + \int_{\Omega} U_{\theta\alpha}^{*}\left(\mathbf{x}', \mathbf{X}\right) q_\alpha\left(\mathbf{X}\right) d\Omega +
$$

$$
\int_{\Omega} U_{\theta\alpha}(\mathbf{x}', \mathbf{X}) \Lambda_{\theta\alpha}^{m} \ddot{u}_\alpha(\mathbf{X}) d\Omega + \int_{\Omega} U_{\theta\alpha}(\mathbf{x}', \mathbf{X}) \Lambda_{\theta\alpha}^{bm} \ddot{w}_\alpha(\mathbf{X}) d\Omega \quad (6.21)
$$

where $\Lambda^m_{\alpha\beta} = I_0\delta_{\alpha\beta}$; the fundamental displacement solutions $U^*_{\alpha\beta}(\mathbf{x}',\mathbf{x})$ and $T^*_{\alpha\beta}(\mathbf{x}',\mathbf{x})$ are given by the expressions [1, 19]:

$$U^*_{\alpha\beta} = \frac{1}{4\pi B(1-\nu)}\left[(3-\nu)\ln\left(\frac{1}{r}\right)\delta_{\theta\alpha} + (1+\nu)r_{,\theta}r_{,\alpha}\right] \tag{6.22}$$

$$T^*_{\theta\alpha} = -\frac{1}{4\pi r}\left\{r_{,n}[(1-\nu)\delta_{\theta\alpha} + 2(1+\nu)r_{,\theta}r_{,\alpha}] + (1-\nu)\left[n_\theta r_{,\alpha} - n_\alpha r_{,\theta}\right]\right\} \tag{6.23}$$

$U^*_{\alpha\beta}(\mathbf{x}',\mathbf{x})$ presents terms with weak singularity, which is treated through logarithmic integration, and $T^*_{\alpha\beta}(\mathbf{x}',\mathbf{x})$ presents strong singularity, which is treated through rigid body considerations in a similar way to how it was presented in the previous chapter. That is, for a traction-free shell $t_\alpha = 0$, $q_\alpha = 0$, and for an arbitrary displacement, C, on the neutral surface of the shell, the following conditions apply [3]:

$$u_1 = C, \quad u_2 = 0, \quad w_1 = 0, \quad w_2 = 0, \quad w_3 = 0$$
$$u_2 = C, \quad u_1 = 0, \quad w_1 = 0, \quad w_2 = 0, \quad w_3 = 0$$

The following equality is obtained by replacing this in Eq. (6.21):

$$c_{\theta\alpha}\left(\mathbf{x}'\right) + \fint_\Gamma T^{(i)*}_{\theta\alpha}\left(\mathbf{x}',\mathbf{x}\right)d\Gamma = 0 \tag{6.24}$$

since $u = C$. Through this relation, the singular terms on the diagonal of the matrix $H_{\alpha\beta}$ can be indirectly calculated. Equations (6.17) and (6.21) represent five boundary integral equations, the first two in (6.17) ($i = \alpha = 1, 2$) are for rotations, the third ($i = 3$) is for out-of-plane displacements, and two in (6.21) ($\alpha = 1, 2$) for in-plane displacements, which can be used to solve shear-deformable surface plate bending problems.

6.3 Treatment of Domain Integrals

In this formulation, domain integrals related to distributed forces, curvatures, displacements, and derivatives of displacements appear. The first three can be treated using the DRBEM formulation presented in the previous chapter, while the integrals related to derivatives of displacements are (see [12, 16]):

$$\int_\Omega W^*_{i3}\frac{\partial u_1}{\partial x_1}d\Omega, \quad \int_\Omega W^*_{i3}\frac{\partial u_2}{\partial x_2}d\Omega, \quad \int_\Omega U^*_{\alpha 1}\frac{\partial w_3}{\partial x_1}d\Omega, \quad \int_\Omega U^*_{\alpha 2}\frac{\partial w_3}{\partial x_2}d\Omega \tag{6.25}$$

Domain integrals related to inertial terms are treated as presented in Chap. 5. Considering as radial approximation function $f_m(r) = 1 + r_m$ (see Fig. 6.2), the particular solutions of Eq. (6.4) are obtained as

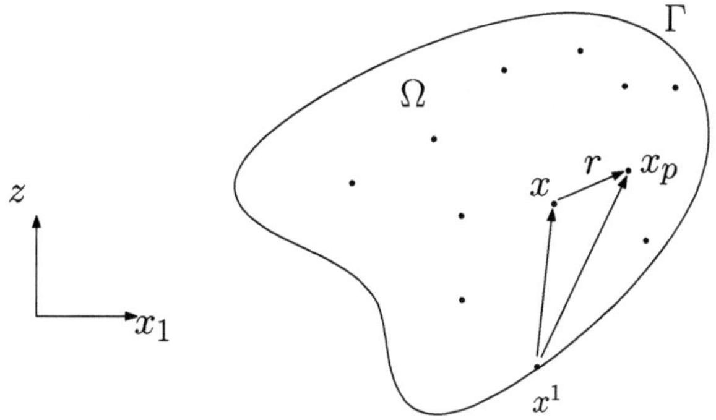

Fig. 6.2 Dual reciprocity boundary element method

$$L_{\alpha k}^{p,adj}\hat{w}_{mk}^{\alpha} = 0, \quad L_{3k}^{p,adj}\hat{w}_{mk}^{\alpha} = \frac{\partial f_m(r)}{\partial x_\alpha} = \frac{x_\alpha}{r_m} \tag{6.26}$$

where $L_{\alpha k}^{p,adj}$ and $L_{3k}^{p,adj}$ are the adjoint operators of the operator L^p. Applying the same procedure used in the previous chapter to transform domain integrals into boundary integrals using DRBEM, we obtain

$$\int_\Omega W_{i3}^* \frac{\partial u_1}{\partial x_1} d\Omega \approx \sum_{m=1}^M \Big[c_{ik}^p(\mathbf{x}')\hat{w}_{mk}^1(\mathbf{x}') - \int_\Gamma U_{ik}^*(\mathbf{x}',\mathbf{x})\hat{p}_{mk}^1(\mathbf{x})d\Gamma$$
$$+ \fint_\Gamma T_{ik}^*(\mathbf{x}',\mathbf{x})\hat{w}_{mk}^1(\mathbf{x})d\Gamma \Big] \mathbf{F}^{-1}\mathbf{u}_1 \tag{6.27}$$

$$\int_\Omega W_{i3}^* \frac{\partial u_2}{\partial x_2} d\Omega \approx \sum_{m=1}^M \Big[c_{ik}^p(\mathbf{x}')\hat{w}_{mk}^2(\mathbf{x}') - \int_\Gamma U_{ik}^*(\mathbf{x}',\mathbf{x})\hat{p}_{mk}^2(\mathbf{x})d\Gamma$$
$$+ \fint_\Gamma T_{ik}^*(\mathbf{x}',\mathbf{x})\hat{w}_{mk}^2(\mathbf{x})d\Gamma \Big] \mathbf{F}^{-1}\mathbf{u}_2 \tag{6.28}$$

Similarly, for the penultimate integral in (6.21), considering $f_m(r) = 1 + r_m$, particular solutions of the plane stress equations are obtained as

$$L_{1k}^{s,adj}\hat{u}_{m\alpha}^1 = \frac{\partial f_m(r)}{\partial x_1} = \frac{x_1}{r}, \quad L_{2\alpha}^{s,adj}\hat{u}_{m\alpha}^1 = 0 \tag{6.29}$$

In this way,

$$\int_{\Omega} U^*_{\alpha 1}\frac{\partial w_3}{\partial x_1}d\Omega \approx \sum_{m=1}^{M}\left[c^s_{\alpha\beta}(\mathbf{x}')\hat{u}^1_{m\beta}(\mathbf{x}') - \int_{\Gamma} U^*_{\alpha\beta}(\mathbf{x}',\mathbf{x})\hat{t}^1_{m\beta}(\mathbf{x})d\Gamma + \fint_{\Gamma} T^*_{\alpha\beta}(\mathbf{x}',\mathbf{x})\hat{u}^1_{m\beta}(\mathbf{x})d\Gamma\right]\mathbf{F}^{-1}\mathbf{w}_3 \tag{6.30}$$

Finally, for the last of the integrals, we have

$$L^{s,adj}_{1k}\hat{u}^2_{m\alpha} = 0, \quad L^{s,adj}_{2\alpha}\hat{u}^2_{m\alpha} = \frac{\partial f_m(r)}{\partial x_2} = \frac{x_2}{r} \tag{6.31}$$

which is approximated as:

$$\int_{\Omega} U^*_{\alpha 2}\frac{\partial w_3}{\partial x_2}d\Omega \approx \sum_{m=1}^{M}\left[c^s_{\alpha\beta}(\mathbf{x}')\hat{u}^2_{m\beta}(\mathbf{x}') - \int_{\Gamma} U^*_{\alpha\beta}(\mathbf{x}',\mathbf{x})\hat{t}^2_{m\beta}(\mathbf{x})d\Gamma + \fint_{\Gamma} T^*_{\alpha\beta}(\mathbf{x}',\mathbf{x})\hat{u}^2_{m\beta}(\mathbf{x})d\Gamma\right]\mathbf{F}^{-1}\mathbf{w}_3 \tag{6.32}$$

The particular solutions for $\hat{w}^{\alpha}_{mk}$ are given by the expressions:

$$\begin{aligned}
\hat{w}^1_{m1} &= -(3x_1^2 + r^2)\frac{r}{45D}\\
\hat{w}^1_{m2} &= -\frac{x_1x_2r}{15D}\\
\hat{w}^1_{m3} &= -[30 - (1-\nu)\lambda^2r^2]\frac{rx_1}{45(1-\nu)\lambda^2D}
\end{aligned} \tag{6.33}$$

$$\begin{aligned}
\hat{w}^2_{m1} &= -\frac{x_1x_2r}{15D}\\
\hat{w}^2_{m2} &= -(3x_1^2 + r^2)\frac{r}{45D}\\
\hat{w}^2_{m3} &= -[30 - (1-\nu)\lambda^2r^2]\frac{rx_2}{45(1-\nu)\lambda^2D}
\end{aligned} \tag{6.34}$$

The particular solutions for $\hat{M}_{\alpha\beta}$ and $\hat{Q}_{\beta}$ are

$$\begin{aligned}
\hat{M}^1_{m11} &= -\frac{x_1}{15}\left[\nu\left(\frac{x_1^2}{r} + 3r\right) + \left(\frac{x_2^2}{r} + r\right)\right]\\
\hat{M}^1_{m12} &= -(1-\nu)\frac{x_2}{15}\left(\frac{x_1^2}{r} + r\right)\\
\hat{M}^1_{m22} &= -\frac{x_1}{15}\left[\nu\left(\frac{x_1^2}{r} + 3r\right) + \left(\frac{x_2^2}{r} + r\right)\right]
\end{aligned} \tag{6.35}$$

$$\hat{M}^2_{m11} = -\frac{x_2}{15}\left[\nu\left(\frac{x_1^2}{r}+r\right)+\left(\frac{x_2^2}{r}+3r\right)\right]$$
$$\hat{M}^2_{m12} = -(1-\nu)\frac{x_1}{15}\left(\frac{x_2^2}{r}+r\right)$$
$$\hat{M}^2_{m22} = -\frac{x_2}{15}\left[\nu\left(\frac{x_1^2}{r}+r\right)+\left(\frac{x_2^2}{r}+3r\right)\right] \tag{6.36}$$

$$\hat{Q}^1_{m1} = -\frac{1}{3}\left(\frac{x_1^2}{r}+r\right)$$
$$\hat{Q}^1_{m2} = -\frac{1}{3}\frac{x_1x_2}{r} \tag{6.37}$$

$$\hat{Q}^2_{m1} = -\frac{1}{3}\frac{x_1x_2}{r}$$
$$\hat{Q}^2_{m2} = -\frac{1}{3}\left(\frac{x_2^2}{r}+r\right). \tag{6.38}$$

6.4 Boundary Element Discretization

Quadratic isoparametric boundary elements describe the geometry and unknown functions along the boundary. Discontinuous elements are used to avoid difficulties with traction discontinuity between elements. Wen et al. [17] In this way, Eq. (6.17) can be rewritten in a discretized form as

$$c_{ij}(\mathbf{x}')w_j(\mathbf{x}') + \sum_{n=1}^{N_e}\sum_{m=1}^{3} w_j^{nm}\int_{\xi=-1}^{\xi=+1} P_{ij}(\mathbf{x}',\mathbf{x})\Phi^m(\xi)J_n(\xi)d\xi$$
$$= \sum_{n=1}^{N_e}\sum_{m=1}^{3} p_j^{nm}\int_{\xi=-1}^{\xi=+1} W_{ij}(\mathbf{x}',\mathbf{x})\Phi^m(\xi)J_n(\xi)d\xi$$
$$-\sum_{n=1}^{N_e}\sum_{m=1}^{3}\kappa_{\alpha\beta}B\frac{1-\nu}{2}\left(u_\alpha^{nm}n_\beta^{nm}+u_\beta^{nm}n_\alpha^{nm}+\frac{2\nu}{1-\nu}u_\gamma^{nm}n_\gamma^{nm}\delta_{\alpha\beta}\right)$$
$$\times\int_{\xi=-1}^{\xi=+1} W_{i3}(\mathbf{x}',\mathbf{x})\Phi^m(\xi)J_n(\xi)d\xi$$

$$+\sum_{k=1}^{N_c}\kappa_{\alpha\beta}B\frac{1-\nu}{2}u_\alpha^k\int_{\eta=-1}^{\eta=+1}\int_{\xi=-1}^{\xi=+1}W_{i3,\beta}(\mathbf{x}',\mathbf{x})J_k^c(\xi,\eta)d\xi d\eta$$

$$+\sum_{k=1}^{N_c}\kappa_{\alpha\beta}B\frac{1-\nu}{2}u_\beta^k\int_{\eta=-1}^{\eta=+1}\int_{\xi=-1}^{\xi=+1}W_{i3,\alpha}(\mathbf{x}',\mathbf{x})J_k^c(\xi,\eta)d\xi d\eta$$

$$+\sum_{k=1}^{N_c}\kappa_{\alpha\beta}B\nu u_\gamma^k\int_{\eta=-1}^{\eta=+1}\int_{\xi=-1}^{\xi=+1}W_{i3,\gamma}(\mathbf{x}',\mathbf{x})J_k^c(\xi,\eta)d\xi d\eta$$

$$-\sum_{k=1}^{N_c}\kappa_{\alpha\beta}B[(1-\nu)\kappa_{\alpha\beta}+\nu\delta_{\alpha\beta}\kappa_{\gamma\gamma}]w_3^k\int_{\eta=-1}^{\eta=+1}\int_{\xi=-1}^{\xi=+1}W_{i3}(\mathbf{x}',\mathbf{X})J_k^c(\xi,\eta)d\xi d\eta$$

$$+\sum_{n=1}^{NDR}\ddot{\alpha}_l^n(t)\Big[c_{ij}\hat{W}_{lj}^n-\sum_{k=1}^{N_e}\sum_{r=1}^{3}\int_{\xi=-1}^{\xi=+1}W_{ij}(\mathbf{x}',\mathbf{x})\hat{P}_{lj}^n\Phi^r(\xi)J_k(\xi)d\Gamma$$

$$+\sum_{k=1}^{N_e}\sum_{r=1}^{3}\int_{\xi=-1}^{\xi=+1}P_{ij}(\mathbf{x}',\mathbf{x})\hat{W}_{lj}^n\Phi^r(\xi)J_k(\xi)d\Gamma\Big]$$

$$+\sum_{n=1}^{NDR}\ddot{\gamma}_l^n(t)\Big[c_{i\alpha}\hat{W}_{l\alpha}^n-\sum_{k=1}^{N_e}\sum_{r=1}^{3}\int_{\xi=-1}^{\xi=+1}W_{i\alpha}(\mathbf{x}',\mathbf{x})\hat{P}_{l\alpha}^n\Phi^r(\xi)J_k(\xi)d\Gamma$$

$$+\sum_{k=1}^{N_e}\sum_{r=1}^{3}\int_{\xi=-1}^{\xi=+1}P_{i\alpha}(\mathbf{x}',\mathbf{x})\hat{W}_{l\alpha}^n\Phi^r(\xi)J_k(\xi)d\Gamma\Big]$$

$$+\sum_{n=1}^{NDR}\beta_l^n\Big[c_{i\alpha}\hat{W}_{l\alpha}^n-\sum_{k=1}^{N_e}\sum_{r=1}^{3}\int_{\xi=-1}^{\xi=+1}W_{i\alpha}(\mathbf{x}',\mathbf{x})\hat{P}_{l\alpha}^n\Phi^r(\xi)J_k(\xi)d\Gamma$$

$$+\sum_{k=1}^{N_e}\sum_{r=1}^{3}\int_{\xi=-1}^{\xi=+1}P_{i\alpha}(\mathbf{x}',\mathbf{x})\hat{W}_{l\alpha}^n\Phi^r(\xi)J_k(\xi)d\Gamma\Big] \quad (6.39)$$

where β_l are the coefficients of the approximation polynomial: $q_k = f_{kj}\beta_j$. Similarly, Eq. (6.21) can be written as

$$c_{\theta\alpha}(\mathbf{x}')u_\alpha(\mathbf{x}')+\sum_{n=1}^{N_e}\sum_{m=1}^{3}u_\alpha^{nm}\int_{\xi=-1}^{\xi=+1}T_{\theta\alpha}(\mathbf{x}',\mathbf{x})\Phi^m(\xi)J_n(\xi)d\xi$$

$$= -\sum_{k=1}^{N_c} B[\kappa_{\alpha\beta}(1-\nu) + \nu\delta_{\alpha\beta}\kappa_{\gamma\gamma}]w_3^k \int_{\eta=-1}^{\eta=+1}\int_{\xi=-1}^{\xi=+1} U_{\theta\alpha,\beta}(\mathbf{x}', \mathbf{X}) J_k^c(\xi,\eta) d\xi d\eta$$

$$+ \sum_{n=1}^{NDR} \ddot{\gamma}_l^n(t) \Big[c_{i\alpha} \hat{W}_{l\alpha}^n - \sum_{k=1}^{N_e}\sum_{r=1}^{3} \int_{\xi=-1}^{\xi=+1} W_{i\alpha}(\mathbf{x}', \mathbf{x}) \hat{P}_{l\alpha}^n \Phi^r(\xi) J_k(\xi) d\Gamma$$

$$+ \sum_{n=1}^{NDR} \ddot{\chi}_l^n(t) \Big[c_{\theta\alpha} \hat{U}_{l\alpha}^n - \sum_{k=1}^{N_e}\sum_{r=1}^{3} \int_{\xi=-1}^{\xi=+1} U_{\theta\alpha}(\mathbf{x}', \mathbf{x}) \hat{T}_{l\alpha}^n \Phi^r(\xi) J_k(\xi) d\Gamma$$

$$+ \sum_{k=1}^{N_e}\sum_{r=1}^{3} \int_{\xi=-1}^{\xi=+1} T_{\theta k}(\mathbf{x}', \mathbf{x}) \hat{U}_{l\alpha}^n \Phi^r(\xi) J_k(\xi) d\Gamma \Big] + \tag{6.40}$$

In these equations, N_e and N_c are the number of boundary elements and internal cells, respectively; n_α are the components of the boundary normal vector; J represents the Jacobian of the transformation $d\Gamma = J(\xi)d\xi$ for boundary elements; J_r^c is the Jacobian of the transformation $d\Omega_c = J(\xi,\eta)d\xi d\eta$ for cells; ξ and η are the local coordinates; Φ^r with $(r = 1, 2, 3)$ are the quadratic form functions.

Applying these equations at each placement point will give the following linear system of equations:

$$\begin{bmatrix} \mathbf{H}^b & \mathbf{H}^u \\ \mathbf{H}^w & \mathbf{H}^p \end{bmatrix} \begin{Bmatrix} \mathbf{w} \\ \mathbf{u} \end{Bmatrix} = \begin{bmatrix} \mathbf{G}^b & \mathbf{0} \\ \mathbf{0} & \mathbf{G}^p \end{bmatrix} \begin{Bmatrix} \mathbf{p} \\ \mathbf{t} \end{Bmatrix} \tag{6.41}$$

where $\mathbf{u} = \{u_1, u_2\}^T$, $\mathbf{w} = \{w_1, w_2, w_3\}^T$, $\mathbf{p} = \{p_1, p_2, p_3\}^T$ and $\mathbf{t} = \{t_1, t_2\}^T$ are displacement and tension vectors for plane stress and plate bending formulations, respectively; $\ddot{\mathbf{u}} = \{\ddot{u}_1, \ddot{u}_2\}^T$, $\ddot{\mathbf{w}} = \{\ddot{w}_1, \ddot{w}_2, \ddot{w}_3\}^T$ are the in-plane acceleration and bending vectors; $\mathbf{M}^b$ and $\mathbf{M}^p$ are mass matrices for plane stress and plate bending formulations; $\mathbf{M}^u$ and $\mathbf{M}^w$ are coupling mass matrices between the bending and membrane responses; $\mathbf{H}^p$, $\mathbf{H}^p$, $\mathbf{G}^b$, and $\mathbf{G}^b$ are boundary element influence matrices for plane stress and plate flexure formulations, respectively, and $\mathbf{M}^u$, $\mathbf{M}^w$, $\mathbf{H}^u$, $\mathbf{H}^w$ are matrices containing coupled terms between the shell flexure and plane stress formulations.

Equations (6.41) can be rewritten in compact form as

$$\mathbf{H}_s \mathbf{w}_s = \mathbf{G}_s \mathbf{p}_s + \mathbf{Q}_s \mathbf{q}_s \tag{6.42}$$

where $\ddot{\mathbf{w}}_s = \{\ddot{\mathbf{w}}, \ddot{\mathbf{u}}\}^T$, $\mathbf{w}_s = \{\mathbf{w}, \mathbf{u}\}^T$, $\mathbf{p}_s = \{\mathbf{p}, \mathbf{t}\}^T$, where the subscript s denotes the plate. The boundary integral equations discussed above contain integrands with several different singularity orders. These singular integrals are treated separately based on their singularity order [3, 17]. All regular integrals are evaluated numerically using the standard Gauss quadrature formula. The influence matrix $\mathbf{G}_s$ contains weakly singular integrals, which are treated using the nonlinear Telles coordinate transformation method [9]. However, an appropriate number of element subdivisions should be used with the Telles transformation for higher numerical accuracy. In this work, four element subdivisions are used. The $\mathbf{M}_s$ matrix combines strong and weakly singular integrals. The influence matrix $\mathbf{H}_s$ contains strongly singular integrals; in this work, these integrals are computed indirectly by considering generalized rigid body motions.

6.5 Internal Stress Resultants

The resulting bending moment and shear force are calculated using the equations presented in Sect. 5.3, while the resulting internal membrane force is calculated using the following expression [2]:

$$
\begin{aligned}
N_{\alpha\beta}\left(\mathbf{X}'\right) &= \int_{\Gamma} U^{*}_{\alpha\beta\gamma}\left(\mathbf{X}',\mathbf{x}\right) t_{\gamma}(\mathbf{x})\, d\Gamma - \int_{\Gamma} T^{(i)*}_{\alpha\beta\gamma}\left(\mathbf{X}',\mathbf{x}\right) u_{\gamma}(\mathbf{x})\, d\Gamma \\
&+ \int_{\Omega} U^{*}_{\alpha\beta\gamma}\left(\mathbf{X}',\mathbf{x}\right) B\left[(1-\nu)\kappa_{\alpha\gamma} + \nu\delta_{\alpha\gamma}\kappa_{\phi\phi}\right] w_3(\mathbf{x})\, n_{\gamma}(\mathbf{x})\, d\Omega \\
&= \int_{\Gamma} U^{*}_{\alpha\beta\gamma}\left(\mathbf{X}',\mathbf{X}\right) q_{\gamma}(\mathbf{x}) + B\left[(1-\nu)\kappa_{\alpha\beta} + \nu\delta_{\alpha\beta}\kappa_{\phi\phi}\right] d\Gamma
\end{aligned}
\tag{6.43}
$$

where the kernels $U^{*}_{\alpha\beta\gamma}$ and $T^{*}_{\alpha\beta\gamma}$ are given by

$$
U^{*}_{\alpha\beta\gamma} = \frac{1}{4\pi r(1-\nu)}\left\{(1-2\nu)\left\{\delta_{\alpha\beta} r_{,\gamma} + \delta_{\beta\gamma} r_{,\beta} - \delta_{\beta\gamma} r_{,\alpha}\right\} + 2 r_{,\alpha} r_{,\beta} r_{,\gamma}\right\}
\tag{6.44}
$$

$$
T^{*}_{\alpha\beta\gamma} = \frac{2G}{4\pi r^{2}(1-\nu)}\left\{2 r_{,n}\left[(1-2\nu)\delta_{\beta\gamma} r_{,\alpha} + \nu\left(\delta_{\beta\alpha} r_{,\gamma} + \delta_{\gamma\alpha} r_{,\beta}\right) - \gamma r_{,\alpha} r_{,\beta} r_{,\gamma}\right]\right.
\tag{6.45}
$$

$$
\begin{aligned}
&+ 2\nu\left(n_{\beta} r_{,\gamma} r_{,\alpha} + n_{\gamma} r_{,\beta} r_{,\alpha}\right) \\
&\left. + (1-2\nu)\left(2 n_{\alpha} r_{,\beta} r_{,\gamma} + n_{\gamma}\delta_{\beta\alpha} + n_{\beta}\delta_{\gamma\alpha}\right) - (1-4\nu)\, n_{\alpha}\delta_{\beta\gamma}\right\}
\end{aligned}
\tag{6.46}
$$

6.6 Numerical Examples

6.6.1 *Analysis of a Cylindrical Shallow Shell*

Consider a cylindrical shallow shell as shown in Fig. 6.3, with geometric properties are as follows: $a = 5.0, h/a = 0.02, a/R = 0.05, \kappa_{11} = 1/R = 0.05, \kappa_{22} = 0$, with elastic constants $E = 210$ MPa, $\nu = 0.3$ and mass density $\rho = 500$ Kgm/m^3. Pressure load for this example is given by

$$P(t) = 10 < t > -290 < t - 0.1 > +10 < t - 0.2 > \tag{6.47}$$

A convergence solution for the central deflection of the plate is obtained for a time step of 1×10^{-2} s. Figure 6.3 show the history of central deflection w_3 and displacements u_2 for different time-steps together with the finite element solution. Again, the finite element solution was obtained using a finite element in-house code.

Analysis of a Square Shallow Spherical Shell

In this example a square shallow spherical shell as shown in Fig. 6.4 is analyzed. An uniform load of 1 N/m^2 is considered. The properties of the shell are $a = b = 1$, $h/b = 0.1$, $a/R = 0.5$, $k_{11} = k_{22} = 0.5$, $R_1/R_2 = 1$. Elastic constants are $E = 210000$ MPa, $\nu = 0.3$ and mass density ρ=5 $\times 10^{-4}$ Kgm/m^3. Simply supported boundary conditions at four edge were considered for this problem.

A convergence analysis for the central deflection versus time step was carried out using a BEM model with 108 DRBEM internal collocation points. The history of the central deflection for time interval from 0 to 0.5 s for different time steps is shown in

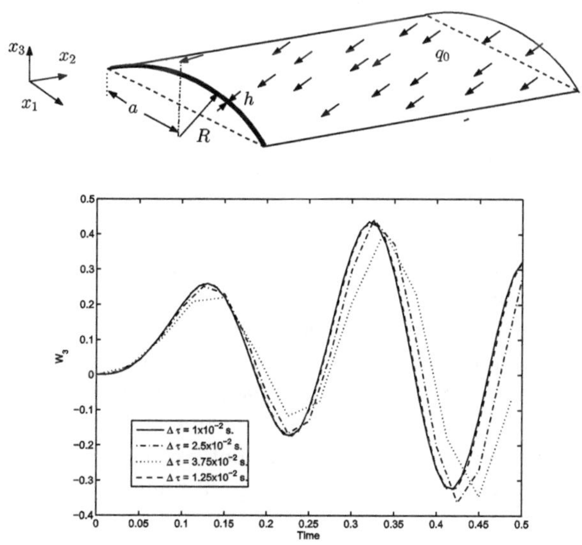

Fig. 6.3 Rectangular shallow shell geometry. Below: Central transversal displacement for cylindrical shallow shell versus time step

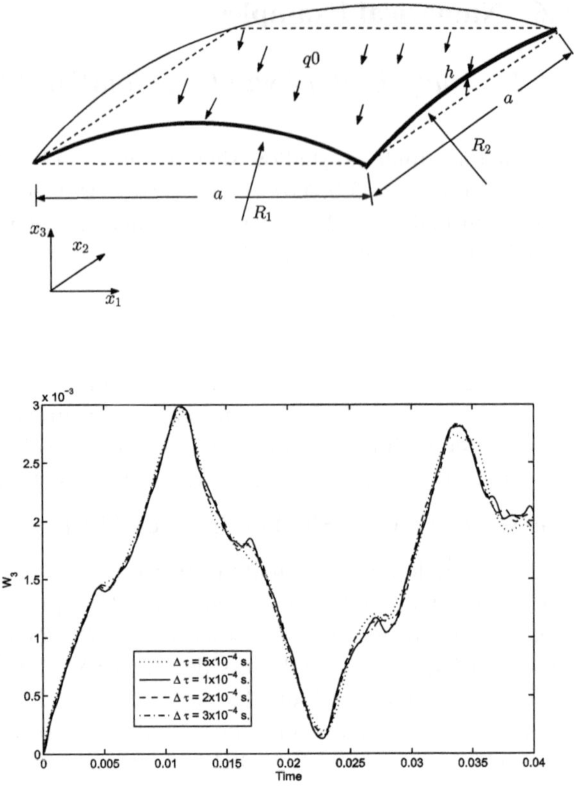

Fig. 6.4 Square shallow spherical shell geometry. Below: Convergence of central transversal displacement for spherical shallow shell under dynamic concentrated load versus time step

Fig. 6.4. A convergence solution for the central deflection of the plate is obtained for a time step of 1×10^{-2} s. As can be seen DRBEM provides results with very good agreement when compared with the finite element solution.

Analysis of a Cylindrical Shallow Shell

In this example a cylindrical square shallow as shown in Fig. 6.4 is analyzed. The geometric properties of shell are as follows: $a/b = 0.75$, $h/a = 0.0043$, $a/R = 0.1$, $\kappa_{11} = 1/R_{11} = 0.033$, $= \kappa_{22} = 1/R_{22} = 0$ with elastic constants $E = 61670$ MPa, $\nu = 0.3$ and mass density ρ=2.375 $\times 10^{-3}$ Kgm/m^3. Two kinds of boundary conditions were considered here: simply supported (SSSS) and fully clamped (CCCC) at four edges.

For harmonic analysis, a uniform load of 1 N/m^2 is considered as the applied load. Figure 6.5 shows the convergence analysis for the FRS in the range from 500 to 2000 Hz for SSSS boundary condition and from 500 to 3000 Hz for CCCC condition. FRS were obtained plotting deflection at the center of the shell versus frequency of excitation. The boundary element convergent solution for the first two resonance frequencies was obtained using 28 boundary elements with a total of 525 DRM points.

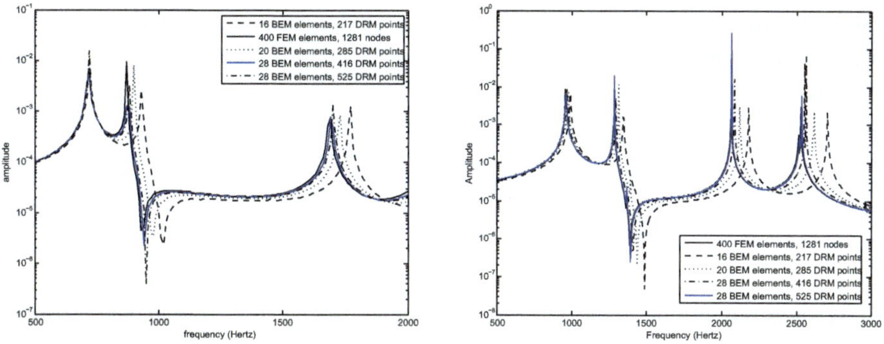

Fig. 6.5 Harmonic analysis of a circular shallow shell. Left: Harmonic response for ssss boundary conditions. Right: Harmonic response for cccc boundary conditions

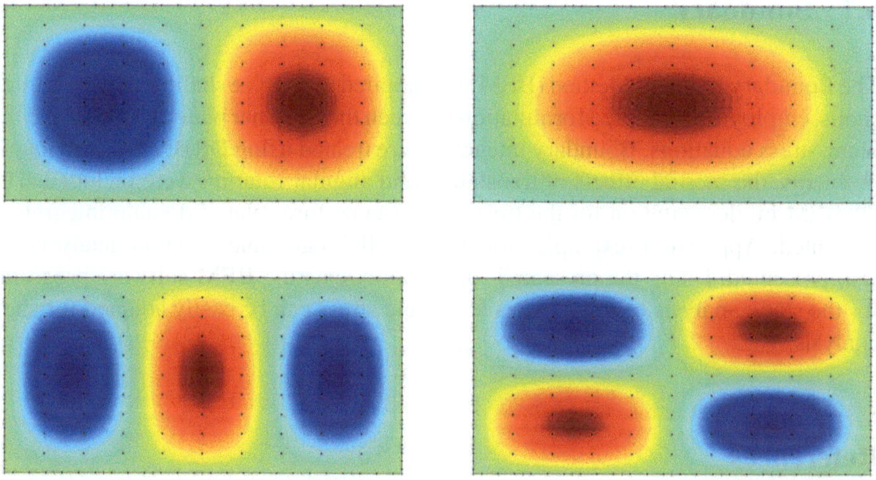

Fig. 6.6 First six modal shapes for a simply supported cylindrical shallow shell. Upper left: first mode 547.818 Hz. Upper Right: second mode 698.316 Hz. Below left: third mode 846.153 Hz. Below right: fourth mode 1104.788 Hz

A finite element convergent solution for the FRS was found with a mesh of 400 elements and 1281 nodes. An inspection of Fig. 8 demonstrates a good correlation for the first and second resonance frequencies, between convergent boundary element and finite element solutions for both SSSS and CCCC conditions. For the third resonance frequency a convergent behavior in BEM solution is observed; however, mesh refinement is needed for the SSSS condition.

For modal analysis, a mesh of 28 boundary elements and 525 DRM points was used. Table 6.1 shows the first six natural vibration frequencies parameters. Figure 6.6 shows the first four natural bending modes. Again, results obtained using

Table 6.1 Frequency parameter $\lambda = \omega a\sqrt{\rho/E}$ for an SSSS cylindrical shallow shell

Mode sequence number				
Method	λ_1	λ_2	λ_3	λ_4
BEM	0.0515	0.0656	0.0795	0.1038
FEM	0.0533	0.0665	0.0804	0.1050
Reddy [8]	0.0514	0.0657	0.0792	0.1038

BEM show good agreement with those obtained using the FEM and with results reported in [4, 8].

6.7 Summary

The Boundary Element Method (BEM) was applied to analyze shallow isotropic thick shells, employing a coupled formulation between the Reissner plate equations and the plane stress equations. Boundary integrals were treated using DRBEM. For dynamic problems, the inertial effects of translation and angular rotation were included. The DRBEM implementation for the treatment of curvature-related domain integrals is presented. Application examples for thick shells' static and dynamic analysis are also presented. Using the presented formulation, existing BEM software programs for the static analysis of thick plates and plane stress can be easily adapted for shell analysis.

References

1. Brebbia CA, Wrobel LC (1996) Boundary element techniques in engineering. Springer
2. Dirgantara T (2002) Boundary element analysis of cracks in shear deformable plates and shells. WIT Press, Southampton
3. Dirgantara T, Aliabadi MH (1999) New boundary element formulation for shear deformable shells analysis. Int. J. Num. Anal. Eng. 45(9):1257–1275
4. Lim CW, Liew KM (1995) A high order theory for vibration of shear deformable cylindrical shallow shells. Int J Mech Sci 37(3):277–295
5. Lu P, Haung M (1992) Boundary element analysis of shallow shells involving shear deformation. Int J Solids Struct 29:1273–1282
6. Newtonm DA, Tottenham H (1968) Boundary value problems in thin shallow shells of arbitrary plan form. J Eng Math 2:211–233
7. Providakis CP, Beskos DE (1991) Free and forced vibration of shallow shells by boundary and interior elements. Comput Method Appl Mech Eng 91:55–74
8. Reddy JN (2004) Mechanics of laminated composite plates and shells: theory and analysis, 2nd edn. CRC Press, New York
9. Telles JCF (1987) A self-adaptive coordinate transformation for efficient numerical evaluation of general boundary element integrals. International Journal for Numerical Methods in Engineering 24:959–973

10. Tosaka N, Miyake S (1983) tA boundary integral equation formulation for elastic shallow shell bending problems. In: Boundary elements. Springer
11. Tottenham H (1980) The boundary element method for plates and shells: development in BEM, vol 10
12. Useche J, Albuquerque EL (2015) Transient dynamic analysis of shear deformable shallow shells. Eng Struct 87(15):1–7
13. Useche J, Harnish C (2016) A boundary element method formulation for modal analysis of doubly curved thick shallow shells. Appl Math Model 40(5–6):3591–3600
14. Useche J, Medina C (2018) Boundary element analysis of laminated composite shear deformable shallow shells. Compos Struct J 199:24–37
15. Wen PH, Aliabadi MH (2000) Application of dual reciprocity method to plates and shells. Eng Anal Bound Elem 24:583–590
16. Wen PH, Aliabadi MH, Young A (2000) Application of dual reciprocity method to plates and shells. Eng Anal Bound Elem 24(7–8):583–590
17. Wen PH, Aliabadi MH, Young A (2000) Plane stress and plate bending coupling in BEM analysis of shallow shells. Int J Numer Methods Eng 48:1107–1125
18. Wen PH, Aliabadi MH, Young A (2006) Post buckling analysis of reissner plates by the boundary element method. J Strain Anal Eng Des 41:239–252
19. Wrobel LC, Aliabadi MH (2002) The boundary element method. Volume 2: Applications in solid and structures. Wiley, New York
20. Zhang JD, Atluri SN (1986) A boundary/interior element method for quasi static and transient response analysis of shallow shells. Comput Struct 24:213–223

Chapter 7
Boundary Element Analysis of Composite Thin Plates and Shells

Abstract This chapter presents the boundary element formulation of thin plates. The development of the boundary integral formulation, part of the governing equations given in the last chapter, is also presented. This integral form presents the discretization of the integrals using the boundary element method and the application of the Dual Reciprocity Boundary Element Method to transform the domain integrals into boundary integrals. At the end of the chapter, application examples for static and dynamic problems are presented.

7.1 Introduction

Mukherjee and Banerjee [11] first applied the (BEM) method to analyze thin composite plates in the early 1980s; they were among the pioneers who extended BEM to the analysis of thin plates made of composite materials. Their pioneering work presents the application of BEM to analyze anisotropic thin plates, laying the groundwork for subsequent studies in composite materials. Fenner and Berger [8] developed BEM formulations for bending problems in anisotropic thin plates. Dominguez [7] applied it to the dynamic analysis of thin composite plates. Brebbia and Wrobel [5] addressed the BEM analysis of vibration of thin composite plates. Albuquerque and Sollero [2] presented the application of BEM to the analysis of thin composite plates. Albuquerque et al. [3] present a BEM formulation for the modal analysis of plates.

This chapter presents the boundary element method applied to analyzing laminated composite plates and shells under Kirchhoff's theory. The boundary integral formulation for these equations is established, and fundamental solutions for displacements, bending moments, and shear forces are shown. The radial integration method and its application to the treatment of domain integrals are presented. Numerical examples and preliminary conclusions are drawn.

J. Useche Vivero, *The Boundary Element Method for Engineering Analysis of Plates and Shells*, Solid Mechanics and Its Applications 281,
https://doi.org/10.1007/978-3-031-93050-8_7

7.2 Equilibrium Equations in Terms of Displacements

Consider an orthotropic elastic plate of uniform thickness h, mass density ρ, occupying an area Ω in the x_1x_2 plane, bounded by the boundary $\Gamma = \Gamma_w \bigcup \Gamma_p$ with $\Gamma = \Gamma_w \bigcap \Gamma_p \equiv 0$ (Fig. 7.1). The flexural response of the plate was modeled using the thin plate theory. The governing equations, as presented in Chap. 3, are given by [14]:

$$
\begin{aligned}
&B_{11}\frac{\partial^3 u_1}{\partial x_1^3} + B_{12}\frac{\partial^3 u_2}{\partial x_1^2 \partial x_2} + B_{16}\left(\frac{\partial^3 u_2}{\partial x_2^2 \partial x_1} + \frac{\partial^3 u_2}{\partial x_1^3}\right) - D_{11}\frac{\partial^4 w_3}{\partial x_1^4} \\
&- 2D_{12}\frac{\partial^4 w_3}{\partial x_1^2 \partial x_2^2} - 4D_{16}\frac{\partial^4 w_3}{\partial x_1^3 \partial x_2} \\
&+ 2B_{16}\frac{\partial^3 u_1}{\partial x_1^2 \partial x_2} + 2B_{26}\frac{\partial^3 u_2}{\partial x_1 \partial x_2} + 2B_{66}\left(\frac{\partial^3 u_1}{\partial x_1 \partial x_2^2} + \frac{\partial^3 u_2}{\partial x_1^2 \partial x_2}\right) \\
&- 2D_{26}\frac{\partial^4 w_3}{\partial x_1 \partial x_2^3} - 4D_{66}\frac{\partial^4 w_3}{\partial x_1^2 \partial x_2^2} \\
&+ B_{12}\frac{\partial^3 u_1}{\partial x_1 \partial x_2^2} + B_{22}\frac{\partial^3 u_2}{\partial x_3} + B_{26}\frac{\partial^3 u_1}{\partial x_2^3} - D_{22}\frac{\partial^4 w_3}{\partial x_2^4} - 2D_{26}\frac{\partial^4 w_3}{\partial x_1 \partial x_2^3} + q(x_1, x_2) = \\
&- I_0\frac{\partial^2 w_3^0}{\partial t^2} + I_2\frac{\partial^2}{\partial t^2}\left(\frac{\partial w_1^0}{\partial x_1} - \frac{\partial w_2^0}{\partial x_2}\right) - I_1\frac{\partial^2}{\partial t^2}\left(\frac{\partial u_1^0}{\partial x_1} - \frac{\partial u_2^0}{\partial x_2}\right)
\end{aligned}
\tag{7.1}
$$

with the boundary conditions:

$$
\begin{aligned}
u_n = u_n^*, \quad u_s = u_s^*, \quad w_3 = w_3^*, \quad &\text{in } \Gamma_w \\
V_n = Q_n + M_{ns,s} = V_n^*, \quad M_{nn} = M_{nn}^*, \quad &\text{in } \Gamma_t.
\end{aligned}
\tag{7.2}
$$

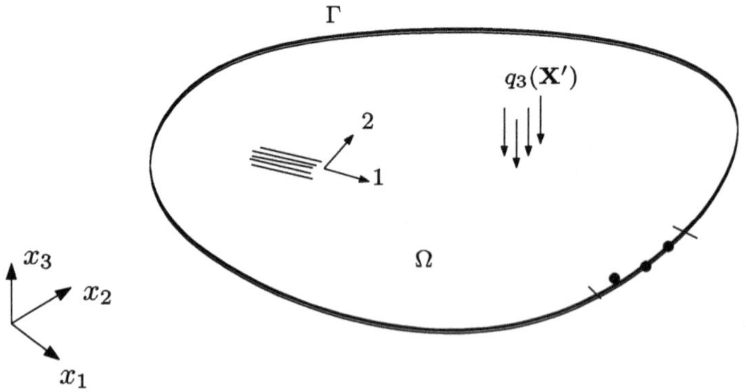

Fig. 7.1 Thin shell geometry

7.2.1 Symmetrical Laminates

For special symmetrical laminates (0° and 90°), $A_{\alpha 6} = 0$, $D_{\alpha 6} = 0$ and $B_{ij} = 0$, the bending equilibrium equation reduces to

$$D_{11}\frac{\partial^4 w_3}{\partial x_1^4} + 2D_{12}\frac{\partial^4 w_3}{\partial x_1^2 \partial x_2^2} + 4D_{66}\frac{\partial^4 w_3}{\partial x_1^2 \partial x_2^2} + D_{22}\frac{\partial^4 w_3}{\partial x_2^4} - q(x_1, x_2) = I_0\frac{\partial^2 w_3^0}{\partial t^2}. \tag{7.3}$$

7.3 Fundamental Solutions

7.3.1 Symmetrical Laminates

The transversal displacement plate bending fundamental solution is computed by placing the non-homogeneous term of the differential equation (7.1) equal to a concentrated force given by a Dirac delta function $\delta(\mathbf{x}', \mathbf{x})$. As shown by [15], the transversal displacement fundamental solution is given by

$$W^*(\rho, \theta) = \frac{1}{8\pi}\left\{\frac{(d_1 - d_2)^2 - (e_1^2 - e_2^2)}{GHe_1}R_1(\rho, \theta)\right.$$
$$\left. + \frac{(d_1 - d_2)^2 + (e_1^2 - e_2^2)}{GHe_2}R_2(\rho, \theta) + \frac{4(d_1 - d_2)}{GH}\left[S_1(\rho, \theta) - S_2(\rho, \theta)\right]\right\} \tag{7.4}$$

where

$$\rho = [(x_1 - x_1')^2 + (x_2 - x_2')^2]^{1/2}, \quad \theta = \arctan\frac{x_2 - x_2'}{x_1 - x_1'} \tag{7.5}$$

$$G = (d_1 - d_2)^2 + (e_1 + e_2)^2, \quad H = (d_1 - d_2)^2 + (e_1 - e_2)^2 \tag{7.6}$$

$$R_i = \rho^2\left[(\cos\theta + d_i \sin\theta)^2 - e_i^2 \sin^2\theta\right]$$
$$\left\{\log\left[\frac{\rho^2}{a^2}\left((\cos\theta + d_i \sin\theta)^2 + e_i^2 \sin^2\theta\right)\right] - 3\right\}$$
$$- 4\rho^2 e_i \sin\theta\,(\cos\theta + d_i \sin\theta)\arctan\frac{e_i \sin\theta}{\cos\theta + d_i \sin\theta} \tag{7.7}$$

and

$$S_i = \rho^2 e_i \sin\theta\,(\cos\theta + d_i \sin\theta)\left\{\log\left[\frac{\rho^2}{a^2}\left((\cos\theta + d_i \sin\theta)^2 + e_i^2 \sin^2\theta\right)\right] - 3\right\}$$

$$+\rho^2\left[(\cos\theta + d_i\sin\theta)^2 - e_i^2\sin^2\theta\right]\arctan\frac{e_i\sin\theta}{\cos\theta + d_i\sin\theta} \tag{7.8}$$

The repeated index i in terms of R_i and S_i does not imply summation. The coefficient a is an arbitrary constant taken as $a = 1$. Other fundamental solutions are given by

$$M_n^* = -\left(f_1\frac{\partial^2 W^*}{\partial x^2} + f_2\frac{\partial^2 W^*}{\partial x\partial y} + f_3\frac{\partial^2 W^*}{\partial y^2}\right), \tag{7.9}$$

$$R_{c_i}^* = -\left(g_1\frac{\partial^2 W^*}{\partial x^2} + g_2\frac{\partial^2 W^*}{\partial x\partial y} + g_3\frac{\partial^2 W^*}{\partial y^2}\right), \tag{7.10}$$

$$V_n^* = -\left(h_1\frac{\partial^3 W^*}{\partial x^3} + h_2\frac{\partial^3 W^*}{\partial x^2\partial y} + h_3\frac{\partial^3 W^*}{\partial x\partial y^2} + h_4\frac{\partial^3 W^*}{\partial y^3}\right) - \frac{1}{\bar{R}}\left(h_5\frac{\partial^2 W^*}{\partial x^2} + h_6\frac{\partial^2 W^*}{\partial x\partial y} + h_7\frac{\partial^2 W^*}{\partial y^2}\right) \tag{7.11}$$

where $\bar{R}$ is the radius of curvature at a smooth point of the Γ boundary. Other constants can be obtained from [2]. As can be seen, the derivatives of R_i and S_i present weak singularities, which are treated by logarithmic integration, Telles, and element subdivision (see [16]). The terms with strong singularities are treated by rigid body motion considerations, as already seen, and the hypersingular ones by Taylor series expansion.

7.3.2 Boundary Element Equations

The boundary integral equations for a thin plate are presented in Chap. 4 and rewritten here [2]:

$$cw(\mathbf{x}) + \int_\Gamma \left(V^*(\mathbf{x}',\mathbf{x})w + M_n^*(\mathbf{x}',\mathbf{x})\frac{\partial w}{\partial n}\right)d\Gamma + \sum_{i=1}^{N_c} R_{c_i}W_{c_i}^* = \int_\Gamma \left(W^*(\mathbf{x}',\mathbf{x})V + \frac{\partial W^*(\mathbf{x}',\mathbf{x})}{\partial n}M_n\right)d\Gamma + \int_\Omega W^*(\mathbf{x}',\mathbf{x})q(\mathbf{x})d\Omega + \sum_{i=1}^{N_c} R_{c_i}^* w_{c_i} \tag{7.12}$$

and

$$\frac{1}{2}\frac{\partial w(\mathbf{x}')}{\partial n_1} + \int_\Gamma \left[\frac{\partial V^*(\mathbf{x}',\mathbf{x})}{\partial n_1}w(\mathbf{x}) - \frac{\partial m_n^*(\mathbf{x}',\mathbf{x})}{\partial n_1}\frac{\partial w}{\partial n}(\mathbf{x})\right]d\Gamma +$$

$$\sum_{i=1}^{N_c} \frac{\partial R^*_{c_i}(\mathbf{x}', \mathbf{x})}{\partial n_1} w_{c_i}(\mathbf{x}) = \int_\Gamma \left\{ V_n(\mathbf{x}) \frac{\partial W^*}{\partial n_1}(\mathbf{x}', \mathbf{x}) - m_n(\mathbf{x}) \frac{\partial}{\partial n_1} \left[\frac{\partial W^*(\mathbf{x}', \mathbf{x})}{\partial n} \right] \right\} d\Gamma + \sum_{i=1}^{N_c} R_{c_i}(\mathbf{x}) \frac{\partial W^*_{c_i}(\mathbf{x}', \mathbf{x})}{\partial n_1} + \int_\Omega q(\mathbf{x}) \frac{\partial W^*(\mathbf{x}', \mathbf{x})}{\partial n_1} d\Omega \quad (7.13)$$

In these equations, $\frac{\partial()}{\partial n}$ is the derivative in the direction of the external vector $\mathbf{n}$ which is normal to the Γ; m_n and V_n are, respectively, the normal bending moment and the Kirchhoff equivalent shear force at the Γ; R_c is the thin plate reaction at the corners; w_{ci} is the transverse displacement at the corners; q_3 is the domain force in the transverse direction.

7.3.3 Unsymmetrical Laminates

Fundamental solutions for unsymmetrical laminates were proposed by [10] and are presented in this section. These essential solutions are much more complex than those for symmetric laminates. They require a special treatment of singularities, which are dealt with by separating the integrals into their symmetric and non-symmetric parts. These are then evaluated explicitly, thus avoiding numerical complications. However, the procedure is rather cumbersome, and the study of this treatment is left to the reader (see [10]). The fundamental solutions of asymmetric plates have been obtained as

$$\begin{aligned}
U^*_{ij}(\mathbf{x}', \mathbf{x}) &= 2Re\{\mathbf{h}^T_j(\mathbf{x})\mathbf{A}\mathbf{f}_i(\mathbf{x}', \mathbf{x})\}, \quad i = 1, 2, 3, 4, \quad j = 1, 2, 4 \\
U^*_{i3}(\mathbf{x}', \mathbf{x}) &= -2Re\{\mathbf{i}^T_3 \mathbf{A}\tilde{\mathbf{f}}_i(\mathbf{x}', \mathbf{x})\} \\
P^*_{i1}(\mathbf{x}', \mathbf{x}) &= -2Re\{\mathbf{n}^T_p(\mathbf{x})\mathbf{B} < \mu_\alpha > \mathbf{f}'_i(\mathbf{x}', \mathbf{x})\} \\
P^*_{i2}(\mathbf{x}', \mathbf{x}) &= 2Re\{\mathbf{n}^T_p(\mathbf{x})\mathbf{B}\mathbf{f}'_i(\mathbf{x}', \mathbf{x})\} \\
P^*_{i3}(\mathbf{x}', \mathbf{x}) &= 2Re\{\mathbf{s}^T_b(\mathbf{x})\mathbf{B} < s'_1 + s'_2\mu_\alpha > \mathbf{f}'_i(\mathbf{x}', \mathbf{x})\} \\
&\quad +2Re\{\mathbf{s}^T_b(\mathbf{x})\mathbf{B} < n'_2 - n'_1\mu_\alpha > \mathbf{f}'_i(\mathbf{x}', \mathbf{x})\} \\
&\quad +2Re\{\mathbf{s}^T_b(\mathbf{x})\mathbf{B} < (n'_2 - n'_1\mu_\alpha)^2 > \mathbf{f}'_i(\mathbf{x}', \mathbf{x})\} \\
P^*_{i4}(\mathbf{x}', \mathbf{x}) &= 2Re\{\mathbf{n}^T_b(\mathbf{x})\mathbf{B} < n'_2 - n'_1\mu_\alpha > \mathbf{f}'_i(\mathbf{x}', \mathbf{x})\} \\
P^*_{ic}(\mathbf{x}', \mathbf{x}_k) &= 2Re\{\mathbf{n}^T_c(\mathbf{x})\mathbf{B}\mathbf{f}'_i(\mathbf{x}', \mathbf{x}) + \mathbf{s}_c(\mathbf{x})_k\mathbf{B} < \mu_\alpha > \mathbf{f}'_i(\mathbf{x}', \mathbf{x})\}, \quad i = 1, 2, 3, 4
\end{aligned} \quad (7.14)$$

where $\mathbf{f}(\mathbf{x}', \mathbf{x})$ is given by

$$\mathbf{f}_i(\mathbf{x}', \mathbf{x}) = \frac{1}{2\pi i} \langle \mathrm{In}(z_\alpha - \hat{z}_\alpha) \rangle \mathbf{A}^T \mathbf{h}_i(\mathbf{x}'), \quad i = 1, 2, 4$$

$$\mathbf{f}_3(\mathbf{x}', \mathbf{x}) = \frac{1}{2\pi i}\langle(z_\alpha - \hat{z}_\alpha)\left[\ln(z_\alpha - \hat{z}_\alpha) - 1\right]\rangle \mathbf{A}^T \mathbf{i}_3 \tag{7.15}$$

With:

$$\mathbf{n}_p = \begin{Bmatrix} n_1 \\ n_2 \\ 0 \\ 0 \end{Bmatrix}, \quad \mathbf{n}_b = \begin{Bmatrix} 0 \\ 0 \\ n_1 \\ n_2 \end{Bmatrix}, \quad \mathbf{s}_b = \begin{Bmatrix} 0 \\ 0 \\ s_1 \\ s_2 \end{Bmatrix}, \quad \mathbf{s}_b' = \begin{Bmatrix} 0 \\ 0 \\ s_1' \\ s_2' \end{Bmatrix},$$

$$\mathbf{i}_1 = \begin{Bmatrix} 1 \\ 0 \\ 0 \\ 0 \end{Bmatrix}, \quad \mathbf{i}_2 = \begin{Bmatrix} 0 \\ 1 \\ 0 \\ 0 \end{Bmatrix}, \quad \mathbf{i}_3 = \begin{Bmatrix} 0 \\ 0 \\ 1 \\ 0 \end{Bmatrix},$$

$$\mathbf{n}_t = \begin{Bmatrix} 0 \\ 0 \\ \frac{1}{2} - n_1^2 \\ -n_1 n_2 \end{Bmatrix}, \quad \mathbf{s}_t = \begin{Bmatrix} 0 \\ 0 \\ -n_1 n_2 \\ n_1^2 - \frac{1}{2} \end{Bmatrix}, \tag{7.16}$$

$$\begin{gathered} \mathbf{n}_c = \mathbf{n}_t^+ - \mathbf{n}_t^-, \mathbf{s}_c = \mathbf{s}_t^+ - \mathbf{s}_t^-, \\ \mathbf{h}_1(\mathbf{x}') = \mathbf{h}_1(\mathbf{x}) = \mathbf{i}_1, \mathbf{h}_2(\mathbf{x}') = \mathbf{h}_2(\mathbf{x}) = \mathbf{i}_2, \\ \mathbf{h}_4(\mathbf{x}') = \mathbf{n}_b(\mathbf{x}'), \mathbf{h}_4(\mathbf{x}) = \mathbf{n}_b(\mathbf{x}), \\ z_\alpha = x_1 + \mu_\alpha x_2, \hat{z}_\alpha = x_1' + \mu_\alpha x_2' \end{gathered} \tag{7.17}$$

In these equations, Re represents the real part of a complex number; the angle bracket represents the diagonal matrix whose components vary according to the subscript $\alpha = 1, 2, 3, 4$, that is, $<f_\alpha> = diag[f_1, f_2, f_3, f_4]$; μ_α, with $\alpha = 1, 2, 3, 4$, are the material eigenvalues of the Stroh-type formalism, which have been assumed to be distinct and have a positive imaginary part; **A** and **B** are the material eigenvector matrices; (n_1, n_2), (s_1, s_2), (s_1', s_2'), (x_1, x_2), and (x_1', x_2') are, respectively, the components of the normal direction **n**, the tangential direction **s**, the tangential change $\mathbf{s}'$, the field point **x** and the origin point $\mathbf{x}'$; the subscripts p, b, t and c denote, respectively, the values related to the in-plane, bending, torsion, and corner responses. The superscripts T, $+$, and $-$ signify the transpose and evaluation of the matrix immediately in front of and behind the corner, respectively. A prime $(\cdot')$ and an overlaid tilde $(\tilde{\cdot})$ mean differentiation and integration, respectively.

7.3.4 Boundary Element Equations

The boundary integral equations are obtained in a similar way to that presented in Sect. 4.3. Thus,

$$c_{ij}w_j(\mathbf{x}') = \int_\Gamma P^*_{ij}(\mathbf{x},\mathbf{x})w_j\mathbf{x})d\Gamma + \sum_{k=1}^{N_c} R_{ic}(\mathbf{x}_k)w_3(\mathbf{x}_k) + \int_\Gamma W^*_{ij}(\mathbf{x},\mathbf{x})p_j(\mathbf{x})d\Gamma + \int_\Omega W^*_{ij}(\mathbf{x},\mathbf{x})q_j(\mathbf{x})d\Omega + \sum_{k=1}^{N_c} W_{i3}(\mathbf{x}_k)R_c(\mathbf{x}_k) \tag{7.18}$$

However, unlike the equations for symmetrical laminates, the displacement and force vectors are defined as follows: w_α with $\alpha = 1, 2$ represents membrane displacements; w_3 represents deflection, and w_i, $i = 4, 5$ represents rotations about the normal and tangent axes to the boundary. Similarly, where p_α with $\alpha = 1, 2$ represents in-plane tractions; p_3 represents transverse shear force at the boundary, and p_4 represents bending moment tangent to the plate boundary. q_α with $\alpha = 1, 2$ represents tensile forces per unit area, q_3 is transverse force per unit area, and $q_4 = m_l$ is moment distributed in an arbitrary direction l.

For plates with general laminates, explicit solutions of c_{ij} and its alternative expressions are obtained by imposing five rigid body motions as presented in the work of [10].

7.4 Domain Integral Treatment

7.4.1 Radial Integration Method

In order to transform domain integrals related with distributed pressure and inertial terms into boundary integrals, the Radial Integration Method—RIM [9], was used as presented in [2, 3, 6] where body forces are approximated with the use of radial basis functions.

Consider the domain term $p(t)$ in Eq. (7.13) approximated over the domain as a sum of the M products between radial basis functions f_m and unknown coefficients $\gamma^m(t)$, that is (see Fig. 7.2).

$$p(t) \approx \sum_{m=1}^{M} f^m \gamma^m(t) \tag{7.19}$$

The domain integral of Eq. (7.13) on the right-hand side can be written as

$$\int_\Omega W^*_{i3} p(t) d\Omega = \sum_{m=1}^{M} \gamma^m(t) \int_\Omega W^*_{i3}(\mathbf{x}',\mathbf{x}) f^m d\Omega$$

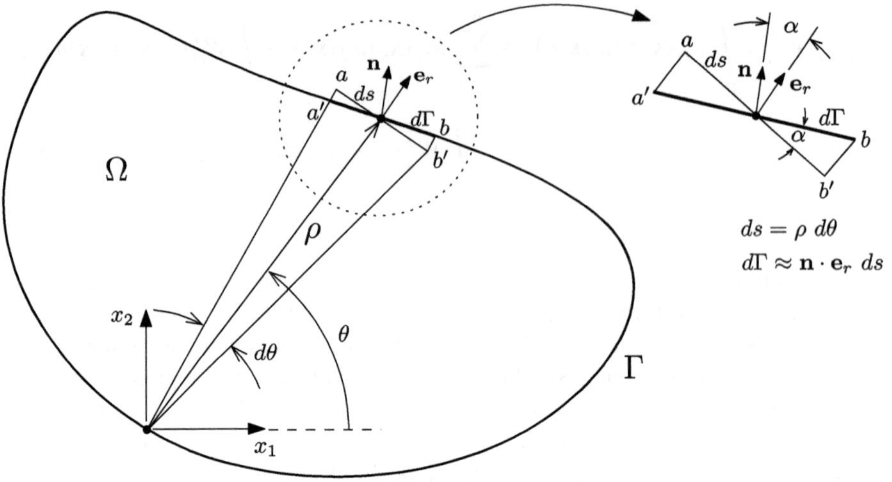

Fig. 7.2 Transformation of domain integral into boundary integral

$$= \sum_{m=1}^{M} \gamma^m(t) \int_{\theta} \int_{0}^{r} W_{i3}^*(\mathbf{x}', \mathbf{x}) f^m \rho d\rho d\theta \tag{7.20}$$

where r is the value of ρ in a point of the boundary Γ as shown in Fig. 7.3. Defining E_{i3}^m as the following integral:

$$E_{i3}^m(\mathbf{x}') = \int_{0}^{r} W_{i3}^*(\mathbf{x}', \mathbf{x}) f^m \rho d\rho \tag{7.21}$$

we can re-write the integral in (7.20) as

$$\sum_{m=1}^{M} \gamma^m(t) \int_{\Omega} W_{i3}^*(\mathbf{x}', \mathbf{x}) f^m d\Omega = \sum_{m=1}^{M} \gamma^m(t) \int_{\theta} E_{i3}^m(\mathbf{x}') d\theta \tag{7.22}$$

Considering an infinitesimal angle $d\theta$, the relation between the arch length $rd\theta$ and the infinitesimal boundary length $d\Gamma$ can be written as

$$d\theta = \frac{\mathbf{n} \cdot \mathbf{r}}{r} d\Gamma \tag{7.23}$$

Substituting Eq. (7.23) into Eq. (7.22) we have

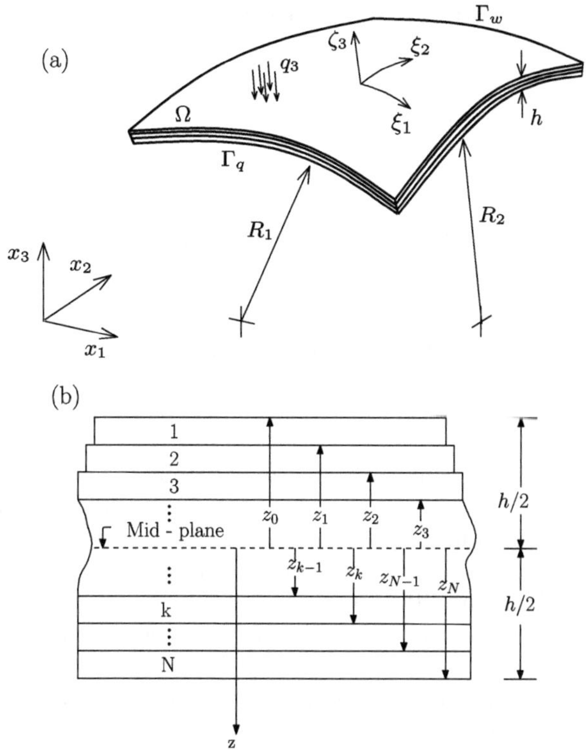

Fig. 7.3 Laminated thin shell geometry

$$\sum_{m=1}^{M} \gamma^m(t) \int_{\Omega} W_{i3}^*(\mathbf{x}', \mathbf{x}) f^m d\Omega = \sum_{m=1}^{M} \gamma^m(t) \int_{\Gamma} \frac{E_{i3}^m(\mathbf{x}')}{r} \mathbf{n} \cdot \mathbf{r} d\Gamma \tag{7.24}$$

In this way, the domain integral on the right-hand side of Eq. (7.13) is transformed to a boundary integral as

$$\int_{\Omega} W_{i3}^*(\mathbf{x}', \mathbf{x}) p(\mathbf{x}, t) d\Omega = \sum_{m=1}^{M} \gamma^m(t) \int_{\Gamma} \frac{E_{i3}^m(\mathbf{x}')}{r} \mathbf{n} \cdot \mathbf{r} d\Gamma \tag{7.25}$$

In this work, the approximation functions f_m are taken as $f_m = 1 + r_m$, where r_m represents distance between the center of the radial basis function and the integration point as presented in [3]. Coefficients $\gamma^m(t)$ are related to $p^m(t)$ through expression:

$$p^m(t) = F_{lk} \gamma^k(t), \quad l = 1, 2 \ldots, M \tag{7.26}$$

where F_{ik} is a matrix of coefficients obtained by taking the value of $p(t)$ at different RIM points.

7.5 Numerical Examples

In order to assess the accuracy of the proposed formulation some numerical problems are analyzed and their results compared with some results available in literature.

7.5.1 *Orthotropic Simply Supported Square Plate*

The first problem is a square plate of side length $a = 1$ m and thickness $h = 0.01$ m. The material is orthotropic with properties $E_x = 2.068 \times 10^{11}$ Pa, $E_y = E_x/15$, $\nu_{xy} = 0.3$, $G_{xy} = 6.055 \times 10^8$ Pa. The plate is under a uniformly distributed load $q = 1 \times 10^4$ Pa. Simply supported boundary condiciones along its four edges are used.

The problem is solved using different meshes and the results for transversal displacements at point A and at point B are compared with series solution for point A and for point B given by $w_{se.} = 8.1258 \times 10^{-3}$ m and $w_{se.} = 4.5211 \times 10^{-3}$ m, respectively. Table 7.1 shows transversal displacements computed by the present BEM using different meshes and their respective errors compared to [17] series solution. It can be seen that a very poor agreement is obtained when 12 elements (3 elements per side) are used. However, the convergence to the series solutions is obtained as the number of elements is increased. When 48 boundary elements are used, transversal displacements in both points present errors below 1% when compared with series solutions.

7.5.2 *Cross-Ply Laminate Graphite-Epoxy Composite Square Plate with Simply Supported Edges*

The second problem that has been analyzed in this work is a nine layer ply simply supported laminate $[0°/90°/0°/90°/0°]_s$ of side length $a = 1$ m under a uniformly distributed load $q = 6.9 \times 10^3$ Pa. The properties of each layer of a high modulus graphite-epoxy composite material used in this analysis are $E_x = 2.07 \times 10^9$ Pa,

Table 7.1 Accuracy of transversal displacements obtained by BEM for the orthotropic square plate with simply supported edges under uniformly distributed loads

Number of elements	Transversal displacements (10^{-6} m)		Errors (%)	
	Point A	Point B	Point A	Point B
12	−0.7985	−0.4415	1.7310	2.3545
24	−0.8014	−0.4430	1.3806	2.0100
48	−0.8081	−0.4481	0.5513	0.8875

$E_y = 5.17 \times 10^9$ Pa, $G_{xy} = 3.10 \times 10^9$ Pa, and $\nu_{xy} = 0.25$. The total thickness of the laminate is taken as $h = 0.01$ m. All layers have equal thickness. A series solution for the transversal displacement in the center of the plate is given by

$$\frac{w_{an} \cdot E_{22} h^3}{q a^4} \times 10^3 = 4.4718 \tag{7.27}$$

The center point transversal displacement obtained by the proposed formulation using a mesh of 22 boundary elements per side is compared in Table 7.2 with the finite element solution, and with the analytical solution presented. As it can be seen, it is obtained the same accuracy of the finite element results.

7.6 Analysis of Symmetric Laminated Composite Thin Shallow Shells

Albuquerque et al. [1] presented the symmetrically laminated thin-shell formulation, which extends the BEM formulation for laminated thin composite plates. The formulation follows the strategy of coupling the plate-bending and plane-yielding components, and the curvature effects are treated as system forces.

Consider a shell of an anisotropic elastic material whose mid surface is described by $z = z(x_1, x_2)$ as shown in Chap. 3. The base plane of the shell is defined by a Ω domain in the x_1, x_2 plane whose boundary is given by Γ (see Fig. 7.3). The equilibrium equations for the shell can be written as [14]:

$$A_{11}\frac{\partial^2 u_1}{\partial x_1^2} + A_{12}\frac{\partial^2 u_2}{\partial x_1 \partial x_2} + A_{66}\left(\frac{\partial^2 u_1}{\partial x_2^2} + \frac{\partial^2 u_2}{\partial x_1 \partial x_2}\right) + A_{16}\frac{\partial^2 u_1}{\partial x_1 \partial x_2} + A_{26}\frac{\partial^2 u_2}{\partial x_2} + A_{66}\left(\frac{\partial^3 u_1}{\partial x_1 \partial x_2} + \frac{\partial^2 u_2}{\partial x_2^2}\right) = 0$$

$$A_{66}\left(\frac{\partial^2 u_2}{\partial x_1^2} + \frac{\partial^2 u_1}{\partial x_1 \partial x_2}\right) + A_{22}\frac{\partial^2 u_2}{\partial x_2^2} + A_{12}\frac{\partial^2 u_1}{\partial x_1 \partial x_2} + A_{12}\frac{\partial^2 u_1}{\partial x_1 \partial x_2} + A_{22}\frac{\partial^2 u_2}{\partial x_2} + A_{26}\left(\frac{\partial^3 u_2}{\partial x_1 \partial x_2} + \frac{\partial^2 u_1}{\partial x_2^2}\right) = 0$$

Table 7.2 Accuracy of transversal displacement obtained by BEM and FEM for the cross-ply laminate graphite-epoxy composite square plate with simply supported edges under uniformly distributed loads

Numerical methods	Transversal displacements and errors	
	$wE_{22}h^3/(qa^4) \times 10^3$	Errors (%)
BEM	4.4507	0.47
FEM	4.4508	0.47

$$D_{11}\frac{\partial^4 w}{\partial x_1^4}+4D_{16}\frac{\partial^4 w}{\partial x_1^3\partial x_2}+2\left(D_{12}+D_{66}\right)\frac{\partial^4 w}{\partial x_1^2\partial x_2^2}+4D_{26}\frac{\partial^4 w}{\partial x_1^2\partial x_2^3}$$
$$+D_{22}\frac{\partial^4 w}{\partial x_2^4}+\frac{N_{ij}}{R_{ij}}=q_3, \tag{7.28}$$

where $R_{ij}=-1/\kappa_{ij}$ are the radii of curvature of the undeformed shell. The boundary element formulation of isotropic shells is due to [6], who derived it using Betti's reciprocity theorem and Green's theorem. In this formulation, the membrane and bending effects are coupled by domain integrals, similar to the thick-shell formulation. This integral formulation can be extended to symmetric laminated composite shells by changing the constitutive equations to their respective anisotropic ones. Therefore, the integral equations of anisotropic plane elasticity (membrane equations) are given by [1]:

$$c_{ij}u_j+\int_\Gamma T^*_{ik}(\mathbf{x}',\mathbf{x})u_k(\mathbf{x})d\Gamma=\int_\Gamma U^*_{ik}(\mathbf{x}',\mathbf{x})t_k(\mathbf{x})d\Gamma$$
$$+\int_\Omega \kappa_{kj}wU^*_{ikj}(\mathbf{X}',\mathbf{x})d\Omega+\int_\Omega u^*_{ik}(\mathbf{X}',\mathbf{x})q_k(\mathbf{x})d\Omega$$

The constants κ_{kj} depend on the radii of curvature R_{kj} of the shell and are given by

$$\begin{Bmatrix}\kappa_{11}\\ \kappa_{22}\\ \kappa_{12}\end{Bmatrix}=\begin{bmatrix}A_{11} & A_{12} & A_{16}\\ A_{12} & A_{22} & A_{26}\\ A_{16} & A_{26} & A_{66}\end{bmatrix}\begin{Bmatrix}\frac{1}{R_{11}}\\ \frac{1}{R_{22}}\\ \frac{2}{R_{12}}\end{Bmatrix}, \tag{7.29}$$

where $A_{11}, A_{22}, A_{66}, A_{12}, A_{16},$ and A_{26} are the extensional stiffnesses of the anisotropic plate.

The integral bending equation for the plate is given by

$$Kw(\mathbf{x}')+\int_\Gamma\left[V^*_n(\mathbf{x}',\mathbf{x})w(\mathbf{x})-m^*_n(\mathbf{x}',\mathbf{x})\frac{\partial w(\mathbf{x})}{\partial n}\right]d\Gamma(\mathbf{x})+\sum_{i=1}^{N_c}R^*_{c_i}(\mathbf{x}',\mathbf{x})w_{ci}(\mathbf{x})$$
$$=\sum_{i=1}^{N_c}R^*_{c_i}(\mathbf{x})w_{ci}(\mathbf{x}',\mathbf{x})+\int_\Omega q_3(\mathbf{X})W^*(\mathbf{X}',\mathbf{x})d\Omega$$
$$+\int_\Gamma\left[V_n(\mathbf{x})W^*(\mathbf{x}',\mathbf{x})-m_n(\mathbf{x})\frac{\partial W^*}{\partial n}(\mathbf{x}',\mathbf{x})\right]d\Gamma(\mathbf{x})$$
$$+\int_\Gamma \kappa_{ij}n_ju_i(\mathbf{x})W^*(\mathbf{x}',\mathbf{x})d\Gamma(\mathbf{x})+\int_\Omega\frac{\kappa_{ij}}{\rho_{ij}}W^*(\mathbf{X}',\mathbf{x})w(\mathbf{x})d\Omega$$

$$+\int_{\Omega}\left[\kappa_{ij}(\mathbf{X})W^*(\mathbf{X}',\mathbf{x})\right]_{,j}u_i(\mathbf{X})d\Omega, \tag{7.30}$$

and

$$K\frac{\partial w}{\partial m}(\mathbf{x}')+\int_{\Gamma}\left[\frac{\partial V_n^*(\mathbf{x}',\mathbf{x})}{\partial m}w(\mathbf{x})-\frac{\partial m_n^*(\mathbf{x}',\mathbf{x})}{\partial m}\frac{\partial w(\mathbf{x})}{\partial n}\right]d\Gamma(\mathbf{x})$$

$$+\sum_{i=1}^{N_c}\frac{\partial R_{c_i}^*(\mathbf{x}',\mathbf{x})}{\partial m}w_{ci}(\mathbf{x})$$

$$=\sum_{i=1}^{N_c}\frac{\partial R_{c_i}(\mathbf{x})^*}{\partial m}w_{ci}(\mathbf{x}',\mathbf{x})+\int_{\Omega}q_3(\mathbf{X})\frac{\partial W^*(\mathbf{X}',\mathbf{x})}{\partial m}d\Omega$$

$$+\int_{\Gamma}\left[V_n(\mathbf{x})\frac{\partial W^*(\mathbf{x}',\mathbf{x})}{\partial m}-m_n(\mathbf{x})\frac{\partial^2 W^*}{\partial n\partial m}(\mathbf{x}',\mathbf{x})\right]d\Gamma(\mathbf{x})$$

$$+\int_{\Gamma}\kappa_{ij}n_ju_i(\mathbf{x})\frac{\partial W^*(\mathbf{x}',\mathbf{x})}{\partial m}d\Gamma(\mathbf{x})+\int_{\Omega}\frac{\kappa_{ij}}{\rho_{ij}}\frac{\partial W^*(\mathbf{X}',\mathbf{x})}{\partial m}w(\mathbf{X})d\Omega \tag{7.31}$$

$$+\int_{\Omega}\left[\kappa_{ij}(\mathbf{X})\frac{\partial W^*(\mathbf{X}',\mathbf{x})}{\partial m}\right]_{,j}u_i(\mathbf{X})d\Omega. \tag{7.32}$$

Domain integrals are treated by the Radial Integration Method [9], as presented in [4], which can analyze thin shells using collocation points at the boundary only. No other formulations for thin-laminated shells using the boundary element method are reported in the literature. Nor is the use of BEM for the dynamic analysis of thin-laminated shells using this formulation reported.

7.7 Numerical Examples

7.7.1 Cylindrical Shell with Simply Supported Edges

Consider a cross ply symmetrically-laminated cylindrical shell under uniform distributed load $q_3 = 1$ with $\kappa_{11} = 1/50$, $\kappa_{11} = 0$, $a/h = 10$ and $a/b = 1$. The shell is simply supported at edges (see Fig. 7.4). Material properties used are: $E_{11} = 25E_{22}$, Ψ and $\nu_{12} = 0.25$. The laminated has eight layers $[90°/0°/0°/90°]_s$ with thickness $h/8 = 0.125$.

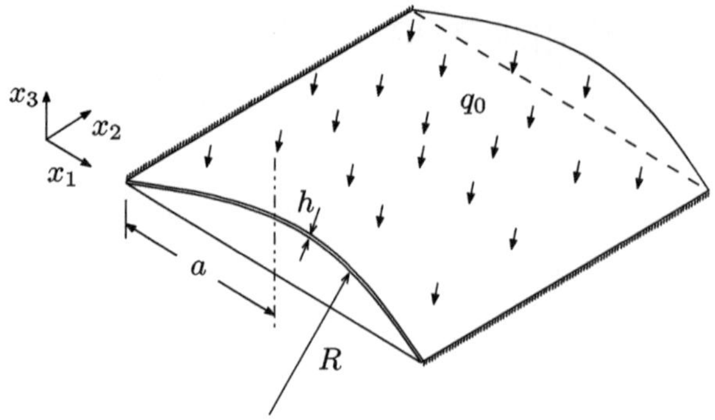

Fig. 7.4 Simply supported cylindrical shell

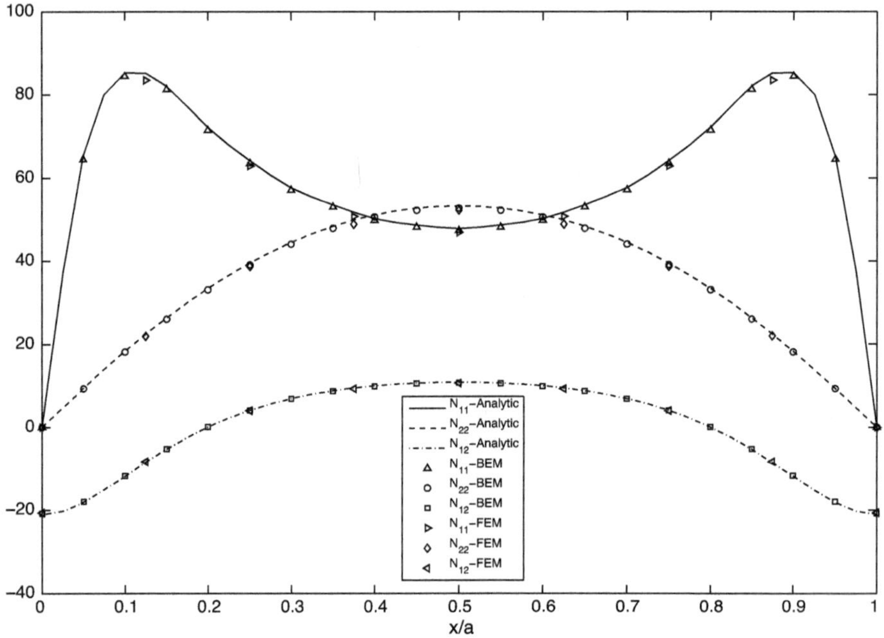

Fig. 7.5 Normal forces along x_2 axis for cylindrical shell

The shell was discretized with sixteen elements and 72 uniformly distributed domain RIM points. Figures 7.5 through 7.6 show the results for internal forces and bending moments along the central lines $x_2 = b/2$ and $x_1 = a/2$. Good BEM results were obtained when compared with the finite element solutions.

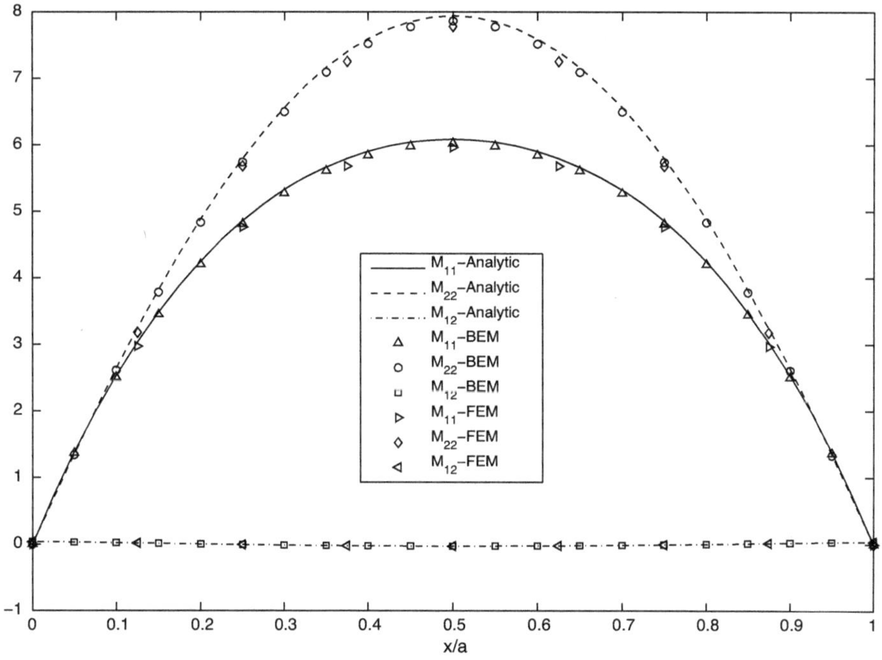

Fig. 7.6 Bending moments along x_2 axis for cylindrical shell

References

1. Albuquerque EL, Aliabadi MH (2008) A boundary element formulation for boundary only analysis of thin shallow shells. CMES—Comput Model Eng Sci 29:63–73
2. Albuquerque EL, Sollero P, Venturini W, Aliabadi MH (2006) Boundary element analysis of anisotropic Kirchhoff plates. Int J Solids Struct 43:4029–4046
3. Albuquerque EL, Sollero P, Portilho de Paiva W (2007) The radial integration method applied to dynamic problems of anisotropic plates. Commun Numer Methods Eng 23:805–818
4. Albuquerque EL, Aliabadi MH (2008) A boundary element formulation for boundary only analysis of thin shallow shells. CMES J 29:63–73
5. Brebbia CA, Wrobel LC (1996) Boundary element techniques in engineering. Springer
6. Portilho de Paiva W, Sollero P, Albuquerque EL (2011) Modal analysis of anisotropic plates using the boundary element method. Eng Anal Bound Elem 35:1248–1255
7. Dominguez J (1993) Boundary elements in dynamics. Computational mechanics, New York
8. Fenner RT, Berger JR (1986) Boundary integral equation solutions for anisotropic plate bending problems. Int J Numer Methods Eng 23(4):687–698
9. Gao X (2002) The radial integration method for evaluation of domain integrals with boundary only discretization. Eng Anal Bound Elem 26:905–916
10. Hwu C (2012) Boundary element formulation for the coupled stretching-bending analysis of thin laminated plates. Eng Anal Bound Elem 36:1027–1039
11. Mukherjee S, Banerjee PK (1982) Boundary element analysis of anisotropic plates. Int J Numer Methods Eng 18(5):661–677
12. Narvaez A, Useche J (2020) The kriging integration method applied to the boundary element analysis of Poisson problems. Eng Anal Bound Elem 121:1–20

13. Narvaez A, Useche J (2022) A new radial basis integration method applied to the boundary element analysis of 2d scalar wave equations. Eng Anal Bound Elem 136:77–92
14. Reddy JN (2004) Mechanics of laminated composite plates and shells: theory and analysis, 2nd edn. CRC Press, New York
15. Shi G, Bezine G (1988) A general boundary integral formulation for the anisotropic plate bending problems. J Compos Mater 22:694–716
16. Telles JCF (1987) A self-adaptive coordinate transformation for efficient numerical evaluation of general boundary element integrals. Int J Numer Methods Eng 24:959–973
17. Timoshenko SP, Woinowsky-Krieger S (1964) The theory of plates and shells. Springer

Chapter 8
Boundary Element Analysis of Shear Deformable Composite Plates

Abstract This chapter presents the development of the application of the Boundary Element Method for the analysis of thick composite laminated plates. The fundamental solutions proposed for symmetric and asymmetric laminates, are used. The application of the Radial Integration Method for transforming domain integrals to boundary integrals is presented for the treatment of integrals related to distributed forces and inertial terms in the governing equation.

8.1 Introduction

The specialized literature shows that little work has been done on analyzing thick plates. One of the first works was published by Wang [9], who presented a BEM formulation for orthotropic shear deformable plates, taking into account the transverse shear deformation. Useche et al. [8] use this formulation for the harmonic analysis of shear-deformable orthotropic cracked plates, where the Radial Integration Method was used. Useche and Alvarez [7] presented a BEM approach for the elastodynamic analysis of thick multilayer composite plates, considering the dynamic stress analysis using the Reissner plate theory. Hsu et al. [3] propose a formulation based on the Boundary Element Method using a new fundamental solution for the governing equation of asymmetrically shear deformable composite plates.

This chapter presents the development of the application of the Boundary Element Method for the analysis of thick composite laminated plates. The fundamental solutions proposed by Wang [9] and Hsu et al. [3] for symmetric and asymmetric laminates, respectively, are used. The application of the Radial Integration Method for transforming domain integrals to boundary integrals is presented for the treatment of integrals related to distributed forces and inertial terms in the governing equation.

J. Useche Vivero, *The Boundary Element Method for Engineering Analysis of Plates and Shells*, Solid Mechanics and Its Applications 281,
https://doi.org/10.1007/978-3-031-93050-8_8

8.2 Symmetric Laminated Plate

8.2.1 *Equilibrium Equations in Terms of Displacements*

Consider a laminated plate of constant thickness h and bounded by the planes $+h/2 \leq x_3 \leq +h/2$. The midsurface of the laminate coincides with the $x_3 = 0$ surface (see Fig. 8.1). The laminate's fiber orientation angles in each layer are defined by the angle θ concerning the x_1-axis. From first-order shear deformation theory, the governing equations for shells with symmetric laminates are given by Reddy [4], Useche and Alvarez [7], Useche et al. [8]:

$$D_{11}\frac{\partial^2 w_1}{\partial x_1^2} + D_{12}\frac{\partial^2 w_2}{\partial x_2 \partial x_1} + D_{16}\left(\frac{\partial^2 u_1}{\partial x_2 \partial x_1} + \frac{\partial^2 w_2}{\partial x_1^2}\right) + D_{16}\frac{\partial^2 w_1}{\partial x_1 \partial x_2} + D_{26}\frac{\partial^2 w_2}{\partial x_2^2} + D_{66}\left(\frac{\partial^2 w_1}{\partial x_2^2} + \frac{\partial^2 w_2}{\partial x_2 \partial x_1}\right) - KA_{55}\left(\frac{\partial w_3}{\partial x_1} + w_1\right) - KA_{45}\left(\frac{\partial w_3}{\partial x_2} + w_2\right) = I_2\frac{\partial^2 w_1}{\partial t^2} + I_1\frac{\partial^2 u_1}{\partial t^2} \tag{8.1}$$

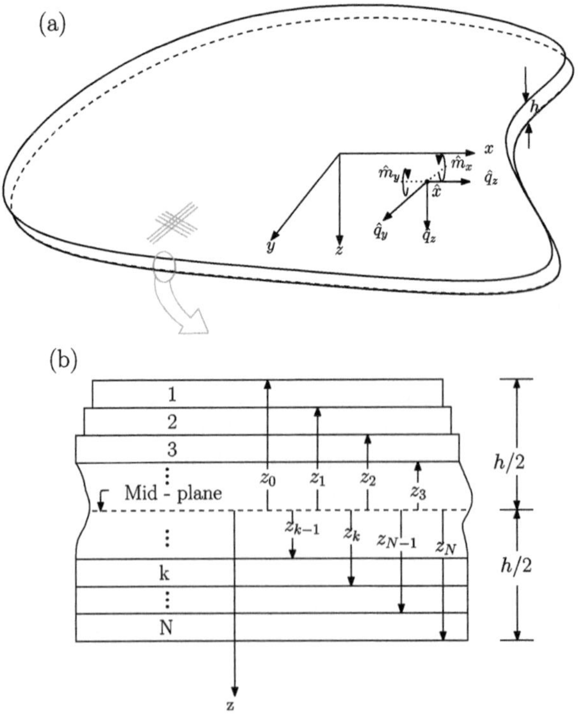

Fig. 8.1 Composite thick plate

$$D_{16}\frac{\partial^2 w_1}{\partial x_1^2} + D_{26}\frac{\partial^2 w_2}{\partial x_2 \partial x_1} + D_{66}\left(\frac{\partial^2 w_1}{\partial x_2 \partial x_1} + \frac{\partial^2 w_2}{\partial x_1^2}\right) + D_{12}\frac{\partial^2 w_1}{\partial x_1 \partial x_2} + D_{22}\frac{\partial^2 w_2}{\partial x_2^2} +$$
$$D_{26}\left(\frac{\partial^2 w_1}{\partial x_2^2} + \frac{\partial^2 w_2}{\partial x_2 \partial x_1}\right) - KA_{45}\left(\frac{\partial w_3}{\partial x_1} + w_1\right) - KA_{44}\left(\frac{\partial w_3}{\partial x_2} + w_2\right)$$
$$= I_2\frac{\partial^2 w_2}{\partial t^2} + I_1\frac{\partial^2 u_2}{\partial t^2} \tag{8.2}$$

$$KA_{55}\left(\frac{\partial^2 w_3}{\partial x_1^2} + \frac{\partial^2 w_1}{\partial x_1}\right) + KA_{45}\left(\frac{\partial^2 w_3}{\partial x_2 \partial x_1} + \frac{\partial w_2}{\partial x_1}\right) + KA_{45}\left(\frac{\partial^2 w_3}{\partial x_2 \partial x_1} + \frac{\partial w_1}{\partial x_2}\right) +$$
$$KA_{44}\left(\frac{\partial^2 w_3}{\partial x_2^2} + \frac{\partial w_2}{\partial x_2}\right) = I_0\frac{\partial^2 w_3}{\partial t^2} \tag{8.3}$$

8.2.2 Fundamental Solutions for Symmetrical Laminates

The fundamental solutions for symmetrical laminates start from considering the solution to the differential equation (See Wang [9]):

$$£^*_{ij} U^*_{kj}(\mathbf{x}', \mathbf{x}) = \delta(\mathbf{x}', \mathbf{x})\delta_{ki} \tag{8.4}$$

The solution to this equation is obtained using the plane wave decomposition method, to convert these equations to a system of ordinary differential equations (see Wang [9]), whose solution allows the fundamental solutions for symmetric laminates, presented below

$$U^*_{\alpha\beta}(\mathbf{x}', \mathbf{x}) = a_{\alpha\beta} D^4 \alpha\beta\phi(\rho) - C\omega_\alpha\omega_\beta D^2\phi(\rho)$$
$$U^*_{\alpha 3}(\mathbf{x}', \mathbf{x}) = -U^*_{3\alpha} = f_\alpha D^3\phi(\rho) - C\omega_\alpha D\phi(\rho)$$
$$U^*_{33}(\mathbf{x}', \mathbf{x}) = AD^4\phi(\rho) - BD^2\phi(\rho) + C\phi(\rho) \tag{8.5}$$

where $\rho = \cos(\theta)(x_1 - x_1') + \sin(\theta)(x_2 - x_2')$ and

$$\phi(\rho) = C_f\left(p^2\rho^2 \ln|\rho| + 2\ln|\rho| + 3 + e^{p\rho}\int_\rho^\infty \frac{e^{-p\sigma}}{\sigma}d\sigma - e^{-p\rho}\int_{-\infty}^\rho \frac{e^{-p\sigma}}{\sigma}d\sigma\right) \tag{8.6}$$

with $C_f = 1/8\pi^2 p^4 a$. Derivatives of $D\phi(\rho)$ are given by $D\phi(\rho) = d^k\phi(\rho)/d\rho^k$, $(k = 1, 2 \ldots)$ and

$$a_{11} = D_{66}A_{55}\omega_1^4 + D_{22}A_{44}\omega_2^4 + (D_{66}A_{44} + D_{22}A_{55} + 4D_{26}A_{45})\omega_1^2\omega_2^2$$
$$+ 2(D_{26}A_{44} + A_{45}D_{22})\omega_1\omega_2^3 + 2(D66A45 + A55D26)\omega_2\omega_1^3$$

$$
\begin{aligned}
a_{12} = a_{21} &= -D_{16}A_{55}\omega_1^4 - D_{26}A_{44}\omega_2^4 - (D_{16}A_{44} + D_{26}A_{55} + 2A_{45}(D_{12} + D_{66}))\omega_1^2\omega_2^2 \\
&- (2D_{26}A_{45} + A_{44}(D_{12} + D_{66}))\omega_1\omega_2^3 - (2D_{16}A_{45} + A_{55}(D_{12} + D_{66}))\omega_2\omega_1^3 \\
a_{22} &= D_{11}A_{55}\omega_{14} + D_{66}A_{44}\omega_{24} + (D_{11}A_{44} + D_{66}A_{55} + 4D_{16}A_{45})\omega_1^2\omega_2^2 + \\
&2(D_{16}A_{44} + A_{45}D_{66})\omega_1\omega_2^3 + 2(D_{11}A_{45} + A_{55}D_{16})\omega_2\omega_1^3 \\
f_1 &= (D_{16}A_{55} - D_{66}A_{55})\omega_1^3 + (D_{26}A_{44} - D_{22}A_{45})\omega_2^3 + (D_{16}A_{44} + D_{12}A_{45} - 2D_{26}A_{55})\omega_1^2\omega_2 \\
&+ ((D_{12} + D_{66})A_{44} - D_{22}A_{55} - D_{26}A_{45})\omega_1\omega_2^2 \\
f_2 &= (D_{16}A_{55} - D_{11}A_{55})\omega_1^3 + (D_{26}A_{45} - D_{66}A_{44})\omega_2^3 + (D_{12}A_{45} + D_{26}A_{55} - 2D_{16}A_{44})\omega_1^2\omega_2 \\
&+ ((D_{12} + D_{66})A_{55} - D_{11}A_{44} - D_{16}A_{45})\omega_1\omega_2^2 \\
A &= (D_{11}D_{66} - D_{16}^2)\omega^4 + (D_{22}D_{66} - D_{26}^2)\omega_2^4 + (D_{11}D_{22} + 2D_{16}D_{26} - D_{12}^2 - 2D_{12}D_{66})\omega_1^2\omega_2^2 \\
&+ 2(D_{11}D_{26} - D_{16}D_{12})\omega_1^3\omega_2 + 2(D_{16}D_{22} - D_{12}D_{26})\omega_1\omega_2^3 \\
B &= (A_{44}D_{11} - A_{55}D_{66} + 2A_{45}D_{16})\omega_1^2 + (A_{44}D_{66} + A_{55}D_{22} - 2A_{45}D_{26})\omega_2^2 \\
&+ 2(D_{16}A_{44} + A_{55}D_{26} - A_{45}(D_{12} + D_{26}))\omega_1\omega_2 \\
C &= A_{44}A_{55} - A_{45}^2
\end{aligned}
\tag{8.7}
$$

The *phi* function is given by

$$\phi(\rho) = C_f(p^2\rho^2 \ln|\rho| + 2\ln|\rho| + 3 + A_0(p\rho)) \tag{8.8}$$

The derivatives of $\phi(\rho)$ are given by

$$
\begin{aligned}
&D\phi(\rho) = C_f(2p^2\rho\ln|\rho| + p2\rho + pA1(p\rho)), \quad D^2\phi(\rho) = C_f(2p^2\ln|\rho| + 3p^2\rho + p^2A_0(p\rho)) \\
&D^3\phi(\rho) = C_f p^3 A_1(p\rho), \quad D^4\phi(\rho) = C_f p^4 A_0(p\rho) \\
&D^5\phi(\rho) = C_f\left(p^5 A_1(p\rho) - \frac{2p^4}{\rho}\right), \quad D^6\phi(\rho) = C_f\left(p^6 A_0(p\rho) + \frac{2p^4}{\rho^2}\right)
\end{aligned}
\tag{8.9}
$$

The functions $A_0(p\rho)$ and $A_1(p\rho)$ can be expressed in terms of exponential integrals $E_i(p\rho)$ and $E_1(p\rho)$ as follows. For $\rho > 0$:

$$A_0(p\rho) = e^{(p\rho)}E_1(p\rho) - e^{(-p\rho)}E_i(p\rho), \quad A_1(p\rho) = e^{(p\rho)}E_1(p\rho) + e^{(-p\rho)}E_i(p\rho) \tag{8.10}$$

For $\rho < 0$:

$$A_0(p\rho) = -e^{|p\rho|}E_1(|p\rho|) + e^{|-p\rho|}E_i(|p\rho|), \quad A_1(p\rho) = -e^{|p\rho|}E_1(|p\rho|) - e^{|-p\rho|}E_i(|p\rho|) \tag{8.11}$$

The fundamental solution P_{ij}^* is defined as

$$P_{ij}^*(\mathbf{x}', \mathbf{x}) = \int_0^{2\pi} \hat{P}_{ij}\, d\theta \tag{8.12}$$

where

$$\hat{P}_{i1} = \hat{M}_{i1}l + \hat{M}_{i6}m, \quad \hat{P}_{i2} = \hat{M}_{i6}l + \hat{M}_{i2}m, \quad \hat{P}_{i3} = \hat{Q}_{i1}l + \hat{Q}_{i2}m$$

where l and m are the direction cosines of the vector normal to the boundary in the element. The terms $\hat{M}_{i1}$, $\hat{M}_{i2}$, $\hat{M}_{i6}$, $\hat{Q}_{i1}$, and $\hat{Q}_{i1}$ are calculated as

$$\begin{bmatrix} \hat{M}_{11} & \hat{M}_{12} & \hat{M}_{16} \\ \hat{M}_{21} & \hat{M}_{22} & \hat{M}_{26} \\ \hat{M}_{31} & \hat{M}_{32} & \hat{M}_{36} \end{bmatrix} = \begin{bmatrix} \hat{\kappa}_{11} & \hat{\kappa}_{12} & \hat{\kappa}_{16} \\ \hat{\kappa}_{21} & \hat{\kappa}_{22} & \hat{\kappa}_{26} \\ \hat{\kappa}_{31} & \hat{\kappa}_{32} & \hat{\kappa}_{36} \end{bmatrix} \begin{bmatrix} D_{11} & D_{12} & D_{16} \\ D_{21} & D_{22} & D_{26} \\ D_{31} & D_{32} & D_{36} \end{bmatrix} \tag{8.13}$$

and

$$\begin{bmatrix} \hat{Q}_{11} & \hat{Q}_{12} \\ \hat{Q}_{21} & \hat{Q}_{22} \\ \hat{Q}_{31} & \hat{Q}_{32} \end{bmatrix} = \begin{bmatrix} \hat{\varepsilon}_{11} & \hat{\varepsilon}_{12} \\ \hat{\varepsilon}_{21} & \hat{\varepsilon}_{22} \\ \hat{\varepsilon}_{31} & \hat{\varepsilon}_{32} \end{bmatrix} \begin{bmatrix} A_{55} & A_{45} \\ A_{45} & A_{44} \end{bmatrix} \tag{8.14}$$

where

$$\begin{aligned}
&\hat{\kappa}_{11} = a_{11}\omega_1^3 D^5\phi(\rho) - C\omega_1^3 D^3\phi(\rho), && \hat{\kappa}_{12} = a_{12}\omega_2 D^5\phi(\rho) - C\omega_1\omega_2^2 - D^3\phi(\rho) \\
&\hat{\kappa}_{16} = (a_{11}\omega_2 + a_{12}\omega_1)D^5\phi(\rho) - 2C\omega_2\omega_1^2 D^3\phi(\rho), && \hat{\kappa}_{21} = a_{21}\omega_1 D^5\phi(\rho) - C\omega_1^2\omega_2 D^3\phi(\rho) \\
&\hat{\kappa}_{22} = a_{22}\omega_2 D^5\phi(\rho) - C\omega_2^3 D^3\phi(\rho), && \hat{\kappa}_{26} = (a_{21}\omega_2 + a_{22}\omega_1)D^5\phi(\rho) - 2C\omega_1\omega_2^2 D^3\phi(\rho) \\
&\hat{\kappa}_{31} = -f_1\omega_1 D^4\phi(\rho) - 2C\omega_1^2 D^2\phi(\rho), && \hat{\kappa}_{32} = -f_2\omega_2 D^4\phi(\rho) - 2C\omega_2^2 D^2\phi(\rho) \\
&\hat{\kappa}_{36} = -(f_1\omega_2 + f_2\omega_1)D^4\phi(\rho) - 2C\omega_1\omega_2 D^2\phi(\rho), && \hat{\varepsilon}_{14} = -(f_1\omega_2 + a_{12})D^4\phi(\rho) \\
&\hat{\varepsilon}_{15} = -(f_1\omega_1 + a_{11})D^4\phi(\rho), && \hat{\varepsilon}_{24} = -(f_2\omega_2 + a_{22})D^4\phi(\rho) \\
&\hat{\varepsilon}_{25} = -(f_2\omega_1 + a_{21})D^4\phi(\rho), && \hat{\varepsilon}_{34} = A\omega_2 D^5\phi(\rho) - (B\omega_2 + f_2)D^3\phi(\rho) \\
&\hat{\varepsilon}_{35} = A\omega_1 D^5\phi(\rho) - (B\omega_1 + f_1)D^3\phi(\rho) &&
\end{aligned} \tag{8.15}$$

To treat the resulting integrals Υ_1, shown in Eq. (8.12), their solution must be carried out in four intervals in the range $0 \leq \theta \leq 2\pi$:

$$\Upsilon_1 = \int_0^{2\pi} F_1(\theta) D^k \phi(\rho) d\theta, \; k = 1, \ldots, 5 \tag{8.16}$$

$$\left[\theta_0, \theta_0 + \frac{\pi}{2}\right], \quad \left[\theta_0 + \frac{\pi}{2}, \theta_0 + \pi\right], \quad \left[\theta_0 + \pi, \theta_0 + \frac{3\pi}{2}\right]$$
$$\left[\theta_0 + \pi, \theta_0 + \frac{3\pi}{2}\right], \quad \left[\theta_0 + \frac{3\pi}{2}, \theta_0 + 2\pi\right]$$

The value of θ_0 is calculated with the following formula: $\theta_0 = \tan^{-1}\left(-\frac{x-\epsilon}{y-\eta}\right)$. The following box presents the computational implementation of the fundamental solutions presented here.

Kernels calculation Matlab function

```
function [He,Ge,FCij] = KernelP(R,RX,RY,n,e)
%Propiedades del material
%----------------------------------------------------------------------------
global CTE AT D K_Amatriz NPGSF TOL
%OJO revisar los factores de corrección cortante
A44=AT(1,1); A45=AT(1,2); A55=AT(2,2); %Factor agregado en la definición
D11=D(1,1); D12=D(1,2); D16=D(1,3); D22= D(2,2); D26=D(2,3); D66=D(3,3);
%----------------------------------------------------------------------------
[eg wg] = gaussNodes(NPGSF,TOL);
n_GPoints=length(eg(:,1));%CANTIDAD DE PUNTOS DE GAUSS
%------- FUNCIONES DE FORMA DISCONTINUAS PARA ELEMENTOS CUADRATICOS--------
phi1 = e*(9/8*e-3/4);
phi2 = 1-9/4*e^2;
phi3 = e*(9/8*e+3/4);
PHI=[phi1   0      0      phi2   0      0      phi3   0      0;...
     0      phi1   0      0      phi2   0      0      phi3   0;...
     0      0      phi1   0      0      phi2   0      0      phi3];
%----------------------------------------------------------------------------
%INICIALIZACIÓN DE LA SOLUCIÒN FUNDAMENTAL A SER INTEGRADA
Uast=zeros(3,3); Past=zeros(3,3);
theta0=atan2(-(RX),RY);
%INTEGRACIÓN
%----------------------------------------------------------------------------
for I=1:2
    y1=theta0;
    y2=theta0+pi/2;
    if(I==1||I==3)
        y_f=y1;
    else
        y_f=y2;
    end
    eet = (y1+y2-2*y_f)/(y1-y2);
for i=1:n_GPoints; %Solución de la integral de 0 a 2pi de la sol fund
 [et,Jt] = telles(eg(i),eet);
 theta=theta0+(et+1)*pi/4;
 Jac=pi/4;
 w1 = cos(theta); w2 = sin(theta);
 rho = w1*RX + w2*RY;
%----------------------------------------------------------------------------
d11=D11*w1^2 + 2*D16*w1*w2 + D66*w2^2;
d12=D16*w1^2 + (D12+D66)*w1*w2 + D26*w2^2;
d13=A45*w2 + A55*w1;
% d23=A44*w2 + A45*w1;
% d22=D66*w1^2 + 2*D26*w1*w2 + D22*w2^2;
% d33=A55*w1^2 + 2*A45*w1*w2 + A44*w2^2;
%----------------------------------------------------------------------------
a11=D66*A55*w1^4 + D22*A44*w2^4 + (D66*A44 + D22*A55 + 4*D26*A45)*...
    w1^2*w2^2 + 2*(D26*A44 + A45*D22)*w1*w2^3 + 2*(D66*A45+A55*D26)*...
    w2*w1^3;
a12=-D16*A55*w1^4 - D26*A44*w2^4 - (D16*A44+D26*A55 + 2*A45*(D12 + D66))*...
w1^2*w2^2 - (2*D26*A45 + A44*(D12 + D66))*w1*w2^3 - (2*D16*A45 + A55*(D12 + ...
D66))*w2*w1^3;
a21=a12;
a22=D11*A55*w1^4 + D66*A44*w2^4 + (D11*A44 + D66*A55 + 4*D16*A45)*w1^2*w2^2+...
    2*(D16*A44+A45*D66)*w1*w2^3 + 2*(D11*A45 + A55*D16)*w2*w1^3;
f1=(D16*A55-D66*A55)*w1^3 + (D26*A44 - D22*A45)*w2^3 + (D16*A44 + D12*A45 -...
    2*D26*A55)*w1^2*w2 + ((D12 + D66)*A44 - D22*A55 - D26*A45)*w1*w2^2;
f2=(D16*A55-D11*A45)*w1^3 + (D26*A45 - D66*A44)*w2^3 + (D12*A45 + D26*A55 -...
    2*D16*A44)*w1*w2^2 + ((D12 + D66)*A55 - D11*A44 - D16*A45)*w1^2*w2;
A=(D11*D66-D16^2)*w1^4 + (D22*D66 - D26^2)*w2^4 + (D11*D22 + 2*D16*D26 -...
    D12^2 - 2*D12*D66)*w1^2*w2^2 + 2*(D11*D26 - D16*D12)*w1^3*w2 +...
    2*(D16*D22 - D12*D26)*w1*w2^3;
```

```
B=(A44*D11 + A55*D66 - 2*A45*D16)*w1^2 + (A44*D66 +A55*D22 - 2*A45*D26)*w2^2 + ...
    2*(D16*A44 + A55*D26 - A45*(D12 + D26))*w1*w2;
C=A44*A55 - A45^2;
%---------------------------------------------------------------------------
a=a11*d11 + a12*d12;
b=a11*A55 + C*d11*w1^2 + a12*A45 + C*d12*w1*w2 + f1*d13;
p=sqrt(b/a);
Cf= 1/(8*pi^2*p^4*a);
%CALCULA LA INTEGRAL EXPONENCIAL
Ei = real(-expint(-p*rho));
E1=expint(p*rho);
%---------------------------------------------------------------------------
if(rho>0)
    A0=exp(p*rho)*E1-exp(-p*rho)*Ei;
    A1=exp(p*rho)*E1+exp(-p*rho)*Ei;
else
    A0=-exp(-abs(p*rho))*Ei + exp(abs(p*rho))*E1;
    A1=-exp(-abs(p*rho))*Ei - exp(abs(p*rho))*E1;
end
phi=Cf*(p^2*rho^2*log(abs(rho)) + 2*log(abs(rho)) + 3 + A0);
dphidrho=Cf*(2*p^2*rho*log(abs(rho)) + p^2*rho + p*A1);
d2phidrho2=Cf*(2*p^2*log(abs(rho)) + 3*p^2 + p^2*A0);
d3phidrho3=Cf*p^3*A1;
d4phidrho4=Cf*p^4*A0;
d5phidrho5=Cf*(p^5*A1 - ((2*p^4)/rho));
d6phidrho6=Cf*(p^6*A0 + ((2*p^4)/(rho^2)));
%---------------------------------------------------------------------------
%SOLUCIÓN FUNDAMENTAL U
Utilde(1,1)= a11*d4phidrho4 - (C*w1^2)*d2phidrho2;
Utilde(1,2)= a12*d4phidrho4 - (C*w1*w2)*d2phidrho2;
Utilde(2,1)= a21*d4phidrho4 - (C*w2*w1)*d2phidrho2;
Utilde(2,2)= a22*d4phidrho4 - (C*w2^2)*d2phidrho2;
Utilde(1,3)= f1*d3phidrho3 + (C*w1)*dphidrho;
Utilde(2,3)= f2*d3phidrho3 + (C*w2)*dphidrho;
Utilde(3,1)= - f1*d3phidrho3 - (C*w1)*dphidrho;
Utilde(3,2)= - f2*d3phidrho3 - (C*w2)*dphidrho;
Utilde(3,3)= A*d4phidrho4 -B*d2phidrho2 +C*phi;
%---------------------------------------------------------------------------
%SOLUCIÓN FUNDAMENTAL P
k11=a11*w1*d5phidrho5 - C*w1^3*d3phidrho3;
k12=a12*w2*d5phidrho5 - C*w1*w2^2*d3phidrho3;
k16=(a11*w2 + a12*w1)*d5phidrho5 - 2*C*w2*w1^2*d3phidrho3;
k21=a21*w1*d5phidrho5 - C*w1^2*w2*d3phidrho3;
k22=a22*w2*d5phidrho5 - C*w2^3*d3phidrho3;
k26=(a21*w2 + a22*w1)*d5phidrho5 - 2*C*w1*w2^2*d3phidrho3;
k31= -f1*w1*d4phidrho4 - C*w1^2*d2phidrho2;
k32= -f2*w2*d4phidrho4 - C*w2^2*d2phidrho2;
k36= -(f1*w2 + f2*w1)*d4phidrho4 - 2*C*w1*w2*d2phidrho2;
e14= (f1*w2 + a12)*d4phidrho4;
e15= (f1*w1 + a11)*d4phidrho4;
e24= (f2*w2 + a22)*d4phidrho4;
e25= (f2*w1 + a21)*d4phidrho4;
e34= A*w2*d5phidrho5 - (B*w2 + f2)*d3phidrho3;
e35= A*w1*d5phidrho5 - (B*w1 + f1)*d3phidrho3;
k_ij=[ k11 k12 k16;...
       k21 k22 k26;...
       k31 k32 k36];
e_ij=[ e15 e14;...
       e25 e24;...
       e35 e34];
%Asegurarse que 'D' solo corresponda a la matriz de rigidez a flexión.
M= k_ij*D;
Q= e_ij*[A55 A45;A45 A44];
%---------------------------------------------------------------------------
```

```
l = n(1); m = n(2);
Ptilde(1,1)=M(1,1)*l + M(1,3)*m; Ptilde(1,2)=M(1,3)*l + M(1,2)*m;
Ptilde(1,3)=Q(1,1)*l + Q(1,2)*m; Ptilde(2,1)=M(2,1)*l + M(2,3)*m;
Ptilde(2,2)=M(2,3)*l + M(2,2)*m; Ptilde(2,3)=Q(2,1)*l + Q(2,2)*m;
Ptilde(3,1)=M(3,1)*l + M(3,3)*m; Ptilde(3,2)=M(3,3)*l + M(3,2)*m;
Ptilde(3,3)=Q(3,1)*l + Q(3,2)*m;
%---------------------------------------------------------------------------
% Utilde
% Ptilde
% pause;
Uast=Uast + 2*Utilde*wg(i)*Jac*Jt;
Past=Past + 2*Ptilde*wg(i)*Jac*Jt;
end
    theta0=theta0+pi/2;
end
U11=Uast(1,1); U12=Uast(1,2); U13=Uast(1,3);
U21=Uast(2,1); U22=Uast(2,2); U23=Uast(2,3);
U31=Uast(3,1); U32=Uast(3,2); U33=Uast(3,3);
%-------------------
P11=Past(1,1); P12=Past(1,2); P13=Past(1,3);
P21=Past(2,1); P22=Past(2,2); P23=Past(2,3);
P31=Past(3,1); P32=Past(3,2); P33=Past(3,3);

Ge = [U11 U12 U13;U21 U22 U23;U31 U32 U33]*PHI;
He = [P11 P12 P13;P21 P22 P23;P31 P32 P33]*PHI;
%----- ARGUMENTO DE LA INTEGRAL Cij
HS11 = P11 - RX*P13; HS12 = P12 - RY*P13; HS21 = P21 - RX*P23;
HS22 = P22 - RY*P23; HS31 = P31 - RX*P33; HS32 = P32 - RY*P33;
HS13 = P13; HS23 = P23; HS33 = P33;
FCij = [HS11 HS12 HS13;HS21 HS22 HS23;HS31 HS32 HS33];
return
```

8.3 Unsymmetric Laminates

The fundamental solution for thick plates laminated with asymmetric composites was recently proposed by Hsu et al. [3]. The fundamental solution is based on plane wave decomposition. The equations for asymmetric laminates for thick plates are given by

$$
\begin{aligned}
& B_{11}\frac{\partial^2 u_1}{\partial x_1^2} + B_{12}\frac{\partial^2 u_2}{\partial x_2 \partial x_1} + B_{16}\left(\frac{\partial^2 u_1}{\partial x_2 \partial x_1} + \frac{\partial^2 u_2}{\partial x_1^2}\right) + D_{11}\frac{\partial^2 w_1}{\partial x_1^2} + D_{12}\frac{\partial^2 w_2}{\partial x_2 \partial x_1} \\
& + D_{16}\left(\frac{\partial^2 u_1}{\partial x_2 \partial x_1} + \frac{\partial^2 w_2}{\partial x_1^2}\right) + B_{16}\frac{\partial^2 u_1}{\partial x_2 \partial x_1} + B_{26}\frac{\partial^2 u_2}{\partial x_2^2} + B_{66}\left(\frac{\partial^2 u_1}{\partial x_2^2} + \frac{\partial^2 u_2}{\partial x_1 \partial x_2}\right) \\
& \qquad + D_{16}\frac{\partial^2 w_1}{\partial x_1 \partial x_2} + D_{26}\frac{\partial^2 w_2}{\partial x_2^2} + D_{66}\left(\frac{\partial^2 w_1}{\partial x_2^2} + \frac{\partial^2 w_2}{\partial x_2 \partial x_1}\right) \\
& \qquad - K A_{55}\left(\frac{\partial w_3}{\partial x_1} + w_1\right) - K A_{45}\left(\frac{\partial w_3}{\partial x_2} + w_2\right) \\
& \qquad = I_2\frac{\partial^2 w_1}{\partial t^2} + I_1\frac{\partial^2 u_1}{\partial t^2}
\end{aligned}
\tag{8.17}
$$

$$B_{16}\frac{\partial^2 u_1}{\partial x_1^2} + B_{26}\frac{\partial^2 u_2}{\partial x_2 \partial x_1} + B_{66}\left(\frac{\partial^2 u_1}{\partial x_2 \partial x_1} + \frac{\partial^2 u_2}{\partial x_1^2}\right) + D_{16}\frac{\partial^2 w_1}{\partial x_1^2} + D_{26}\frac{\partial^2 w_2}{\partial x_2 \partial x_1}$$
$$+D_{66}\left(\frac{\partial^2 w_1}{\partial x_2 \partial x_1} + \frac{\partial^2 w_2}{\partial x_1^2}\right) + B_{12}\frac{\partial^2 u_1}{\partial x_2 \partial x_1} + B_{22}\frac{\partial^2 u_2}{\partial x_2^2} + B_{26}\left(\frac{\partial^2 u_1}{\partial x_2^2} + \frac{\partial^2 u_2}{\partial x_1 \partial x_2}\right)$$
$$+D_{12}\frac{\partial^2 w_1}{\partial x_1 \partial x_2} + D_{22}\frac{\partial^2 w_2}{\partial x_2^2} + D_{26}\left(\frac{\partial^2 w_1}{\partial x_2^2} + \frac{\partial^2 w_2}{\partial x_2 \partial x_1}\right)$$
$$-KA_{45}\left(\frac{\partial w_3}{\partial x_1} + w_1\right) - KA_{44}\left(\frac{\partial w_3}{\partial x_2} + w_2\right) =$$
$$I_2\frac{\partial^2 w_2}{\partial t^2} + I_1\frac{\partial^2 u_2}{\partial t^2} \tag{8.18}$$

$$KA_{55}\left(\frac{\partial^2 w_3}{\partial x_1^2} + \frac{\partial^2 w_1}{\partial x_1}\right) + KA_{45}\left(\frac{\partial^2 w_3}{\partial x_2 \partial x_1} + \frac{\partial w_2}{\partial x_1}\right) + KA_{45}\left(\frac{\partial^2 w_3}{\partial x_2 \partial x_1} + \frac{\partial w_1}{\partial x_2}\right) +$$
$$KA_{44}\left(\frac{\partial^2 w_3}{\partial x_2^2} + \frac{\partial w_2}{\partial x_2}\right) = I_0\frac{\partial^2 w_3}{\partial t^2} \tag{8.19}$$

In a compact form, we have

$$L^*\mathbf{v} + \mathbf{p} = \mathbf{0} \tag{8.20}$$

where $\mathbf{v} = \{\mathbf{u}, \boldsymbol{\beta}, w\}^T$; $\mathbf{u} = \{u_1, u_2\}^T$, $\boldsymbol{\beta} = \{w_1, w_2\}^T$; $\mathbf{p} = \{\mathbf{f}, \boldsymbol{m}, q_3\}^T$; $\mathbf{f} = \{f_1, f_2\}^T$, $\mathbf{m} = \{m_1, m_2\}^T$. The operator L^* is given by

$$L^* = \begin{bmatrix} L^T\mathbf{A}L & L^T\mathbf{B}L & \mathbf{0} \\ L^T\mathbf{B}L & L^T\mathbf{A}L - \mathbf{A}_t & -\mathbf{A}_t\nabla \\ \mathbf{0} & \nabla^T\mathbf{A}_t & \nabla^T\mathbf{A}_t\nabla \end{bmatrix} \tag{8.21}$$

With:

$$L = \begin{bmatrix} ()_{,1} & 0 \\ 0 & ()_{,2} \\ ()_{,2} & ()_{,1} \end{bmatrix}, \nabla = \begin{Bmatrix} ()_{,1} \\ ()_{,2} \end{Bmatrix} \tag{8.22}$$

8.3.1 Fundamental Solutions

Recently, the fundamental solution for unsymmetric shear deformable plates has been proposed using the plane wave decomposition method of the governing equations. Consider a shear deformable laminated composite plate with infinite dimension subjected to concentrated forces: $\mathbf{p} = \hat{\mathbf{p}}\delta(\mathbf{x}', \mathbf{x})$, where $\hat{\mathbf{p}} = \{\hat{f}_1, \hat{f}_2, \hat{m}_1, \hat{m}_2, \hat{q}_3\}^T$ acting at the source: $\mathbf{x}'$. The fundamental solution to this problem is obtained by solving the equation (see Hsu et al. [3], Hsu and Hwu [2]):

$$L^*_{ij} W^*_{ij} + \hat{p}_j \delta(\mathbf{x}', \mathbf{x}) = 0 \tag{8.23}$$

In the solution of this equation by plane wave decomposition, the integral transformations are used (Fig. 8.2):

$$\delta(\mathbf{x}', \mathbf{x}) = -\frac{1}{4\pi^2} \int_0^{2\pi} \rho^{-2} d\theta, \quad W^*_{ij} = \int_0^{2\pi} \hat{v}_{ij}(\rho) d\theta \tag{8.24}$$

where $\rho = (x_1 - x_1') \cos(\theta) + (x_2 - x_2') \sin(\theta) = (\mathbf{x} - \mathbf{x}')\boldsymbol{\omega}$. The governing equation then depends on $\hat{\mathbf{v}}(\rho)$, which is obtained by the development presented in Hsu and Hwu, where the fundamental solution W^*_{ij}, of displacement in the x_i direction at a field point $\mathbf{x}$ for a concentrated unit force $\hat{p}$ applied in the x_j direction is given by the expression:

$$\mathbf{W}^*(\mathbf{x}', \mathbf{x}) = \frac{1}{4\pi^2} \left[\int_0^{2\pi} \mathbf{I}_c \mathbf{Z} \mathbf{F}(\rho) \mathbf{Z}^{-1} d\theta \right] \hat{\mathbf{f}} \tag{8.25}$$

where $\hat{\mathbf{f}} = \{\mathbf{0}, \mathbf{L}_2^{-1}\hat{\mathbf{p}}\}^T$; $\mathbf{I}_c = [\mathbf{I}, \mathbf{0}]^T$ y $\mathbf{I}_r = [\mathbf{0}, \mathbf{I}]^T$ are constant matrices of 5×10. The matrix $\mathbf{Z}$ is given by

$$\mathbf{Z} = [\mathbf{z}_1, \mathbf{z}_2, \ldots, \mathbf{z}_{10}] \tag{8.26}$$

Fig. 8.2 Polar decomposition

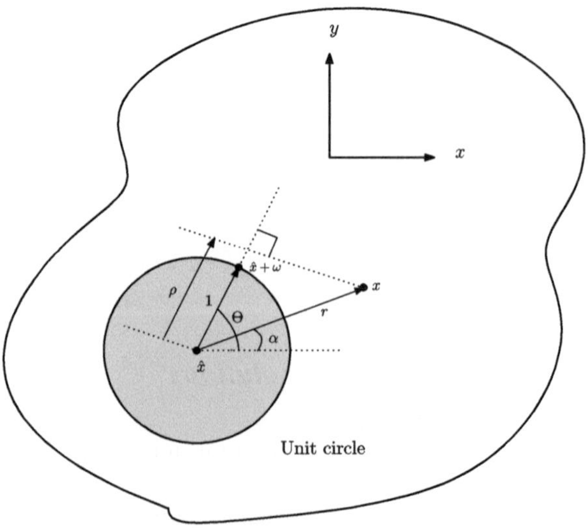

where

$$\mathbf{z}_1 = \{\mathbf{i}_1, \mathbf{0}\}^T, \quad \mathbf{z}_2 = \{\mathbf{i}_1, \mathbf{i}_1\}^T, \quad \mathbf{z}_3 = \{\mathbf{i}_2, \mathbf{0}\}^T, \quad \mathbf{z}_4 = \{\mathbf{i}_2, \mathbf{i}_2\}^T$$
$$\mathbf{z}_5 = \{\mathbf{i}_5, \mathbf{0}\}^T, \quad \mathbf{z}_6 = \{\boldsymbol{\omega}_6, \mathbf{i}_5\}^T, \quad \mathbf{z}_7 = \{\boldsymbol{\omega}_7, \boldsymbol{\omega}_6\}^T, \quad \mathbf{z}_8 = \{\boldsymbol{\omega}_8, \boldsymbol{\omega}_7\}^T,$$
$$\mathbf{z}_9 = \{\mathbf{g}^+, \lambda_d \mathbf{g}^+\}^T, \quad \mathbf{z}_{10} = \{\mathbf{g}^-, -\lambda_d \mathbf{g}^-\}^T \tag{8.27}$$

where $\mathbf{i}_n$, $i = 1, 2, \ldots, 5$ is the 5×1 basis vector whose nth component equals one and the others are zeros. The vector $\mathbf{g}^\pm$ is defined as

$$\mathbf{g}^\pm = \begin{Bmatrix} -\tilde{\mathbf{A}}^{-1}\tilde{\mathbf{B}}\mathbf{d} \\ \mathbf{d} \\ \mp\lambda_d^{-1}\tilde{A}_t^{-1}\boldsymbol{\omega}^T\mathbf{A}_t\mathbf{d} \end{Bmatrix} \tag{8.28}$$

$\lambda_d^2 = tr(\mathbf{D}_0) = d_{11} + d_{22}$, with matrix $\mathbf{D}_0$ defined as

$$\mathbf{D}_0 = \begin{bmatrix} d_{11} & d_{12} \\ d_{12} & d_{22} \end{bmatrix} = \tilde{\mathbf{D}}^*\tilde{\mathbf{A}}_t(\mathbf{I} - \tilde{A}_t^{-1}\boldsymbol{\omega}\boldsymbol{\omega}^T\mathbf{A}_t) \tag{8.29}$$

with,

$$\tilde{\mathbf{D}}^* = (\tilde{\mathbf{D}} - \tilde{\mathbf{B}}\tilde{\mathbf{A}}^{-1}\tilde{\mathbf{B}})^{-1} \tag{8.30}$$

$$\tilde{\mathbf{D}} = \boldsymbol{\Omega}^T\mathbf{D}\boldsymbol{\Omega}, \quad \mathbf{D} = \begin{bmatrix} D_{11} & D_{12} & D_{16} \\ D_{12} & D_{22} & D_{26} \\ D_{16} & D_{26} & D_{66} \end{bmatrix} \tag{8.31}$$

and

$$\mathbf{d} = \begin{cases} \{1, d_{21}/d_{11}\}^T, & when\ d_{11} \neq 0 \\ \{0, 1\}^T, & when\ d_{11} = 0 \end{cases} \tag{8.32}$$

$$\boldsymbol{\omega}_6 = \begin{Bmatrix} \tilde{\mathbf{A}}^{-1}\tilde{\mathbf{B}}\boldsymbol{\omega} \\ -\boldsymbol{\omega} \\ 1 \end{Bmatrix}, \quad \boldsymbol{\omega}_7 = \begin{Bmatrix} \mathbf{0} \\ -\boldsymbol{\omega} \\ 1 \end{Bmatrix}, \quad \boldsymbol{\omega}_8 = \begin{Bmatrix} \mathbf{0} \\ -(\mathbf{I} + \mathbf{A}_t^{-1}\tilde{\mathbf{D}}^{*-1})\boldsymbol{\omega} \\ 0 \end{Bmatrix} \tag{8.33}$$

$$\mathbf{A}_t = K\begin{bmatrix} A_{55} & A_{45} \\ A_{45} & A_{44} \end{bmatrix}, \quad \tilde{A}_t = \boldsymbol{\omega}^T\mathbf{A}_t\boldsymbol{\omega}, \quad \tilde{\mathbf{A}} = \boldsymbol{\Omega}^T\mathbf{A}\boldsymbol{\Omega} \tag{8.34}$$

$$\boldsymbol{\Omega} = \begin{bmatrix} \cos(\theta) & 0 \\ 0 & \sin(\theta) \\ \sin(\theta) & \cos(\theta) \end{bmatrix}, \quad \mathbf{A} = \begin{bmatrix} A_{11} & A_{12} & A_{16} \\ A_{12} & A_{22} & A_{26} \\ A_{16} & A_{26} & A_{66} \end{bmatrix} \tag{8.35}$$

$$\tilde{\mathbf{B}} = \boldsymbol{\Omega}^T \mathbf{B} \boldsymbol{\Omega}, \quad \mathbf{B} = \begin{bmatrix} B_{11} & B_{12} & B_{16} \\ B_{12} & B_{22} & B_{26} \\ B_{16} & B_{26} & B_{66} \end{bmatrix} \tag{8.36}$$

On the other hand, the matrix $\mathbf{F}(\rho)$ is defined as

$$\mathbf{F}(\rho) = [\mathbf{F}_a(\rho), \mathbf{F}_a(\rho), \mathbf{F}_c(\rho), \mathbf{F}_d(\rho)] \tag{8.37}$$

where

$$\mathbf{F}_a(\rho) = \begin{bmatrix} -\rho^{-1} & -\ln|\rho| - 1 \\ 0 & -\rho^{-1} \end{bmatrix} \tag{8.38}$$

$$\mathbf{F}_c(\rho) = \begin{bmatrix} -\rho^{-1} & -\ln|\rho| - 1 & -\rho\ln|\rho| & \rho^2(1 - 2\ln|\rho|)/4 \\ 0 & -\rho^{-1} & -\ln|\rho| - 1 & -\rho\ln|\rho| \\ 0 & 0 & -\rho^{-1} & -\ln|\rho| - 1 \\ 0 & 0 & 0 & -\rho^{-1} \end{bmatrix} \tag{8.39}$$

$$\mathbf{F}_d(\rho) = \begin{bmatrix} F_1(\rho) & 0 \\ 0 & F_2(\rho) \end{bmatrix} \tag{8.40}$$

where

$$F_1(\rho) = e^{\lambda_d \rho} \int \rho^{-2} e^{-\lambda_d \rho} d\rho, \quad F_2(\rho) = -e^{\lambda_d \rho} \int \rho^{-2} e^{\lambda_d \rho} d\rho \tag{8.41}$$

To evaluate numerically and efficiently, these integrals can be expressed as

$$\begin{aligned} F_1(\rho) &= -\rho^{-1} + \lambda_d e^{-\lambda} R_e\{Ei(-\lambda_d \rho)\} \\ F_2(\rho) &= -\rho^{-1} - \lambda_d e^{-\lambda} R_e\{Ei(\lambda_d \rho)\} \end{aligned} \tag{8.42}$$

where E_i is the exponential integral function. The above integrals are evaluated in the Cauchy sense, since the integrand becomes singular for $\lambda_d \rho = 0$. Numerical approximations for evaluating this integral are proposed in Wang [9] and Hsu et al. [3]. However, in Hsu et al. [3], an efficient approximation is proposed for evaluating the above expressions, which avoids numerical instabilities for large values of $|\lambda_d \rho|$.

On the other hand, the fundamental solutions for force resultants **N**, **M**, and **Q** are given by

$$\mathbf{P}^*(\mathbf{x}', \mathbf{x}) = [\mathbf{N}^*, \mathbf{M}^*, \mathbf{Q}^*] \tag{8.43}$$

where

$$\mathbf{N}^*(\mathbf{x}', \mathbf{x}) = \frac{1}{4\pi^2} \left[\int_0^{2\pi} (\mathbf{A}\boldsymbol{\Omega}\mathbf{I}_u + \mathbf{B}\boldsymbol{\Omega}\mathbf{I}_\beta)\mathbf{I}_u \mathbf{Z}\mathbf{F}(\rho)\mathbf{Z}^{-1} d\theta \right] \hat{\mathbf{f}}$$

$$\mathbf{M}^*(\mathbf{x}',\mathbf{x}) = \frac{1}{4\pi^2}\left[\int_0^{2\pi}(\mathbf{B}\boldsymbol{\Omega}\mathbf{I}_u + \mathbf{D}\boldsymbol{\Omega}\mathbf{I}_\beta)\mathbf{I}_u\mathbf{Z}\mathbf{F}(\rho)\mathbf{Z}^{-1}d\theta\right]\hat{\mathbf{f}}$$

$$\mathbf{Q}^*(\mathbf{x}',\mathbf{x}) = \frac{1}{4\pi^2}\left[\int_0^{2\pi}\mathbf{A}_t(\mathbf{I}_\beta\mathbf{I}_c + \boldsymbol{\omega}\mathbf{i}_5^T\mathbf{I}_r)\mathbf{Z}\mathbf{F}(\rho)\mathbf{Z}^{-1}d\theta\right]\hat{\mathbf{f}} \tag{8.44}$$

where $\mathbf{i}_u = \{\mathbf{i}_1^T, \mathbf{i}_2^T\}^T$ y $\mathbf{i}_\beta = \{\mathbf{i}_3^T, \mathbf{i}_4^T\}^T$.

8.3.2 *Numerical Evaluation of Integrals*

On the other hand, the transformation integrals (8.24) present singularities for $\rho = 0$ at $\theta = \theta^*$ and $\theta = \theta^* + \pi$. These can be efficiently evaluated using the Gaussian integration scheme, employing the coordinate transformation. To do so, consider the transformation integral:

$$I = \int_0^{2\pi} g(\rho)h(\theta)d\theta \tag{8.45}$$

with $g(\rho) = \rho^{-1}$, $\ln(\rho)$, $e^{-\lambda}R_e\{Ei(\pm\lambda_d\rho)\}$, etc. This can be evaluated numerically as

$$I \approx \frac{\pi}{2}\sum_{n=1}^{2}\sum_{i=1}^{n_G} w_i^G g(\rho(\zeta(t_i^G)))h(\theta(\zeta(t_i^G)))J(t_i^G) \tag{8.46}$$

t_i^G y w_i^G are the abscissas and weights of the Gauss points, n_G is the number of Gauss points and

$$\begin{aligned} \zeta(t) &= \frac{1}{4}(t+1)^3, \quad J(t) = \frac{3}{4}(t+1)^3, \qquad when\ n = 1 \\ \zeta(t) &= \frac{1}{4}(t-1)^3, \quad J(t) = \frac{3}{4}(t-1)^3, \qquad when\ n = 2 \end{aligned} \tag{8.47}$$

Finally, integrals with strong singularity of the form:

$$I = \int_0^{2\pi} \rho^{-2}h(\theta)d\theta \tag{8.48}$$

are treated by singularity subtraction:

$$I = \int_0^{2\pi} \rho^{-2}h(\theta)d\theta = \int_0^{2\pi} \rho^{-2}[h(\theta) - h(\theta^*)]d\theta + \int_0^{2\pi} \rho^{-2}d\theta \tag{8.49}$$

The first integral on the right-hand side is evaluated by the expression derived above, while the last integral is equal to $-4\pi^2\delta(\mathbf{x}', \mathbf{x})$, according to the definition given in Hsu et al. [3].

8.4 Boundary Element Equations

The derivation of the integral formulation for the governing equations presented in Sects. 8.2.1 and 8.3 follows the same procedure presented in Chap. 5 for isotropic plates. Therefore, the integral formulation for these equations (considering $q_\alpha = 0$ and $q_3 = p$) is given by

$$c_{ik} w_k(\mathbf{x}', t) + \int_\Gamma P^*_{ik}(\mathbf{x}', \mathbf{x}) w_k(\mathbf{x}, t) d\Gamma = \int_\Gamma W^*_{ik}(\mathbf{x}', \mathbf{x}) p_k(\mathbf{x}, t) d\Gamma + \int_\Omega W^*_{i3}(\mathbf{x}', \mathbf{x}) p(\mathbf{x}, t) d\Omega - \int_\Omega W^*_{ik}(\mathbf{x}', \mathbf{x}) \Lambda_{ik} w_{k,tt}(\mathbf{x}, t) d\Omega \tag{8.50}$$

In this equation, W^*_{ik} and P^*_{ik} are fundamental solutions of the orthotropic shear deformable plates as presented in Wang [9]; $c_{ij}(\mathbf{x}')$ are the jump terms arising from the terms of $O(1/r)$ in the kernel P^*_{ij}. Equation (8.50) represents three integral equations, two (i=1,2) for rotations and one for deflection.

In order to transform domain integrals related with distributed pressure and inertial terms into boundary integrals, the Radial Integration Method (see Useche and Alvarez [7]) was used as presented in Chap. 4, where the body forces are approximated with the use of radial basis functions.

Quadratic discontinuous boundary elements were used to approximate w_k, p_k, throughout the elements. In order to model the geometry, continuous quadratic elements were used. Thus, for a given collocation point i, Eqs. (8.50) can be discretized as

$$c_{ik} w_k(\mathbf{x}', t) + \sum_{j=1}^{N_e} \sum_{r=1}^{3} w_k^{jr}(t) \int_{\xi=-1}^{\xi=+1} P^*_{ik}(\mathbf{x}', \mathbf{x}) \Phi^r(\xi) J_j(\xi) d\xi$$
$$= \sum_{j=1}^{N_e} \sum_{r=1}^{3} p_k^{jr}(t) \int_{\xi=-1}^{\xi=+1} W^*_{ik}(\mathbf{x}', \mathbf{x}) \Phi^r(\xi) J_j(\xi) d\xi$$
$$+ \sum_{m=1}^{M} \gamma^m(t) \sum_{j=1}^{N_e} \int_{\xi=-1}^{\xi=+1} \frac{F^m_{i3}(\xi)}{r(\xi)} n_\alpha(\xi) r_\alpha(\xi) J_j(\xi) d\xi$$
$$- \sum_{n=1}^{N} \lambda^n_{,tt}(t) \sum_{j=1}^{N_e} \int_{\xi=-1}^{\xi=+1} \frac{D^n_{ik}(\xi)}{r(\xi)} n_\alpha(\xi) r_\alpha(\xi) J_j(\xi) d\xi \tag{8.51}$$

where J represents the Jacobian of the transformation ($d\Gamma = J(\xi)d\xi$); ξ is the element local coordinate and Φ^r is the element shape function. Defining influence coefficient matrices H_{ij}, G_{ij} and vector q_i for a given element as

$$
\begin{aligned}
H_{ij} &= \frac{1}{2}\delta_{ik} + \int_{\xi=-1}^{\xi=+1} P_{ik}^*(\mathbf{x}', \mathbf{x})\Phi^r(\xi)J_j(\xi)d\xi \\
G_{ij} &= \int_{\xi=-1}^{\xi=+1} W_{ik}^*(\mathbf{x}', \mathbf{x})\Phi^r(\xi)J_j(\xi)d\xi \\
S_{ij}^n &= \int_{\xi=-1}^{\xi=+1} \frac{D_{ik}^n(\xi)}{r(\xi)} n_\alpha(\xi)r_\alpha(\xi)J_j(\xi)d\xi \\
Q_{ij}^m &= \int_{\xi=-1}^{\xi=+1} \frac{F_{i3}^m(\xi)}{r(\xi)} n_\alpha(\xi)r_\alpha(\xi)J_j(\xi)d\xi
\end{aligned}
\tag{8.52}
$$

Equations (8.51) can be rewritten as

$$
\sum_{j=1}^{N_e}\sum_{r=1}^{3} H_{ij} w_k^{jr}(t) - \sum_{j=1}^{N_e}\sum_{r=1}^{3} G_{ij} p_k^{jr}(t) = \sum_{m=1}^{M}\sum_{j=1}^{N_e} Q_{ij}^m \gamma^m(t) - \sum_{n=1}^{N}\sum_{j=1}^{N_e} S_{ij}^m \lambda_{,tt}^m(t) \tag{8.53}
$$

Applying this equation at each collocation point successively, we obtain a set of equations that can be expressed in compact matrix form as Useche et al. [8], Useche and Albuquerque [6]:

$$
\mathbf{M}\ddot{\mathbf{w}} + \mathbf{H}\mathbf{w} = \mathbf{G}\mathbf{p} + \mathbf{f} \tag{8.54}
$$

In this equation, $\mathbf{f} = \mathbf{Q}\mathbf{F}^{-1}\mathbf{q}$ where $\mathbf{F}$ is the matrix of coefficients and $\mathbf{q}$ is a vector containing pressure loads. The bending structural mass matrix, $\mathbf{M}$, is given by Useche and Alvarez [7]:

$$
\mathbf{M} = \mathbf{\Lambda}\mathbf{S}\mathbf{D}^{-1} \tag{8.55}
$$

where $\mathbf{D}$ is the matrix of coefficients and $\mathbf{\Lambda}$ is a tensor containing inertial terms.

The boundary integral equations discussed above contain integrands with several different orders of singularities. These singular integrals are treated separately based on their order of singularity. In this work, all regular integrals are evaluated numerically using standard Gauss quadrature formula. The influence matrix $\mathbf{G}$ contains weakly singular integrals, which are treated using Telles's nonlinear coordinate transformation method. However, for better numerical accuracy, a suitable number of element subdivisions must be used. In this work, four element subdivisions are used. The influençe matrix $\mathbf{H}$ contains strongly singular integrals that, in this work, are computed indirectly by considering generalized rigid body movements as presented in Chap. 5.

8.5 Internal Stress Resultants

The internal stress resultants $N_{\alpha\beta}$, $M_{\alpha\beta}$ and Q_α at domain point $\mathbf{X}'$ are calculated using equations given in Chap. 3. For this, $u_{\alpha,\gamma}$ and $w_{k,\gamma}$ are obtained by differentiating equations (8.50) with respect to the coordinate of the source point $\mathbf{X}'$ (See Wang [9], Reis et al. [5], dos Reis et al. [1]):

$$w_{j,\gamma}(\mathbf{X}',t) = -\int_\Gamma P^*_{ij,\gamma}(\mathbf{X}',\mathbf{x})w_j(\mathbf{x},t)d\Gamma + \int_\Gamma W^*_{ij,\gamma}(\mathbf{X}',\mathbf{x})p_j(\mathbf{x},t)d\Gamma + \int_\Omega W^*_{i3,\gamma}(\mathbf{X}',\mathbf{X})q_3(\mathbf{x})d\Omega \quad (8.56)$$

8.6 Numerical Examples

This section presents several representative numerical examples to validate the formulations presented for the static case. The obtained solutions are compared with analytical and FEM solutions. All presented results for the deflection w are given for the geometry centroid.

8.6.1 Cantilever Composite Plate with Transverse End Load

A composite plate with a rolling sequence [0°/90°/90°/0°] measuring 10 by 5 inches is subjected to a 3 lbf/in end load as shown in Fig. 3.4. The total plate thickness is 1. The material properties used are summarized in Table 3.1. The top view of the plate is

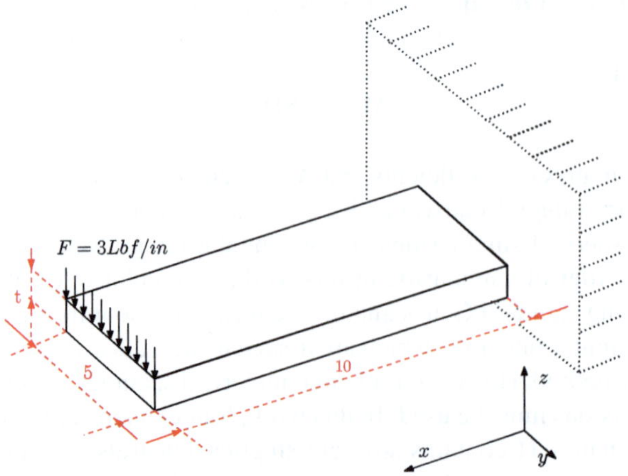

Fig. 8.3 Cantilever plate with force at border

Fig. 8.4 Displacement and rotation for cantilever plate. *Above*: Deflexion along x_3-axis (y = 0). *Middle*: Rotation about x_2-axis. *Below*: Rotation about x_1-axis

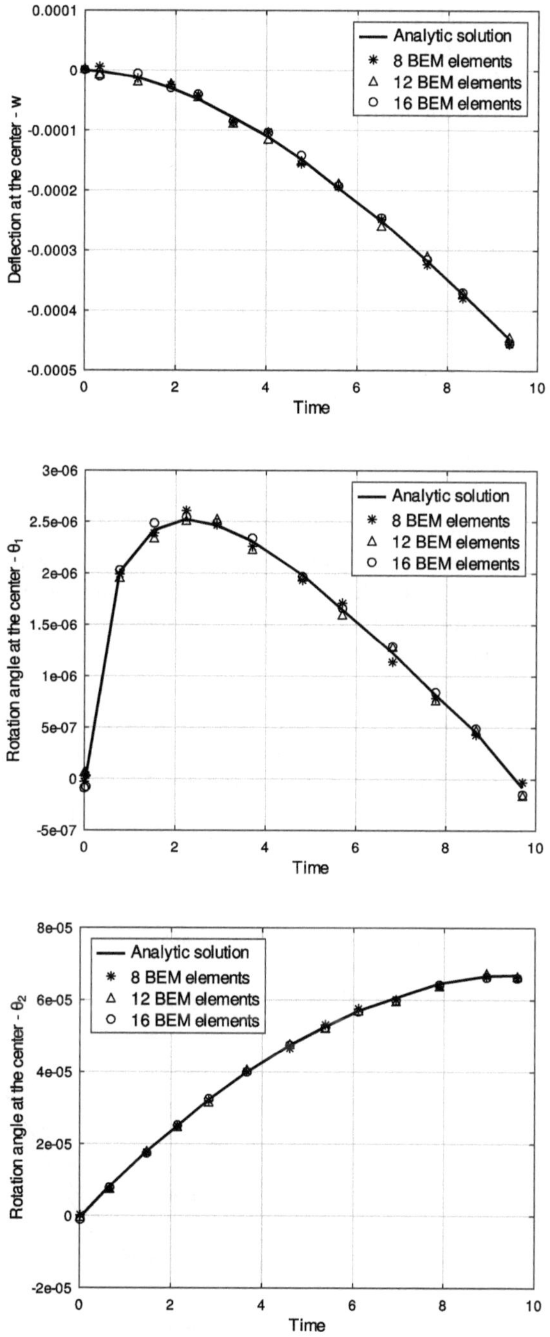

shown in Fig. 3.5. The boundary conditions used for this example are $x = 0,\ w_i = 0,$ $x = L,\ Q_3 = 3$ (Fig. 8.3).

Verifying the results consists of comparing the deflections obtained in the BEM model with analytical solutions and solutions obtained with finite elements. For the indicated loading conditions, the solution for w_i is obtained for 6, 10, 12 and 16 boundary elements as shown in Fig. 8.4 and compared with the analytical solution given in Reddy [4]. In Fig. 8.4, it can be observed that the results obtained with BEM are quite close to the analytical solution with errors below to 1%. In the presented problem, a constant number of 10 Gauss points was used for all meshes.

Simply Supported Plate with Impulsive Pressure Load

A simply supported square plate with side $a = 1.0m$ and a total thickness $h = 0.1m$ and a rolling pattern (0/90/90/0) is subjected to an impulsive pressure load $q(t) = 1.0$. The elastic properties of the material used are: $E11 = 4 \times 106Pa$, $E22 = 0.5E11$, $G12 = G31 = 1 \times 106Pa$, $G32 = 0.5G12$, $v12 = v31 = v23 = 0.5$ and $\rho = 4000kg/m$. Figure 8.5 shows the deflection at the center of the plate as a function of time for different approximation functions ($fm_1 = 1 + r_m + r_m^3$, $fm_2 = r^2 \log(r)$, $fm_3 = 1 - 6(r_m/dA)^2 + 6(r_m/dA)^3 - 3(r_m/dA)^4$ with $fm_3 = 0$ for $r > dA$, where dA is a support radius) with 8-element BEM meshes for each approximation function. It can be seen that with the approximation functions fm_2 and fm_3, a better fit of the solution is obtained with respect to the solution obtained using finite elements. Figure 8.6 shows the convergence of the model for meshes with 12 and 16 elements, using the approximation function fm_3. However, a loss of convergence is observed over time, which suggests that the influence of the time step used δt should be analyzed. Finally, Fig. 8.7 shows the sensitivity of the solution to the passage of time for a model analyzed with 12 elements using the approximation function fm_4.

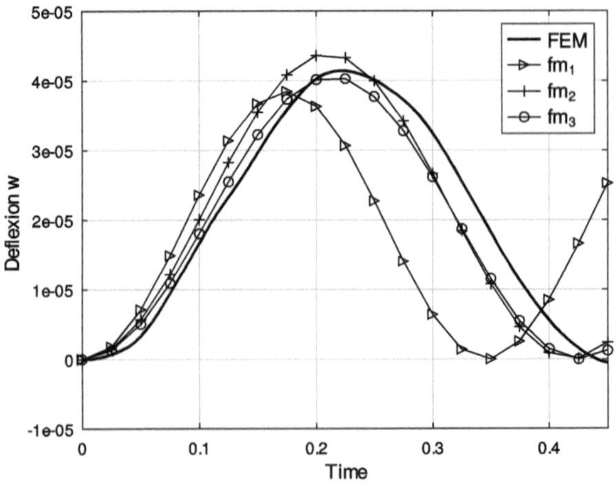

Fig. 8.5 Deflection at the center of the simply supported plate for differents radial-basis functions

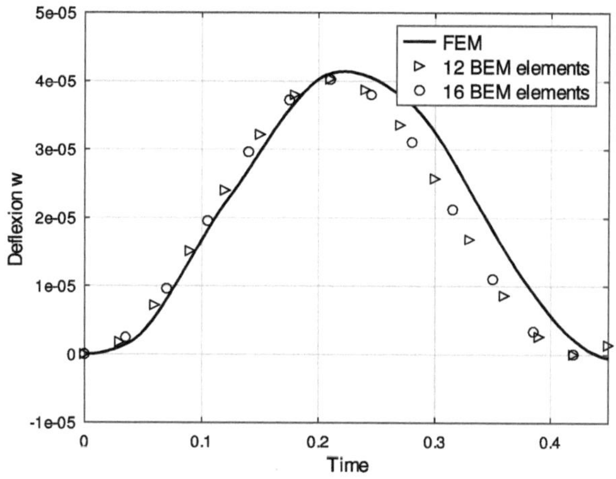

Fig. 8.6 Convergence of deflection at the center of the simply supported plate for two differents boundary element meshes

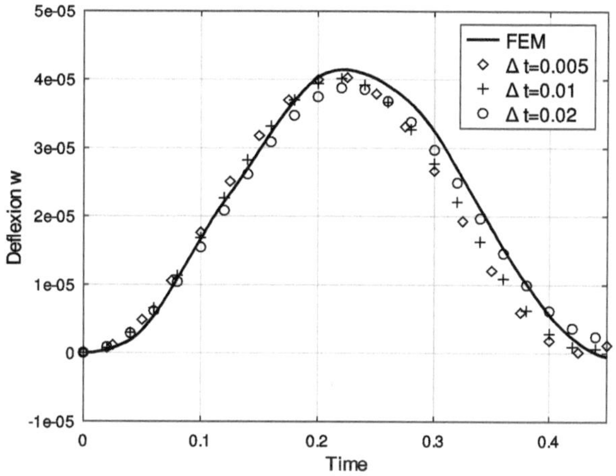

Fig. 8.7 Convergence of deflection at the center of the simply supported plate for two differents boundary element meshes

8.7 Summary

The boundary element formulation for the deflection analysis of thick Laminated plates was presented. The integrals related to distributed loads in the domain were carried to the boundary using the radial integration method. The fundamental solutions proposed by Wang [9] and Hsu, et al. [3] were used for symmetric and asymmetric laminates, respectively. The MATLAB(R) implementation of the RIM is presented. Several representative examples of the static and dynamic cases were presented. Although the RIM presents itself as an appropriate method for these problems, the application of other methodologies is an open research topic, as well as the use of the fundamental solutions Hsu et al. [3] for the dynamic analysis of thick plates with asymmetric laminates.

References

1. dos Reis A, Albuquerque EL, Palermo L Jr (2013) The boundary element method applied to orthotropic shear deformable plates. Eng Anal Bound Elem 37:738–746
2. Hsu C-W, Hwu C (2023) Green's function for thick laminated composite plates with coupled stretching-bending and transverse shear deformation. Compos Struct 320
3. Hsu CW, Mittelstedt C, Hwu C (2024) Boundary element method for bending-extension coupling shear deformable laminated composite plates with multiple stiffeners. Eng Anal Bound Elem
4. Reddy JN (2004) Mechanics of laminated composite plates and shells: theory and analysis, 2nd edn. CRC Press, New York
5. Reis A, Albuquerque EL, Santana AP, Useche J (2012) Computation of moments in orthotropic thick plates by the boundary element method. In: 10th-World congress on computational mechanics, São Paulo, Brasil, July 2012
6. Useche J, Albuquerque EL (2012) Dynamic analysis of shear deformable plates using the dual reciprocity method. Eng Anal Bound Elem 36:627–632
7. Useche J, Alvarez H (2015) Elastodynamic analysis of thick multilayer composite plates by the boundary element method. Comput Model Eng Sci 107(4):277–296
8. Useche J, Albuquerque EL, Sollero P (2012) Harmonic analysis of shear deformable orthotropic cracked plates using the boundary element method. Eng Anal Bound Elem 36(11):1528–1535
9. Wang J (1991) Boundary element method for orthotropic thick plates. Acta Mech Sin 7:258–266

Chapter 9
Boundary Element Analysis of Composite Shear Deformable Shallow Shells

Abstract In this chapter, a boundary element formulation is presented based on the coupling of the boundary element formulation of shear-deformable symmetrically laminated composite plates and the boundary element formulation for anisotropic plane stress analysis. This avoids the use of cumbersome fundamental solutions as given in Wang and Schweizerhorf. This way, an efficient BEM formulation is easily implementable by combining existing computational codes to analyze anisotropic elasticity and thick laminated plates.

9.1 Introduction

Schweizerhof and Wang [4], Wang and Schweizerhof [13] presented the bending analysis of symmetrically laminated shallow shells of moderate thickness based on the first-order shear deformation plate theory. The fundamental solution of the plates used in this work was cumbersome and computationally complex; they used this formulation for the free vibration analysis of shallow rolled anisotropic plates. Later, Useche and Medina [10] used the orthotropic in-plane elasticity BEM formulation coupled to the orthotropic plate BEM formulation proposed by Wang [12], Useche and Alvarez [8] to analyze the static response of symmetric laminated shallow shells.

In this chapter, a boundary element formulation is presented based on the coupling of the boundary element formulation of shear-deformable symmetrically laminated composite plates, as presented in Useche and Medina [10], and the boundary element formulation for anisotropic plane stress analysis. This avoids the use of cumbersome fundamental solutions as given in Wang and Schweizerhorf [13]. This way, an efficient BEM formulation is easily implementable by combining existing computational codes to analyze anisotropic elasticity and thick laminated plates.

J. Useche Vivero, *The Boundary Element Method for Engineering Analysis of Plates and Shells*, Solid Mechanics and Its Applications 281,
https://doi.org/10.1007/978-3-031-93050-8_9

9.2 Laminated Shell Equations

Consider a symmetrically laminated composite shell occupying the Ω space in the $x_1 - x_2$ coordinate system with a total thickness h and defined by a simply curved mean quadratic surface given by (see Fig. 9.1):

$$x_3 = -\frac{1}{2}(\kappa_{11}x_1^2 + \kappa_{22}x_2^2) \tag{9.1}$$

where the curvatures are defined as $\kappa_{11} = 1/R_1$ and $\kappa_{22} = 1/R_2$; R_1 and R_2 are the principal radii of curvature of the shell. The lower surface of the shell is located at $(x_3 - h/2)$ and the upper surface at $(x_3 + h/2)$. The shell is composed of N, especially orthotropic layers. Each layer has principal material coordinates oriented at an angle θ relative to the lamination coordinate x_1. The $x_1 - x_2$ plane is considered the midsurface of the shell, and the x_3 axis is taken normal to this surface. The shell is bounded by the boundary $\Gamma = \Gamma_w \bigcup \Gamma_p$ with $\Gamma_w \bigcap \Gamma_p \equiv 0$. The generalized midsurface displacements are w_i and u_α, where w_α represents in-plane rotations in the x_1 and x_2 directions, w_3 represents a transverse displacement normal to the midsurface, and u_α represents in-plane displacements in the x_1 and x_2 directions. The generalized tractions p_α and t_α denote in-plane moments and in-plane tractions, respectively; p_3 denotes the traction due to the shear resultant.

For this shell, the plane stress equilibrium equations, as a function of the displacements, are given by the equations presented in Chap. 3 (see Useche and Medina [10], Useche and Albuquerque [7], Useche and Harnish [9], Useche [5], Useche and Alvarez [8]):

$$\mathcal{L}^m_{\alpha\beta}u_\alpha + f^*_\alpha = \Lambda^{bm}_{\alpha\beta}\ddot{w}_\beta + \Lambda^m_{\alpha\beta}\ddot{u}_\beta \tag{9.2}$$

where the operator $\mathcal{L}^m_{\alpha\beta}$ is defined as

$$\mathcal{L}^m_{11} = A_{11}\frac{\partial^2}{\partial x_1^2} + A_{66}\frac{\partial^2}{\partial x_2^2} + 2A_{26}\frac{\partial^2}{\partial x_1 x_2}$$

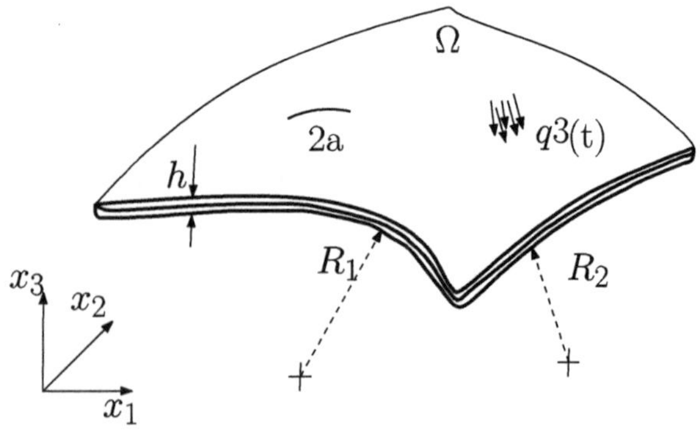

Fig. 9.1 Thick composite shell

$$\mathcal{L}_{12}^m = \mathcal{L}_{21}^m = A_{12}\frac{\partial^2}{\partial x_1 x_2} + A_{16}\frac{\partial^2}{\partial x_1} + A_{26}\frac{\partial^2}{\partial x_2} + A_{66}\frac{\partial^2}{\partial x_1 x_2}$$

$$\mathcal{L}_{22}^m = A_{22}\frac{\partial^2}{\partial x_2^2} + 2A_{26}\frac{\partial^2}{\partial x_1 x_2} + A_{66}\frac{\partial^2}{\partial x_1} \tag{9.3}$$

Considering $\kappa_{\alpha\beta} = 1/R_\alpha \delta_{\alpha\beta}$ as the curvature tensor, the in-plane equivalent distributed load vector f_α^*, can be defined as

$$f_\alpha^* = f_\alpha^b + f_\alpha = A_{\alpha\beta}\kappa_{\beta\beta}w_{3,\alpha} + A_{\beta 6}\kappa_{\beta\beta}(w_{3,\gamma}\delta_{\gamma\gamma} - w_{3,\alpha}) + f_\alpha \tag{9.4}$$

On the other hand, the equilibrium equations for bending are given by the equations presented in Chap. 3:

$$\mathcal{L}_{ik}^b w_k + q_i^* = \Lambda_{ik}^b \ddot{w}_k + \Lambda_{i\alpha}^{bm}\ddot{u}_\alpha \tag{9.5}$$

where the $\mathcal{L}_{ik}^b$ operator is defined as

$$\mathcal{L}_{11}^b = D_{11}\frac{\partial^2}{\partial x_1^2} + 2D_{16}\frac{\partial^2}{\partial x_1 \partial x_2} + D_{66}\frac{\partial^2}{\partial x_2^2} - A_{55}$$

$$\mathcal{L}_{12}^b = \mathcal{L}_{21} = D_{16}\frac{\partial^2}{\partial x_2^2} + (D_{12} + D_{66})\frac{\partial^2}{\partial x_1 \partial x_2} + D_{26}\frac{\partial^2}{\partial x_2^2} - A_{45}$$

$$\mathcal{L}_{13}^b = -\mathcal{L}_{31}^b = -\left(A_{45}\frac{\partial}{\partial x_2} + A_{55}\frac{\partial}{\partial x_1}\right)$$

$$\mathcal{L}_{23}^b = -\mathcal{L}_{32}^b = -\left(A_{44}\frac{\partial}{\partial x_2} + A_{45}\frac{\partial}{\partial x_1}\right)$$

$$\mathcal{L}_{22}^b = D_{66}\frac{\partial^2}{\partial x_1^2} + 2D_{26}\frac{\partial^2}{\partial x_1 \partial x_2} + D_{22}\frac{\partial^2}{\partial x_2^2} - A_{44}$$

$$\mathcal{L}_{33}^b = A_{55}\frac{\partial^2}{\partial x_2^2} + 2A_{45}\frac{\partial^2}{\partial x_1 \partial x_2} + A_{44}\frac{\partial^2}{\partial x_2^2} \tag{9.6}$$

The equivalent moments distributed in the plane q_i^* are defined as $q_\alpha^* = m_\alpha$ and $q_3^* = q_3$. Therefore, the following equations are obtained:

$$q_3^* = q_3 - \{A_{\alpha\beta}\kappa_{\beta\beta}u_{\alpha,\alpha} + A_{\gamma 6}\kappa_{\gamma\gamma}(u_{2,1} + u_{1,2}) + \kappa_{\alpha\alpha}A_{\alpha\beta}\kappa_{\beta\beta}w_3\} \tag{9.7}$$

9.3 Boundary Integral Equations

Since the differential operators $\mathcal{L}_{\alpha\beta}^m$ and $\mathcal{L}_{ik}^b$ are the same for two-dimensional symmetrically anisotropic plate and laminate bending problems, respectively, the BEM formulation for shallow laminated shells can be derived by considering the coupling terms f_α^b and q_3^b as body forces. The boundary integral equations related to the governing equations for in-plane and plate bending are derived using Betti's reciprocal theorem as presented in [1]:

Thus, the boundary integral equations for Eqs. (9.2), without considering in-plane distributed loads, are given by

$$c_{\theta\alpha}(\mathbf{x}')u_\alpha(\mathbf{x}') + \int_\Gamma T_{\theta\alpha}(\mathbf{x}',\mathbf{x})u_\alpha(\mathbf{x})d\Gamma - \int_\Gamma U_{\theta\alpha}(\mathbf{x}',\mathbf{x})t_\alpha(\mathbf{x})d\Omega +$$

$$\int_\Omega U_{\theta\alpha}(\mathbf{x}',\mathbf{X})[A_{\alpha\beta}\kappa_{\beta\beta}w_{3,\alpha}(\mathbf{X}) + A_{\beta 6}\kappa_{\beta\beta}\{w_{3,\gamma}(\mathbf{X})\delta_{\gamma\gamma} - w_{3,\alpha}(\mathbf{X})\}]d\Omega +$$

$$\int_\Gamma U_{\theta\alpha}(\mathbf{x}',\mathbf{x})t_\alpha(\mathbf{x})d\Gamma + \int_\Omega U_{\theta\alpha}(\mathbf{x}',\mathbf{X})\Lambda^m_{\theta\alpha}\ddot{u}_\alpha(\mathbf{X})d\Omega$$

$$+ \int_\Omega U_{\theta\alpha}(\mathbf{x}',\mathbf{X})\Lambda^{bm}_{\theta\alpha}\ddot{w}_\alpha(\mathbf{X})d\Omega \tag{9.8}$$

where $U^*_{\theta\alpha}$ and $T^*_{\theta\alpha}$ are the fundamental displacement and tensile solutions for the anisotropic plane stress problem, respectively. Similarly, the boundary integral equations for the bending equations (9.5) are

$$c_{ij}w_j(\mathbf{x}') + \int_\Gamma P_{ij}(\mathbf{x}',\mathbf{x})w_j(\mathbf{x})d\Gamma = \int_\Gamma W_{ij}(\mathbf{x}',\mathbf{x})p_j(\mathbf{x})d\Gamma$$

$$+ \int_\Omega A_{\alpha\beta}\kappa_{\beta\beta}u_{\alpha,\alpha}(\mathbf{X})W_{i3}(\mathbf{x}',\mathbf{X})d\Omega + \int_\Omega A_{\gamma 6}\kappa_{\gamma\gamma}[u_{2,1} + u_{1,2}]W_{i3}(\mathbf{x}',\mathbf{X})d\Omega$$

$$+ \int_\Omega \kappa_{\alpha\alpha}A_{\alpha\beta}\kappa_{\beta\beta}w_3(\mathbf{X})W_{i3}(\mathbf{x}',\mathbf{X})d\Omega + \int_\Omega W_{i3}(\mathbf{x}',\mathbf{X})q_3(\mathbf{x})d\Omega +$$

$$+ \int_\Omega W_{i3}(\mathbf{x}',\mathbf{X})q_3(\mathbf{x})d\Omega + \int_\Omega W_{ij}(\mathbf{x}',\mathbf{x})\Lambda^b_{ij}\ddot{w}_j(\mathbf{X})d\Omega$$

$$+ \int_\Omega W_{i\alpha}(\mathbf{x}',\mathbf{X})\Lambda^{bm}_{i\alpha}\ddot{u}_\alpha(\mathbf{X})d\Omega \tag{9.9}$$

where W_{ik} and P_{ik} are fundamental solutions for symmetric shear-deformable thick plates. Equations (9.8) and (9.9) represent five integral equations that can be used to solve problems for shallow shear-deformable rolled plates.

9.3.1 Domain Integrals

The domain integrals related to distributed loads and curvature terms appearing in Eqs. (9.8) and (9.9) are treated through the Radial Integration Method (RIM) [2]. The RIM was successfully used for the treatment of domain integrals in the boundary element analysis of thick plate dynamic bending problems Useche et al. [11], Useche and Medina [10], Useche and Alvarez [8], Useche and Harnish [9] and Useche [6]. There are nine domain integrals in Eqs. (9.8) and (9.9). These integrals can be classified into the following groups:

$$\mathcal{I}_1 = \int_{\Omega} A^*_{ij}(\mathbf{x}', \mathbf{X}) v_j d\Omega, \quad \mathcal{I}_2 = \int_{\Omega} A^*_{ij}(\mathbf{x}', \mathbf{X}) v_{j,\gamma} d\Omega, \tag{9.10}$$

where A^*_{ij} and v_j are arbitrary functions defined on Ω. In RIM, arbitrary functions v_j are approximated using radial basis functions, which are defined as

$$v_j = \mathbf{f}(r)\boldsymbol{\alpha} \tag{9.11}$$

where $\mathbf{f}(r) = \{1 + r_1, 1 + r_2, \ldots, 1 + r_m\}$ is the radial basis function vector; $r_m^2 = (x_1 - x_m)^2 + (x_2 - x_m)^2$ and $m = 1, 2, \ldots, M$ are the number of RIM placement points used to approximate v_j at the placement point with coordinates (x_1, x_2). The vector $\boldsymbol{\alpha} = \{\alpha_1, \alpha_2, \ldots, \alpha_M\}^T$ can be obtained by evaluating (9.11) at each RIM placement point:

$$\boldsymbol{\alpha} = \mathbf{F}^{-1}\mathbf{v} \tag{9.12}$$

where $\mathbf{v}$ is a vector of unknowns evaluated at the placement points. Replacing it in Eq. (9.11), the following equations are obtained:

$$v_j = \mathbf{f}(r)\mathbf{F}^{-1}\mathbf{v} \tag{9.13}$$

Derivatives $v_{j,\gamma}$ can be calculated as

$$v_{j,\gamma} = \mathbf{f}_{,\gamma}(r)\mathbf{F}^{-1}\mathbf{v} \tag{9.14}$$

Finally, the integrals $\mathcal{I}_1$ and $\mathcal{I}_2$ can be approximated as

$$\mathcal{I}_1 = \left[\int_{\Gamma} \mathbf{n} \cdot \mathbf{e}_r \left[\int_0^{r(\theta)} A^*_{ij}(\mathbf{x}', \rho, \theta)\mathbf{f}(\rho)\rho d\rho \right] d\Gamma \right] \mathbf{F}^{-1}\mathbf{v} = \mathcal{R}(A^*_{ij}, \mathbf{f})\mathbf{F}^{-1}\mathbf{v} \tag{9.15}$$

$$\mathcal{I}_2 = \left[\int_{\Gamma} \mathbf{n} \cdot \mathbf{e}_r \left[\int_0^{r(\theta)} A^*_{ij}(\mathbf{x}', \rho, \theta)\mathbf{f}_{,\gamma}(\rho)\rho d\rho \right] d\Gamma \right] \mathbf{F}^{-1}\mathbf{v} = \mathcal{R}(A^*_{ij}, \mathbf{f}_{,\gamma})\mathbf{F}^{-1}\mathbf{v} \tag{9.16}$$

Similarly, domain integrals involving inertial terms can be transformed using RIM as Useche [6]:

$$\int_{\Omega} W_{ij}(\mathbf{x}', \mathbf{x})\Lambda^b_{ij}\ddot{w}_j(\mathbf{X})d\Omega + \int_{\Omega} W_{i\alpha}(\mathbf{x}', \mathbf{X})\Lambda^{bm}_{i\alpha}\ddot{u}_\alpha(\mathbf{X})d\Omega \approx \Lambda^b_{ij}\mathcal{R}(W^*_{ij}, \mathbf{f})\mathbf{G}^{-1}\ddot{\mathbf{w}} + \Lambda^{bm}_{i\alpha}\mathcal{R}(W^*_{i\alpha}, \mathbf{f})\mathbf{H}^{-1}\ddot{\mathbf{u}}$$

$$\int_{\Omega} U_{\theta\alpha}(\mathbf{x}', \mathbf{X})\Lambda^{m}_{\theta\alpha}\ddot{u}_{\alpha}(\mathbf{X})d\Omega + \int_{\Omega} U_{\theta\alpha}(\mathbf{x}', \mathbf{X})\Lambda^{bm}_{\theta\alpha}\ddot{w}_{\alpha}(\mathbf{X})d\Omega \approx \Lambda^{bm}_{\theta\alpha}\mathcal{R}(U^{*}_{\theta\alpha}, \mathbf{G})^{-1}\ddot{\mathbf{w}}_{\alpha} + \Lambda^{m}_{\theta\alpha}\mathcal{R}(U^{*}_{\theta\alpha}, \mathbf{H})^{-1}\ddot{\mathbf{u}} \tag{9.17}$$

where **G, H** are matrices containing the radial basis functions used in the approximation of $\ddot{\mathbf{u}}$ and $\ddot{\mathbf{w}}$.

9.4 Boundary Element Approximation

To approximate w_k, p_k, u_α and t_α, discontinuous quadratic boundary elements were used. Continuous quadratic elements approximate the boundary geometry. Thus, for a given placement point i, Eqs. (9.8) and (9.9) can be discretized as

$$c_{\theta\alpha}u_{\theta}(\mathbf{x}', t) + \sum_{j=1}^{N_e}\sum_{r=1}^{3} u_{\theta}^{jr}(t) \int_{\xi=-1}^{\xi=+1} T^{*}_{\theta\alpha}(\mathbf{x}', \mathbf{x})\Phi^{r}(\xi)J_{j}(\xi)d\xi$$
$$= \sum_{j=1}^{N_e}\sum_{r=1}^{3} t_{\theta}^{jr}(t) \int_{\xi=-1}^{\xi=+1} U^{*}_{\theta\alpha}(\mathbf{x}', \mathbf{x})\Phi^{r}(\xi)J_{j}(\xi)d\xi +$$
$$(A_{\alpha\beta} - A_{\beta 6})\kappa_{\beta\beta}\mathcal{R}(U^{*}_{\theta\alpha}, \mathbf{f}_{,\alpha})\mathbf{F}^{-1}\mathbf{w}_3 + A_{\beta 6}\kappa_{\beta\beta}\delta_{\gamma\gamma}\mathcal{R}(U^{*}_{\theta\alpha}, \mathbf{f}_{,\gamma})\mathbf{F}^{-1}\mathbf{w}_3 \tag{9.18}$$

and

$$c_{ik}w_{k}(\mathbf{x}', t) + \sum_{j=1}^{N_e}\sum_{r=1}^{3} w_{k}^{jr}(t) \int_{\xi=-1}^{\xi=+1} P^{*}_{ik}(\mathbf{x}', \mathbf{x})\Phi^{r}(\xi)J_{j}(\xi)d\xi$$
$$= \sum_{j=1}^{N_e}\sum_{r=1}^{3} p_{k}^{jr}(t) \int_{\xi=-1}^{\xi=+1} W^{*}_{ik}(\mathbf{x}', \mathbf{x})\Phi^{r}(\xi)J_{j}(\xi)d\xi + \mathcal{R}(W^{*}_{i3}, \mathbf{f})\mathbf{F}^{-1}\mathbf{q}_3 \tag{9.19}$$

where J represents the Jacobian of the transformation ($d\Gamma = J(\xi)d\xi$); ξ is the local coordinate of the element and Φ^r is the shape function of the element. Applying this equation successively at each placement point, we obtain a set of equations that can be expressed in matrix form as

$$\begin{bmatrix} \mathbf{H}^{m} & \mathbf{H}^{w} \\ \mathbf{H}^{u} & \mathbf{H}^{b} + \mathbf{H}^{\kappa} \end{bmatrix} \begin{Bmatrix} \mathbf{u} \\ \mathbf{w} \end{Bmatrix} = \begin{bmatrix} \mathbf{G}^{m} & \mathbf{0} \\ \mathbf{0} & \mathbf{G}^{b} \end{bmatrix} \begin{Bmatrix} \mathbf{t} \\ \mathbf{q} \end{Bmatrix} + \begin{Bmatrix} \mathbf{0} \\ \mathbf{b} \end{Bmatrix} \tag{9.20}$$

where $\mathbf{u} = \{\mathbf{u}_1, \mathbf{u}_2, \ldots, \mathbf{u}_N\}^T$, $\mathbf{w} = \{\mathbf{w}_1, \mathbf{w}_2, \ldots, \mathbf{w}_N\}^T$, $\mathbf{t} = \{\mathbf{t}_1, \mathbf{t}_2, \ldots, \mathbf{t}_N\}^T$, $\mathbf{q} = \{\mathbf{q}_1, \mathbf{q}_2, \ldots, \mathbf{q}_N\}^T$ y $\mathbf{b} = \{\mathbf{b}_1, \mathbf{b}_2, \ldots, \mathbf{b}_N\}^T$ and N is the number of BEM placement points. $\mathbf{H}^m$ and $\mathbf{G}^m$ are boundary element influence matrices for plane stress elasticity. $\mathbf{H}^b$ and $\mathbf{G}^p$ represent influence matrices for plate bending problems; $\mathbf{H}^u$, $\mathbf{H}^w$, and $\mathbf{H}^\kappa$ are in-plane curvature coupling matrices.

9.5 Stress Calculation

The internal resultant forces $N_{\alpha\beta}$, $M_{\alpha\beta}$, and Q_α are calculated through the constitutive relations presented in Chap. 3 for laminated shells. In this calculation, and similar to what was done in the previous chapter for laminated plates, the derivatives of the displacements can be calculated by differentiating equations (9.8) and (9.9) with respect to $\mathbf{X}'$ ([10]):

$$u_{\alpha,\gamma}(\mathbf{X}',t) + \int_\Gamma T_{\theta\alpha,\gamma}(\mathbf{X}',\mathbf{x})u_\alpha(\mathbf{x},t)d\Gamma + \int_\Omega U_{\theta\alpha,\gamma}(\mathbf{X}',\mathbf{X})f_\alpha(\mathbf{X},t)d\Omega +$$
$$\int_\Omega U_{\theta\alpha,\gamma}(\mathbf{X}',\mathbf{X})[A_{\alpha\beta}\kappa_{\beta\beta}w_{3,\alpha}(\mathbf{X},t) + A_{\beta 6}\kappa_{\beta\beta}\{w_{3,\gamma}(\mathbf{X})\delta_{\gamma\gamma} - w_{3,\alpha}\}(\mathbf{X},t)]d\Omega \tag{9.21}$$

and

$$w_{j,\gamma}(\mathbf{X}',t) + \int_\Gamma P_{ij,\gamma}(\mathbf{X}',\mathbf{x})w_j(\mathbf{x},t)d\Gamma = \int_\Gamma W_{ij,\gamma}(\mathbf{X}',\mathbf{x})p_j(\mathbf{x},t)d\Gamma$$
$$+ \int_\Omega A_{\alpha\beta}\kappa_{\beta\beta}u_{\alpha,\alpha}W_{i3,\gamma}(\mathbf{X}',\mathbf{X})d\Omega + \int_\Omega A_{\theta 6}\kappa_{\theta\theta}[u_{2,1} + u_{1,2}]W_{i3,\gamma}(\mathbf{x}',\mathbf{X})d\Omega$$
$$+ \int_\Omega \kappa_{\alpha\alpha}A_{\alpha\beta}\kappa_{\beta\beta}w_3(\mathbf{X})W_{i3,\gamma}(\mathbf{X}',\mathbf{X})d\Omega + \int_\Omega W_{i3,\gamma}(\mathbf{X}',\mathbf{X})q_3(\mathbf{x},t)d\Omega \tag{9.22}$$

9.6 Numerical Examples

Double curved laminated shell with simply supported edges

Consider now a laminated spherical shell under uniform distributed load $q_3 = 1$ with $\kappa_{11} = \kappa_{22} = 1/100$, $a/h = 100$ and $a = b$. The shell is simply supported at edges (see Fig. 9.2). Material properties are the same as the previous problem. The stacking laminated sequence is $[0°/90°]_s$ with thickness $h/4 = 0.25$.

The shell was discretized with sixteen elements and 98 uniformly distributed domain RIM points. BEM solution is compared with 19-term Navier analytical solution given in Reddy [3]. Figure 9.3 shows the results for internal forces and moments

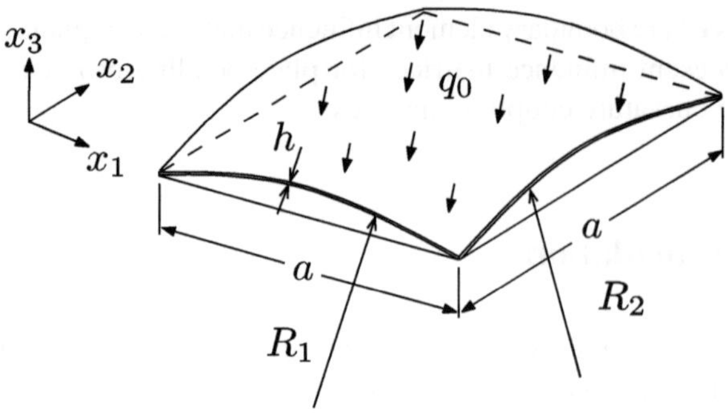

Fig. 9.2 Simply supported double curved shell

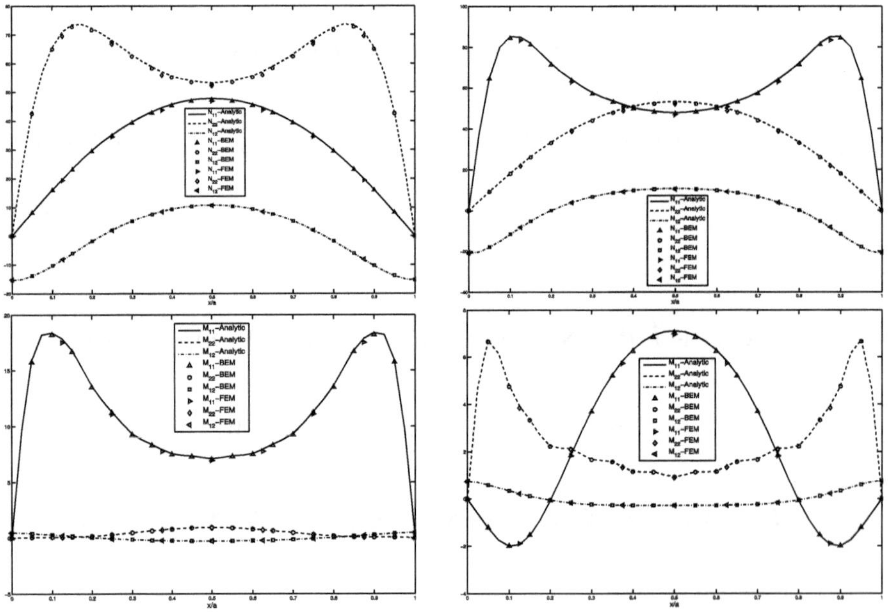

Fig. 9.3 Double curved shell: Normal forces along x_1-axis (Above left) and x_2-axis (Above right). Bending moments along x_1-axis (below left) and x_2-axis (below right)

along the central lines $x_2 = b/2$ and $x_1 = a/2$. Similar results were obtained for both BEM and the analytical solution.

Circular shallow shell with full clamped edges

In this example a fully-clamped laminated circular shell with $a = 5$ is considered (Fig. 9.4). The shell is under a distributed load given by $p(r) = p_{max}(1 + r^2)$ with

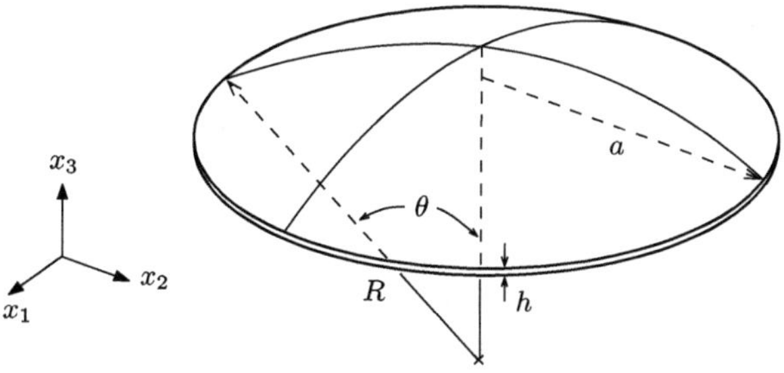

Fig. 9.4 Full clamped circular shell

$r = \sqrt{x_1^2 + x_2^2}$. For this problem, two stacking sequences of laminated are considered: $[90°/0°/90°/90°/0°]_s$ and $[0°/0°/0°/90°/90°]_s$ with thickness $h/10 = 0.01$. The material properties $E_{11} = 25$, $E_{22} = 1.0$, $G_{12} = G_{13} = G_{23} = 0.5$ and $\nu_{12} = 0.25$.

The shell was discretized with sixteen curve-boundary elements and 120 RIM collocation points. In order to validate the BEM solution, a Finite Element Analysis was carried out using the commercial software ANSYS® using the 8-node shear deformable element shell93 with a mesh made of 300 FEM elements and 941 nodes. Tables 9.1 and 9.2 show FEA and boundary element results for internal forces and bending moments for the laminated considered. As it can be seen, good agreement between FEM and BEM results was obtained.

Table 9.1 Normal forces for circular shallow shell for $[90/0/90/90/0]_s$ laminated at $r = a/2$

	N_{11}		N_{22}		N_{12}	
1/R	BEM	FEM	BEM	FEM	BEM	FEM
20	−83.597	−83.744	−111.230	−111.084	−9.279	−9.283
50	−219.667	−219.448	−145.927	−145.439	−4.812	−4.860
80	−264.382	−264.659	−174.137	−174.378	−0.807	−0.898
100	−272.045	−272.570	−179.895	−179.895	−0.873	0.864

Table 9.2 Bending moments for circular shallow shell for $[90/0/90/90/0]_s$ laminated at $r = a/2$

	M_{11}		M_{22}		M_{12}	
1/R	BEM	FEM	BEM	FEM	BEM	FEM
20	−0.191	−0.189	−0.811	−0.803	0.102	0.101
50	−3.471	−3.434	−0.591	−0.585	0.204	0.201
80	−5.710	−5.749	−1.142	−1.140	0.200	0.197
100	−7.003	−6.932	−1.389	−1.375	0.176	0.173

9.7 Summary

In this work, the Boundary Element Method was applied to carry out the stress-analysis of shear deformable shallow laminated composite shallow shells. A boundary element formulation based on a coupled BEM formulation for the classical Reissner plate theory and the two-dimensional plane stress anisotropic elasticity. Effects of shear deformation and rotatory inertia were included in the formulation. The Radial Integration Method was used to treat domain integrals involving distributed forces and curvature terms.

References

1. Albuquerque EL, Sollero P, Aliabadi MH (2002) The boundary element method applied to time dependent problems in anisotropic materials. Int J Solids Struct 39(5):1405–1422
2. Gao XW, Davies TG (2002) Boundary element programming in mechanics. Cambridge University Press, Cambridge
3. Reddy JN (2004) Mechanics of laminated composite plates and shells: theory and analysis, 2nd edn. CRC Press, New York
4. Schweizerhof J, Wang K (1996) Study on free vibration of moderately thick orthotropic laminated shallow shells by boundary-domain elements. Appl Math Modell J 20:579–584
5. Useche J (2014) Vibration analysis of shear deformable shallow shells using the boundary element method. Eng Struct J 62–63:65–74
6. Useche J (2024) Dynamic analysis of cracked thick composite shells by the boundary element method. Eng Anal Bound Elem J 168
7. Useche J, Albuquerque EL (2015) Transient dynamic analysis of shear deformable shallow shells. Eng Struct 87(15):1–7
8. Useche J, Alvarez H (2015) Elastodynamic analysis of thick multilayer composite plates by the boundary element method. Comput Model Eng Sci 107(4):277–296
9. Useche J, Harnish C (2016) A boundary element method formulation for modal analysis of doubly curved thick shallow shells. Appl Math Model 40(5–6):3591–3600
10. Useche J, Medina C (2018) Boundary element analysis of laminated composite shear deformable shallow shells. Compos Struct J 199:24–37
11. Useche J, Albuquerque EL, Sollero P (2012) Harmonic analysis of shear deformable orthotropic cracked plates using the boundary element method. Eng Anal Bound Elem 36(11):1528–1535
12. Wang J (1991) Boundary element method for orthotropic thick plates. Acta Mech Sin 7:258–266
13. Wang J, Schweizerhof K (1995) Computation of fundamental solutions for laminated anisotropic shallow shells. Mech Res Commun 22:393–400

Chapter 10
Boundary Element Analysis of Cracked Isotropic Thick Plates

Abstract This chapter presents the dual boundary element method applied to fracture analysis of plates and shells under static and dynamic loading. The dual boundary element method extends the boundary element method to model the behavior of cracks in solids by introducing the integral boundary equation of displacement and tension on either side of the crack faces. In this way, by using both equations simultaneously, the crack can be modeled without the need to define additional conditions at the fracture, which makes it more efficient than other numerical methods based on domain discretization.

10.1 Introduction

The Dual Boundary Element Method (DBEM) for fracture analysis was proposed by Portela et al. [6] in 1992 and lays the foundation for fracture analysis and crack growth in thick isotropic DBEM plates. Sollero and Aliabadi [7] applied DBEM to fracture analysis in anisotropic plates under plane stress and static loads, and Albuquerque et al. [1] used it for dynamic analysis. Likewise, the DBEM has been used in the analysis of isotropic fractured shells by Dirgantara and Aliabadi [4], Wen et al. [13], Dirgantara and Aliabadi [5], Dirgantara [2], and others. Useche [10] apply DBEM in combination with DRBEM for the analysis of isotropic fractured plates in the time domain.

This chapter presents the dual boundary element method applied to fracture analysis of plates and shells under static and dynamic loading. The dual boundary element method extends the boundary element method to model the behavior of cracks in solids by introducing the integral boundary equation of displacement and tension on either side of the crack faces. In this way, by using both equations simultaneously, the crack can be modeled without the need to define additional conditions at the fracture line, which makes it more efficient than other numerical methods based on domain discretization.

J. Useche Vivero, *The Boundary Element Method for Engineering Analysis of Plates and Shells*, Solid Mechanics and Its Applications 281,
https://doi.org/10.1007/978-3-031-93050-8_10

10.2 The Dual Boundary Element Method

Because the surfaces of a crack are coincident, applying a displacement (or tensile) integral equation at the collocation points on these surfaces will generate a singular system of equations.

The Dual Boundary Element Method (DBEM) is an numerical technique used in particular for solving fracture mechanics problems that extend the Boundary Element Method (BEM) by employing two different boundary integral equations to avoid duplication of equations at placement points on one and the other crack face. This method applies a displacement integral equation to the placement points on one of the crack surfaces and a traction integral equation on the other surface (see Fig. 10.1). This tide allows us to effectively address problems in fracture mechanics.

The dual boundary element method can be applied to the analysis of fractured shells and plates by employing the boundary element formulation seen in Chap. 6, in which the shell problem is reduced to a plane stress and bending problem coupled by curvature terms as domain forces. Therefore, the equations of the dual boundary method for shells are given by the equations of in-plane displacement by bending

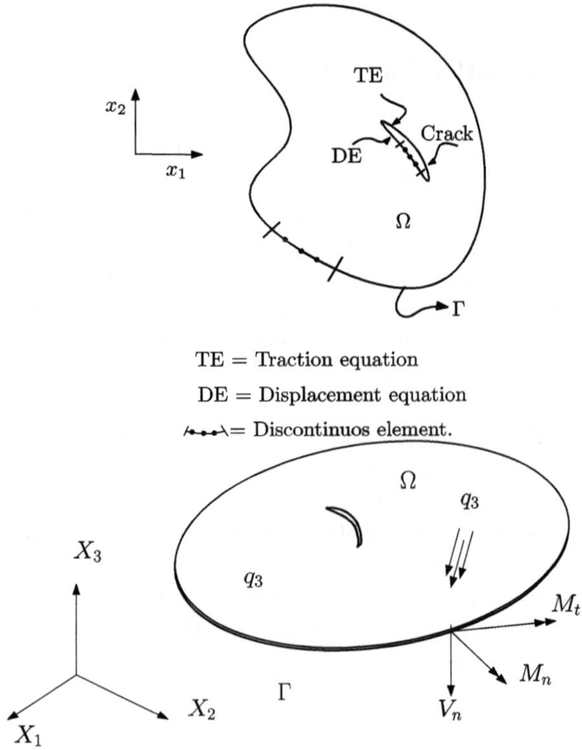

Fig. 10.1 Above: DBEM equations collocations. Below: Isotropic cracked plate

and the equations of in-plane tension and traction by bending moment and transverse shear forces (Dirgantara and Aliabadi [4], Wen et al. [13]).

10.2.1 Boundary Integral Displacement Equations

The integral formulation for the displacements (and rotations) of an isotropic thick shell was given in Chap. 6 and rewritten here:

For plane stress:

$$
\begin{aligned}
& c_{\theta\alpha}\left(\mathbf{x}'\right) u_\theta\left(\mathbf{x}'\right) + \fint_\Gamma T^*_{\theta\alpha}\left(\mathbf{x}', \mathbf{x}\right) u_\alpha\left(\mathbf{x}\right) d\Gamma \\
& + \int_\Gamma U^*_{\theta\alpha}\left(\mathbf{x}', \mathbf{x}\right) B\left[(1-\nu)\kappa_{\alpha\beta} + \nu\delta_{\alpha\beta}\kappa_{\gamma\gamma}\right] w_{3,\beta}\left(\mathbf{x}\right) n_\beta(\mathbf{x}) d\Omega - \\
& \int_\Omega U^*_{\theta\alpha}\left(\mathbf{x}', \mathbf{x}\right) B\left[(1-\nu)\kappa_{\alpha\beta} + \nu\delta_{\alpha\beta}\kappa_{\gamma\gamma}\right] w_3\left(\mathbf{X}\right) d\Omega \\
& = \int_\Gamma U^*_{\theta\alpha}\left(\mathbf{x}', \mathbf{x}\right) t_\alpha\left(\mathbf{x}\right) d\Gamma + \int_\Omega U^*_{\theta\alpha}\left(\mathbf{x}', \mathbf{X}\right) q_\alpha\left(\mathbf{X}\right) d\Omega
\end{aligned}
\tag{10.1}
$$

And for flexion:

$$
\begin{aligned}
& c_{ij}\left(\mathbf{x}'\right) w_j\left(\mathbf{x}'\right) + \fint_\Gamma P^*_{ij}\left(\mathbf{x}', \mathbf{x}\right) w_j\left(\mathbf{x}\right) d\Gamma = \int_\Gamma W^*_{ij}\left(\mathbf{x}', \mathbf{x}\right) p_j\left(\mathbf{x}\right) d\Gamma \\
& - \int_\Gamma \kappa_{\alpha\beta} B\left(\frac{1-\nu}{2}\right)\left[u_\alpha(\mathbf{x}) n_\beta + u_\beta(\mathbf{x}) n_\alpha + \frac{2\nu}{1-\nu} u_\gamma(\mathbf{x}) n_\gamma \delta_{\alpha\beta}\right] W^*_{i3}\left(\mathbf{x}', \mathbf{x}\right) d\Gamma \\
& + \int_\Omega \kappa_{\alpha\beta} B\left(\frac{1-\nu}{2}\right)\left[u_\alpha\left(\mathbf{X}\right) W^*_{i3,\beta}\left(\mathbf{x}', \mathbf{X}\right) + u_\beta\left(\mathbf{X}\right) W^*_{i3,\alpha}\left(\mathbf{x}', \mathbf{X}\right) + \right. \\
& \left. \frac{2\nu}{1-\nu} u_\gamma(\mathbf{X}) n_\gamma W^*_{i3,\gamma}\left(\mathbf{x}', \mathbf{X}\right) \delta_{\alpha\beta}\right] d\Omega \\
& - \int_\Gamma \kappa_{\alpha\beta} B[(1-\nu)\kappa_{\alpha\beta} + \nu\delta_{\alpha\beta}\kappa_{\gamma\gamma}] w_3\left(\mathbf{X}\right) W^*_{i3}\left(\mathbf{x}', \mathbf{X}\right) d\Omega + \int_\Omega W^*_{i3}\left(\mathbf{x}', \mathbf{X}\right) q_3\left(\mathbf{X}\right) d\Omega
\end{aligned}
\tag{10.2}
$$

10.2.2 Boundary Integral Traction Equations

On the other hand, the tensile equations for plane stress bending moments and shear forces are given by (see Dirgantara and Aliabadi [4]):

$$\frac{1}{2}t_\alpha(\mathbf{x}') + n_\beta(\mathbf{x}') \mathop{{=}\!\!\!\!\!\!\int}_{\Gamma} T^*_{\alpha\beta\gamma}(\mathbf{x}',\mathbf{x})\, u_\gamma(\mathbf{x}) d\Gamma + n_\beta \mathop{{-}\!\!\!\!\!\!\int}_{\Gamma} U^*_{\alpha\beta\gamma}(\mathbf{x}',\mathbf{x})$$

$$B[\kappa_{\theta\psi}(1-\nu) + \nu\delta_{\theta\psi}\kappa_{\theta\theta}] w_3(\mathbf{x}) n_\psi d\Gamma$$

$$-n_\beta \int_{\Omega} U^*_{\alpha\beta\gamma}(\mathbf{x}',\mathbf{X})\, B[\kappa_{\theta\psi}(1-\nu) + \nu\delta_{\theta\psi}\kappa_{\theta\theta}] w_{3,\beta}(\mathbf{X}) n_\psi d\Gamma$$

$$= n_\beta(\mathbf{x}') \mathop{{-}\!\!\!\!\!\!\int}_{\Gamma} U^*_{\alpha\beta\gamma}(\mathbf{x}',\mathbf{x})\, t_\gamma d\Gamma + n_\beta(\mathbf{x}') \int_{\Omega} U_{\alpha\beta\gamma}(\mathbf{x}',\mathbf{X})\, B[(1-\nu)\kappa_{\theta\psi} + \nu\delta_{\theta\psi}\kappa_{\phi\phi}] w_3(\mathbf{x})\, d\Omega$$

$$+n_\beta(\mathbf{x}') \int_{\Omega} U_{\alpha\beta\gamma}(\mathbf{x}',\mathbf{X})\, f_k d\Omega \tag{10.3}$$

For bending, similarly to the case of the isotropic plate, the tensile integrals for a shell are obtained by multiplying $M_{\alpha\beta}$ y Q_α, by the normal vector to the boundary n_α the expressions presented in Chap. 6 for the calculation of which is obtained:

$$\frac{1}{2}p_\alpha(\mathbf{x}') + n_\beta(\mathbf{x}') \mathop{{=}\!\!\!\!\!\!\int}_{\Gamma} P^*_{\alpha\beta\gamma}(\mathbf{x}',\mathbf{x})\, w_\gamma(\mathbf{x})\, d\Gamma + n_\beta(\mathbf{x}') \mathop{{-}\!\!\!\!\!\!\int}_{\Gamma} P^*_{\alpha\beta 3}(\mathbf{x}',\mathbf{x})\, w_3(\mathbf{x})\, d\Gamma$$

$$= n_\beta(\mathbf{x}') \mathop{{-}\!\!\!\!\!\!\int}_{\Gamma} W^*_{\alpha\beta\gamma}(\mathbf{x}',\mathbf{x})\, p_\gamma(\mathbf{x})\, d\Gamma + n_\beta(\mathbf{x}') \int_{\Gamma} W^*_{\alpha\beta 3}(\mathbf{x}',\mathbf{x})\, p_3(\mathbf{x})\, d\Gamma -$$

$$n_\beta(\mathbf{x}') \int_{\Omega} \kappa_{\theta\psi} B \frac{1-\nu}{2} \left(u_{\theta,\psi}(\mathbf{X}) + u_{\psi,\theta}(\mathbf{X}) + \frac{2\nu}{1-\nu} u_{\phi,\phi}(\mathbf{X})\, \delta_{\theta\psi} \right) W^*_{\alpha\beta 3}(\mathbf{x}',\mathbf{X})\, d\Omega -$$

$$n_\beta(\mathbf{x}') \int_{\Omega} \kappa_{\theta\psi} B((1-\nu)\kappa_{\theta,\psi} + \nu\delta_{\theta\psi}\kappa_{\phi\phi}) W^*_{\alpha\beta 3}(\mathbf{x}',\mathbf{X})\, d\Omega +$$

$$n_\beta(\mathbf{x}') \int_{\Omega} W^*_{\alpha\beta 3}(\mathbf{x}',\mathbf{X})\, q_3(\mathbf{X})\, d\Omega \tag{10.4}$$

and

$$\frac{1}{2}p_3(\mathbf{x}') + n_\beta(\mathbf{x}') \int\!\!\!\!\!-_{\Gamma} P^*_{3\beta\gamma}(\mathbf{x}', \mathbf{x})\, w_\gamma(\mathbf{x})\, d\Gamma + n_\alpha(\mathbf{x}') \int\!\!\!\!\!=_{\Gamma} P^*_{3\beta 3}(\mathbf{x}', \mathbf{x})\, w_3(\mathbf{x})\, d\Gamma$$
$$= n_\beta(\mathbf{x}') \int_\Gamma W^*_{3\beta\gamma}(\mathbf{x}', \mathbf{x})\, p_\gamma(\mathbf{x}, t)\, d\Gamma + n_\beta(\mathbf{x}') \int_\Gamma W^*_{3\beta 3}(\mathbf{x}', \mathbf{x})\, p_3(\mathbf{x})\, d\Gamma -$$
$$n_\beta(\mathbf{x}') \int_\Omega \kappa_{\theta\psi} B \frac{1-\nu}{2}\left(u_{\theta,\psi}(\mathbf{X}) + u_{\psi,\theta}(\mathbf{X}) + \frac{2\nu}{1-\nu} u_{\phi,\phi}(\mathbf{X})\,\delta_{\theta\psi}\right) W^*_{3\beta 3}(\mathbf{x}', \mathbf{X})\, d\Omega -$$
$$n_\beta(\mathbf{x}') \int_\Omega \kappa_{\theta\psi} B((1-\nu)\kappa_{\theta,\psi} + \nu\delta_{\theta\psi}\kappa_{\phi\phi}) W^*_{3\beta 3}(\mathbf{x}', \mathbf{X})\, d\Omega +$$
$$n_\beta(\mathbf{x}') \int_\Omega W^*_{3\beta 3}(\mathbf{x}', \mathbf{X})\, q_3(\mathbf{X})\, d\Omega \tag{10.5}$$

10.2.3 Traction Bending Equations

The singularity order in the traction integral equations is higher than in the displacement integral equations. Indeed, the matrix **H**, the kernels $P^*_{\alpha\beta\gamma}$ and $P^*_{3\beta\gamma}$ are strongly singular, while the kernels $P^*_{\alpha\beta\gamma}$ and $P^*_{3\beta 3}$ are hypersingular. The shape functions will reduce the singularity order by one for off-diagonal submatrices. This means that the entries of elements in the **H** matrix corresponding to the kernels $P^*_{\alpha\beta 3}$ and $P^*_{3\beta\gamma}$ become smooth, while the elements of the kernels $P^*_{\alpha\beta\gamma}$ and $P^*_{3\beta 3}$ continue to have strong sigularity (Dirgantara [2]).

On the other hand, in the matrix **G**, the off-diagonal submatrices are again smoothed because the shape functions reduce the singularity order. The diagonal matrices, on the other hand, contain the kernels $W^*_{\alpha\beta 3}$ and $W^*_{3\beta\gamma}$ which are weakly singular and $W^*_{\alpha\beta\gamma}$ and $W^*_{3\beta 3}$ have strong singular.

Weak singularity is treated using a nonlinear coordinate transformation as in Telles [9] and element subdivision. Strongly singular and hypersingular integrals are evaluated using Taylor series expansion around the singular point as presented in Dirgantara [2].

The treatment of strong singularities and hypersingularity is carried out using the method of subtraction of singularities based on Taylor series expansion around the singular point (Dirgantara [2]). Subsequently, the singular terms are integrated analytically. For straight elements along the crack, direct analytical integrations most efficiently evaluate strongly singular and hypersingular integrals in membrane integral equations. This method presents no mathematical difficulties in its implementation. However, to keep the present chapter compact, the reader is invited to refer to it in Dirgantara [2].

The displacement integrals require a continuity C^0 to ensure the existence of the integrals in the Cauchy sense. Such a condition is ensured by locating the collocation points on the boundary. On the other hand, the existence of the traction integrals in the Hadamard sense requires that the displacements are at least C^1 functions and the tractions C^0 in order for the integrals in the Cauchy sense to exist. The above continuity requirements are satisfied by locating the collocation points at the internal nodes of discontinuous elements on the crack faces.

The general modeling strategy implemented in this work is closely based on that used by Dirgantara and Aliabadi [3], and can be summarized as follows:

Crack modeling strategy

i Crack boundarys are modeled with discontinuous quadratic plane elements.
ii The displacement equation is applied to the placement points of one of the crack surfaces.
iii The traction equation is applied to place points on the opposite surface.
iv Discontinuous quadratic plane elements are used along the remaining boundary of the structure.

10.3 Multidomain Technique for Crack Modeling

The dual boundary element method requires the use of both the displacement and integral traction equations. In the case of composite shells, presented in Chap. 9, the traction equation involves complex mathematical expressions, making its practical application difficult when dealing with the hyper singularities in the **H** matrix. In this and other types of problems (such as plates made of different materials), the alternative way to deal with fracture problems is the subregion method, also known as the multi-domain method.

For this purpose, consider the case of a cracked plate divided into two different domain Ω_1 and Ω_2 whose boundaries at the interface are Γ_1' and Γ_2' as shown in Fig. 10.2. The discretized boundary equations for each region are given by Useche et al. [12], Useche [11]:

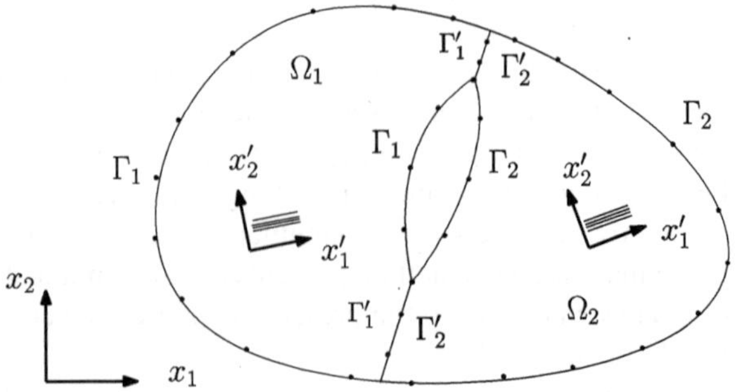

Fig. 10.2 Crack modeling using multi-domain technique

$$\begin{aligned}\mathbf{H}_I\mathbf{w}_I &= \mathbf{G}_I\mathbf{p}_I + \mathbf{q}_I \\ \mathbf{H}_{II}\mathbf{w}_{II} &= \mathbf{G}_{II}\mathbf{p}_{II} + \mathbf{q}_{II}\end{aligned} \tag{10.6}$$

In addition, the displacement vectors and tractions at the interface Γ_I^α, must satisfy the following compatibility and equilibrium conditions:

$$\mathbf{w}_I^1 = \mathbf{w}_I^2, \quad \ddot{\mathbf{w}}_I^1 = \ddot{\mathbf{w}}_I^2, \quad \mathbf{p}_I^1 = -\mathbf{p}_I^2 \tag{10.7}$$

If the displacement and traction vectors for both regions are ordered as a function of Γ_I y Γ_{II}, Γ_I^1, and Γ_{II}^1:

$$\mathbf{w}_I = \{\mathbf{w}_I, \mathbf{w}_I^1, \mathbf{w}_I^2\}^T, \quad \mathbf{p}_I = \{\mathbf{p}_I, \mathbf{p}_I^1, \mathbf{p}_I^2\}^T \tag{10.8}$$

$$\mathbf{w}_{II} = \{\mathbf{w}_{II}, \mathbf{w}_{II}^1, \mathbf{w}_{II}^2\}^T, \quad \mathbf{p}_{II} = \{\mathbf{p}_{II}, \mathbf{p}_{II}^1, \mathbf{p}_{II}^2\}^T \tag{10.9}$$

Equations (10.6) and the coupling and force balance Eqs. (10.7) can be written compactly as

$$\begin{bmatrix} \mathbf{H}_I^1 & \mathbf{H}_{II}^1 & & \\ & & \mathbf{H}_I^2 & \mathbf{H}_{II}^2 \\ \mathbf{C}_I^1 & & -\mathbf{C}_I^2 & \end{bmatrix} \begin{Bmatrix} \ddot{\mathbf{w}}_I^1 \\ \ddot{\mathbf{w}}_{II}^1 \\ \ddot{\mathbf{w}}_I^2 \\ \ddot{\mathbf{w}}_{II}^2 \end{Bmatrix}$$
$$+ \begin{bmatrix} \mathbf{H}_I^1 & \mathbf{H}_{II}^1 & -\mathbf{G}_I^1 & & & \\ & & & \mathbf{H}_I^2 & \mathbf{H}_{II}^2 & -\mathbf{G}_I^2 \\ \mathbf{C}_I^1 & & & -\mathbf{C}_I^2 & & \\ & & \mathbf{E}_I^1 & & & \mathbf{E}_I^2 \end{bmatrix} \begin{Bmatrix} \mathbf{w}_I^1 \\ \mathbf{w}_{II}^1 \\ \mathbf{p}_I^1 \\ \mathbf{w}_I^2 \\ \mathbf{w}_{II}^2 \\ \mathbf{p}_I^2 \end{Bmatrix} = \begin{bmatrix} \mathbf{G}_{II}^1 & \\ & \mathbf{G}_{II}^2 \\ \mathbf{0} & \\ & \mathbf{0} \end{bmatrix} \begin{Bmatrix} \mathbf{p}_{II}^1 \\ \mathbf{p}_{II}^2 \end{Bmatrix} + \begin{Bmatrix} \mathbf{f}^1 \\ \mathbf{f}^2 \end{Bmatrix} \tag{10.10}$$

where $\mathbf{C}_I^\alpha$ and $\mathbf{E}_I^\alpha$ are compatibility matrices related to the Eqs. (10.7). Applying the boundary conditions and solving the system yields unknown displacements and tractions in the plate.

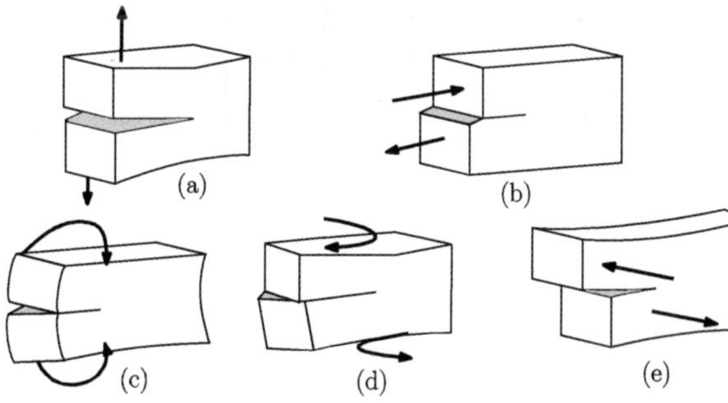

Fig. 10.3 Crack modes: **a** Opening crack-mode I. **b** In-plane shear–mode II. **c** Opening crack–bending mode I. **d** Torsional shear–bending mode II. **d** Out-of-plane shear–bending mode III

10.4 Stress Intensity Factor Evaluation

10.4.1 Crack Tip Opening Displacement–CTOD

For crack problems that combine bending and plane tension, the stress intensity factors can be represented by the superposition of five stress intensity factors (SIF's), two due to membrane loads and three due to bending and shear loads (see Fig. 10.3). In this work, the stress resultant intensity factors for modes I, II, and III are denoted by K_1, K_2, and K_3, respectively. Subscript m is added for stress intensity factors due to membrane loads, and subscript b is added for stress intensity factors due to bending and shear loads.

The displacements on the crack surfaces near the crack tip can be obtained as (Dirgantara [5]):

$$\begin{Bmatrix} \Delta\phi_2 \\ \Delta\phi_1 \\ \Delta w_3 \\ \Delta u_2 \\ \Delta u_1 \end{Bmatrix} = \begin{Bmatrix} \phi_2 \\ \phi_1 \\ w_3 \\ u_2 \\ u_1 \end{Bmatrix}_{\theta=180^\circ} - \begin{Bmatrix} \phi_2 \\ \phi_1 \\ w_3 \\ u_2 \\ u_1 \end{Bmatrix}_{\theta=-180^\circ}$$

$$= \begin{bmatrix} \frac{48}{Eh^3}\sqrt{\frac{2r}{\pi}} & 0 & 0 & 0 & 0 \\ 0 & \frac{48}{Eh^3}\sqrt{\frac{2r}{\pi}} & 0 & 0 & 0 \\ 0 & 0 & \frac{24(1+\upsilon)}{5Eh}\sqrt{\frac{2r}{\pi}} & 0 & 0 \\ 0 & 0 & 0 & \frac{8}{Eh}\sqrt{\frac{2r}{\pi}} & 0 \\ 0 & 0 & 0 & 0 & \frac{8}{Eh}\sqrt{\frac{2r}{\pi}} \end{bmatrix} \begin{Bmatrix} K_{1b} \\ K_{2b} \\ K_{3b} \\ K_{1m} \\ K_{2m} \end{Bmatrix} \tag{10.11}$$

The resultant stress intensity factors can then be written in terms of displacements on the crack surfaces as

$$\{K\} = \frac{1}{\sqrt{r}}\mathbf{C}\{\Delta w\} \tag{10.12}$$

where

$$K = \begin{Bmatrix} K_{1b} \\ K_{2b} \\ K_{3b} \\ K_{1m} \\ K_{2m} \end{Bmatrix}; \ \Delta\{w\} = \begin{Bmatrix} \Delta\phi_2 \\ \Delta\phi_1 \\ w_3 \\ u_2 \\ u_1 \end{Bmatrix} \tag{10.13}$$

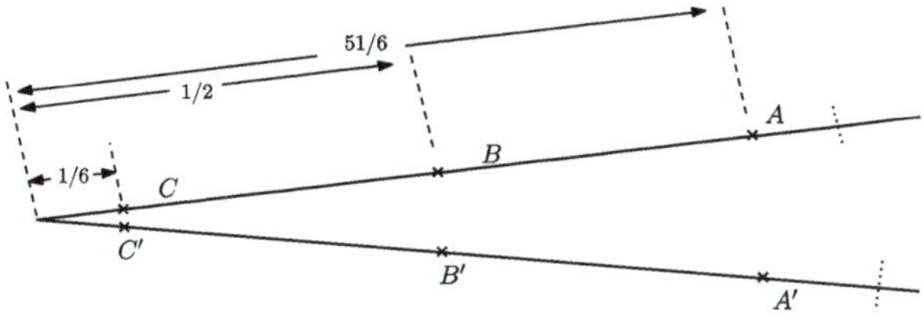

Fig. 10.4 Crack tip element

and

$$\mathbf{C} = \begin{bmatrix} \frac{Eh^3}{48}\sqrt{\frac{\pi}{2}} & 0 & 0 & 0 & 0 \\ 0 & \frac{Eh^3}{48}\sqrt{\frac{\pi}{2}} & 0 & 0 & 0 \\ 0 & 0 & \frac{5Eh(1+\upsilon)}{24(1+\upsilon)}\sqrt{\frac{\pi}{2}} & 0 & 0 \\ 0 & 0 & 0 & \frac{Eh}{8}\sqrt{2\pi} & 0 \\ 0 & 0 & 0 & 0 & \frac{Eh}{8}\sqrt{2\pi} \end{bmatrix} \tag{10.14}$$

When discontinuous elements are used for modeling crack surfaces, then, at crack tip elements, the distance of every node to the crack tip is given in (see Fig. 10.4). Hence,

$$\begin{aligned} \{K\}^{AA'} &= \sqrt{\frac{6}{5l}}\mathbf{C}\left(\{w\}^A - \{w\}^{A'}\right) \\ \{K\}^{BB'} &= \sqrt{\frac{2}{l}}\mathbf{C}\left(\{w\}^B - \{w\}^{B'}\right) \end{aligned} \tag{10.15}$$

Then, the Stress Intensity Factors are extrapolated to the crack tip using relationship:

$$\{K\}^{tip} = \frac{r_{AA'}}{r_{AA'} - r_{BB'}}\left(\{K\}^{BB} - \frac{r_{BB'}}{r_{AA'}}\{K\}^{AA'}\right) \tag{10.16}$$

10.4.2 J-Integral Calculation

The J-integral in the x_δ direction, considering distributed body loads and inertial forces, is defined for plate bending as presented in Sosa and Herrmann [8], Useche [10]:

$$J_{\delta b} = \int_\Gamma (W_b - qw_3)n_\delta \, d\Gamma - \int_\Gamma \left[M_{\alpha\beta}w_{\alpha,\delta} + Q_\beta w_{3,\delta}\right] n_\beta \, d\Gamma \tag{10.17}$$

where $\Gamma = \Gamma_c + \Gamma_p$ is an arbitrary curve surrounding the crack tip as presented in Fig. 10.5; Ω is the area enclosed Γ; n_δ is a unit outward normal to the boundary Γ, and W_b is the strain energy density for plate bending. From above expression, the stress resultant intensity are obtained through the following relationship:

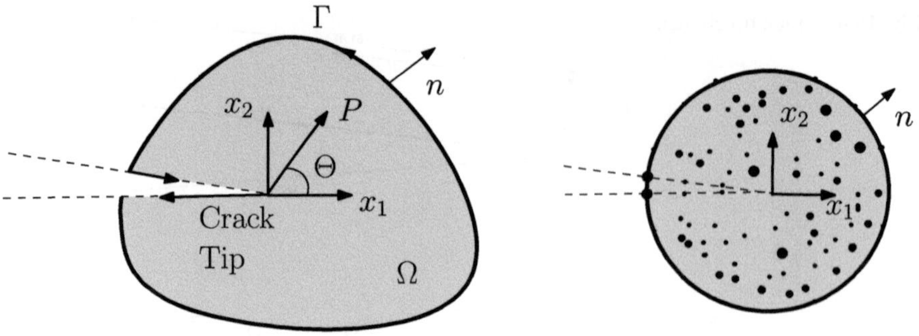

Fig. 10.5 Path-definition for J-Integral calculation

$$J_{1b}^{S} = \frac{12}{Eh^3} K_{1b}^2 \tag{10.18}$$

$$J_{1b}^{AS} = \frac{12}{Eh^3} K_{2b}^2 \left[\frac{10}{h^2(1+\nu)} \left(\frac{\Delta w_3}{\Delta w_1} \right)^2 + 1 \right] \tag{10.19}$$

$$J_{1b}^{AS} = \frac{12(1+\nu)}{10Eh} K_{3b}^2 \left[\frac{h^2(1+\nu)}{10} \left(\frac{\Delta w_1}{\Delta w_3} \right)^2 + 1 \right] \tag{10.20}$$

10.5 Numerical Examples

10.5.1 Plate with a Centre Crack Loaded by Bending and Tension

A rectangular plate with a central crack loaded by edge bending and tension is analyzed (see Fig. 10.6). The properties of the plate are $b/h = 2; c/b = 2; M_o = 1.0;$ $t = 1.0$, $E = 210000$, and $\nu = 0.3$. For DBEM analysis, 32 boundary elements for plate border and 16 discontinuous quadratic elements for each faces of the crack has been used (see Fig. 10.6). Table 10.1 shows SIF for K_{1b} factor for different relations *a*/*b*. The DBEM results show good agreement when compared with those obtained by Dirgantara [2]. Bending deflection distribution is showed in Fig. 10.6.

Clampled square plate with central crack

In this example, a square orthotropic cracked plate as shown in Fig. 10.7 is considered. The plate is fully clamped. The mechanical properties of the plate are the same as in the previous example except $h = 25.4$ mm. For harmonic analysis, a uniform load of 1 N/m^2 is considered as the applied load.

The harmonic response and modal analysis of this plate was analyzed for four different material angles: $\theta = 0°, 15°, 30°$, and $45°$. A convergent mesh of 24 boundary elements and 288 RIM points for both regions was used. A finite element model

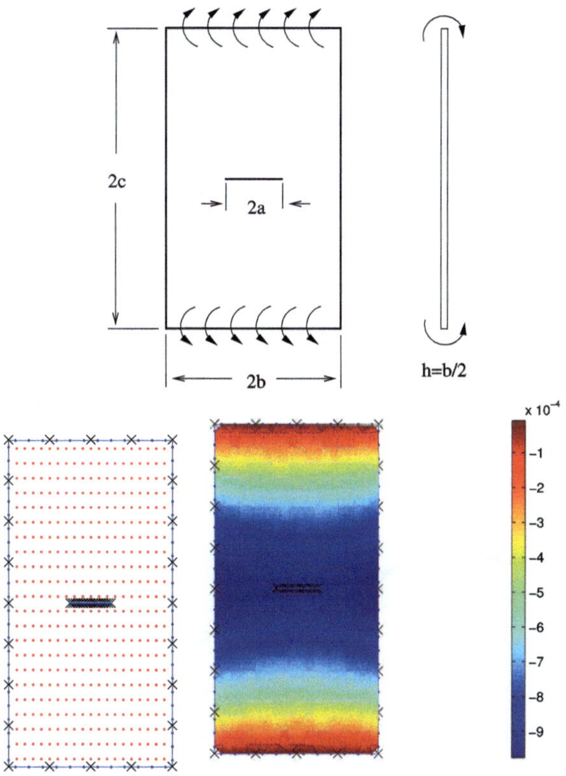

Fig. 10.6 Rectangular plate with central crack. Above: Geometry and loads. Below Left: Dual boundary element model. Right: Displacement solution for deflection

Table 10.1 K_{1b} stress intensity factor for the rectangular plate with central crack

a/b	$K_{1b}/M_o\sqrt{\pi a}$–BEM	$K_{1b}/M_o\sqrt{\pi a}$–Ref. [2]	% error
0.1	0.993	0.995	0.20
0.2	0.992	0.990	0.20
0.4	0.845	0.850	0.59
0.6	0.095	0.100	0.50
0.8	0.134	0.135	0.74

using ANSYS with a mesh of 256 elements and 844 nodes in order to compare BEM solutions. Figure 10.7 demonstrates a good correlation between convergent boundary element and finite element solution for $\theta = 0°$, $15°$, $30°$. A difference between BEM and FEM solutions appears in the fourth resonance frequency for $\theta = 45°$. The first five natural frequencies were calculated for material direction θ from $0°$ to $\theta = 45°$. Again, Fig. 10.7 shows good corelations between boundary element and finite element solutions as function of material direction.

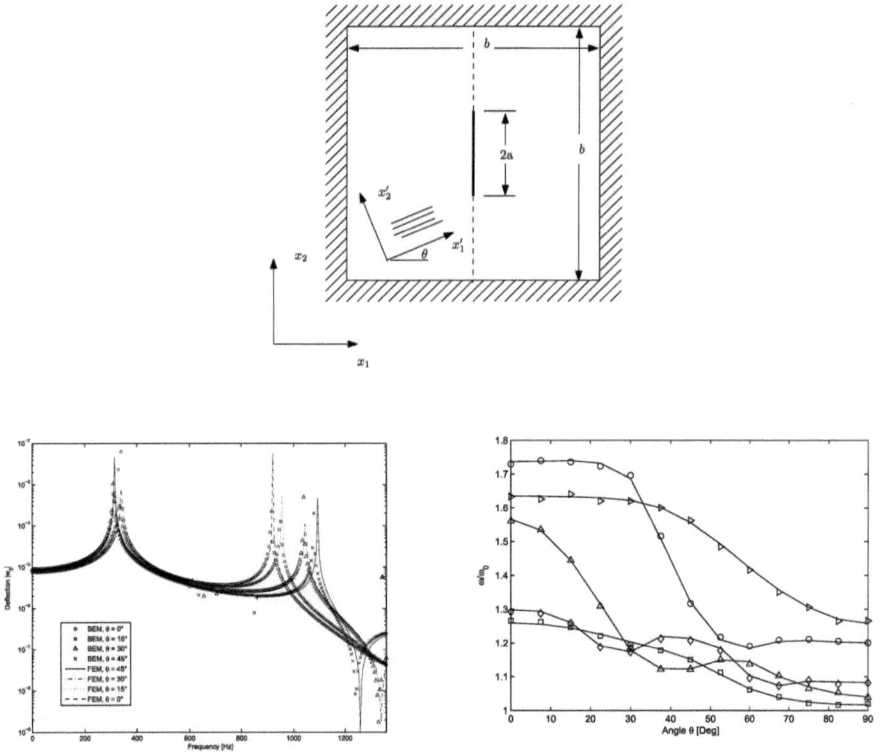

Fig. 10.7 Full clamped plate with central crack. Above: Geometry and boundary conditions. Left: Frequency response spectrum for a square plate with central crack. Right: Natural frequencies for different material orientations

Double curved laminated shell

Now consider a laminated double curved shell with $\kappa_{11} = \kappa_{22} = 1/100, a/h = 100$, as shown in Fig. 10.8. The shell is at the edges. The stacking sequence of the laminated is $[0°/90°]_s$. A time-dependent uniformly distributed pressure over the shell given by ($p_0/E_{22} = 500$ and $t_{max} = 5 \times 10^{-3} s.$) is considered in this example:

$$p(t) = \begin{cases} p_0 t, & t \leq 0.001\ s. \\ p_0, & t \in [0.001, 0.002]\ s. \\ 0.0 & t > 0.002\ s. \end{cases} \tag{10.21}$$

A convergent time-step Δt equal to $1 \times 10^{-4}\ s.$ was used. The shell was discretized with sixteen elements and 9, 18, 32, and 64 uniformly distributed domain RIM internal collocation points. As in the previous case, the BEM solution is compared with the FEM model with 4936 nodes and 1580 shear deformable shell 8-node elements. Figure 10.8 presents good convergent behavior of transverse deflexion as a func-

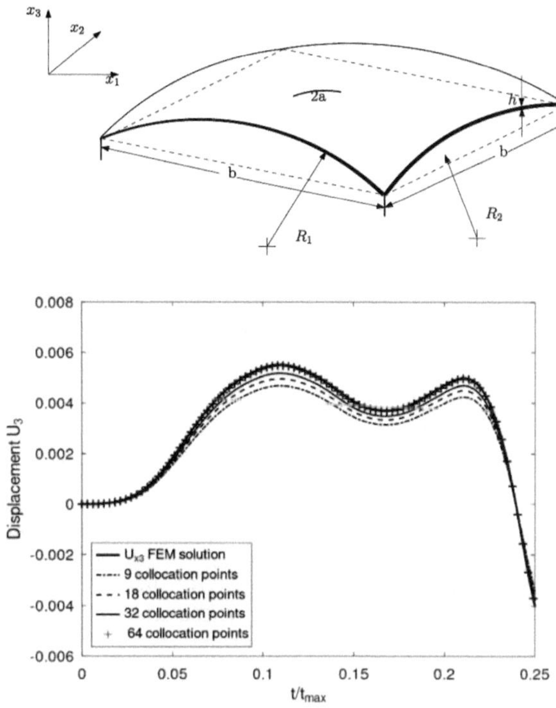

Fig. 10.8 Double curved square laminated shell with encastred edges, curvature radios of R_1 and R_2 and central crack with length of $2a$. Above: BEM convergence solution for the displacement u_3 at crack faces

tion of RIM collocation points when compared with the FEM solution. Variations of 0.01% were obtained for displacements when comparing the BEM convergent curve (64 placement points) and the FEM displacement curve for the considered elapsed time. Using more placement points, which would further reduce the difference between both curves, would increase the computing time from 5.82 to 9.31 min for 75 collocations points.

10.6 Summary

The Dual Boundary Element Method applied to isotropic fracture analysis of plates and shells considering membrane, bending moments, and shear forces were presented. A general methodology for using the Dual Boundary Element Method was presented. The hypersingular equations for plane elasticity and Reissner plate bending were established. The types of singularities appearing in the tensile equations were identified, and the Taylor series expansion methodology for dealing with them was presented. Stress intensification factors were calculated using CTOD and J-integral methodology.

References

1. Albuquerque EL, Sollero P, Aliabadi MH (2002) The boundary element method applied to time dependent problems in anisotropic materials. Int J Solids Struct 39(5):1405–1422
2. Dirgantara T (2002) Boundary element analysis of cracks in shear deformable plates and shells. WIT Press, Southampton
3. Dirgantara T, Aliabadi MH (2000) Crack growth analysis of plates loaded by bending and tension using dual boundary element method. Int J Fract 105(1):27–47
4. Dirgantara T, Aliabadi MH (2001) Dual boundary element formulation for fracture mechanics analysis of shear deformable shells. Int J Solids Struct 38(44–45):7769–7800
5. Dirgantara T, Aliabadi MH (2001) Dual boundary element formulation for fracture mechanics analysis of shear deformable shells. Int J Solids Struct 38:7769–7800
6. Portela A, Aliabadi MH, Rooke DP (1992) The dual boundary element method: Effective implementation for crack problems. Int J Numer Meth Eng 33(6):1269–1287
7. Sollero P, Aliabadi MH (1995) Anisotropic analysis of cracks in composite laminates using the dual boundary element method. Compos Struct 31:229–233
8. Sosa HA, Herrmann G (1979) On invariant integrals for plate bending problems. Int J Fract 15:769–782
9. Telles J (1987) A self-adaptive coordinate transformation for efficient numerical evaluation of general boundary element integrals. Int J Numer Methods Eng 24:959–973
10. Useche J (2020) Fracture dynamic analysis of cracked reissner plates using the boundary element method. Int J Solids Struct 191–192:315–332
11. Useche J (2024) Dynamic analysis of cracked thick composite shells by the boundary element method. Eng Anal Bound Elem J 168
12. Useche J, Albuquerque EL, Sollero P (2012) Harmonic analysis of shear deformable orthotropic cracked plates using the boundary element method. Eng Anal Bound Elem 36(11):1528–1535
13. Wen PH, Aliabadi MH, Young A (2000) Application of dual reciprocity method to plates and shells. Eng Anal Bound Elem 24(7–8):583–590

Chapter 11
Boundary Element Analysis of Thin Walled Spatial Structures

Abstract Plate and shell formulations by the boundary element method can be used to analyze thin-walled space structures. These structures, built by assembling the plate in the space, allow the modeling of structures commonly found in aeronautical, aerospace, naval, and oceanic applications. In this chapter, we present a methodology for the boundary element analysis of thin-walled spatial structures by assembling the discretized boundary equations of their constituent elements. These equations are obtained from the methodologies presented in previous chapters. These equations are coupled mathematically from conditions of continuity of displacements and rotations and equilibrium of forces and moments in contours and coupling surface between the plates. By means of application examples, the methodology is applied to the analysis of thin-walled structures. As a special case and of relevance to the aircraft industry, the application of the methodology to the analysis of fractured plates repaired with composite plates is presented.

11.1 Introduction

Plate and shell formulations by the boundary element method can be used to analyze thin-walled space structures. These structures, built by assembling the plate in the space, allow the modeling of structures commonly found in aeronautical, aerospace, naval.

In this chapter, we present a methodology for the boundary element analysis of thin-walled spatial structures by assembling the discretized boundary equations of their constituent elements: Dirgantara [2], Baiz and Aliabadi [1], Pisa et al. [5]. These equations are obtained from the methodologies presented in previous chapters. These equations are coupled mathematically from conditions of continuity of displacements and rotations and equilibrium of forces and moments in contours and coupling surface between the plates. By means of application examples, the methodology is applied to the analysis of thin-walled structures. As a special case and of relevance to the aircraft industry, the application of the methodology to the analysis of fractured plates repaired with composite plates is presented.

J. Useche Vivero, *The Boundary Element Method for Engineering Analysis of Plates and Shells*, Solid Mechanics and Its Applications 281,
https://doi.org/10.1007/978-3-031-93050-8_11

11.2 Spatial Assembled Formulation for Plates

Consider a structure in space consisting of multiple plates referred to as local coordinate systems, $O^r - x_1^r - x_2^r - x_3^r$. The global coordinate system is defined by $O - x_1 - x_2 - x_3$, as presented in the Fig. 11.1. In plate space as a structure, it implies a coupling between the membrane (in the plane of the plate) and bending responses. Therefore, in the study of these structures, the equations of plane stress and bending of each plate must be considered, which are given by the boundary equations presented in the previous chapters (either for isotropic or laminated materials) and written here again:

$$\int_{\Gamma} N^i_{\alpha\beta} n_\beta U^*_\alpha d\Gamma - \int_{\Omega} N^i_{\alpha\beta} U^*_{\alpha,\beta} d\Omega + \int_{\Omega} N^{ii}_{\alpha\beta,\beta} U^*_\alpha d\Omega + \int_{\Omega} q_\alpha U^*_\alpha d\Omega + \int_{\Omega} U^*_\theta \Lambda^m_{\theta\alpha} \ddot{u}_\alpha d\Omega + \int_{\Omega} U^*_\theta \Lambda^{bm}_{\theta\alpha} \ddot{w}_\alpha d\Omega = 0 \tag{11.1}$$

$$w^r_j(\mathbf{x}') + \int_{\Gamma} P^*_{ij}(\mathbf{x}', \mathbf{x})\, w^r_j(\mathbf{x})\, d\Gamma = \int_{\Gamma} W^*_{ij}(\mathbf{x}', \mathbf{x})\, p^r_j(\mathbf{x})\, d\Gamma + \int_{\Omega} W^*_{ij}(\mathbf{x}', \mathbf{x})\, q^r_j(\mathbf{x})\, d\Omega - \Lambda_{ij} \int_{\Omega} W^*_{ij}(\mathbf{X}', \mathbf{X})\, \ddot{w}^r_j(\mathbf{X})\, d\Omega \tag{11.2}$$

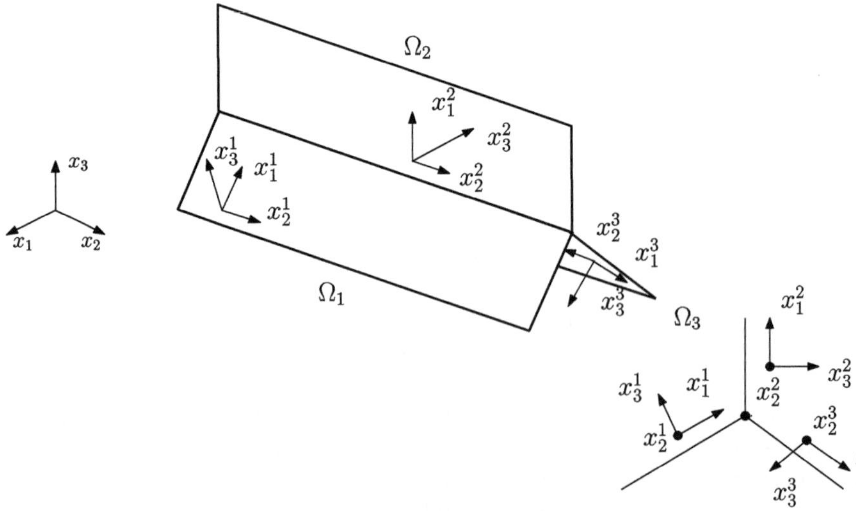

Fig. 11.1 Example

Superscript r intends for *region*. Discretizing the contour, we obtain the discretized form for the above equations, which is written in matrix form as, in discretized matrix form:

$$\mathbf{H}^r_p \ddot{\mathbf{w}}^r + \mathbf{H}^r_p \mathbf{w}^r = \mathbf{G}^r_p \mathbf{p}^r - \mathbf{B}^r_p \mathbf{q}^r \tag{11.3}$$

where, in the above equations, the superindex r intends for the plate r $\ddot{\mathbf{w}}^r = \{\ddot{w}^r_1, \ddot{w}^r_1, \ddot{w}^r_1\}^T$, $\mathbf{w}^r = \{w^r_1, w^r_1, w^r_1\}^T$ $\mathbf{p}^r = \{p^r_1, p^r_2, p^r_3\}^T$ and $\mathbf{q}^r = \{q^r_1, q^r_2, q^r_3\}^T$.

For flat stress:

$$u^r_\theta\left(\mathbf{X}'\right) + \int_\Gamma T^*_{\theta\alpha}\left(\mathbf{X}', \mathbf{x}\right) u^r_\alpha(\mathbf{x})\, d\Gamma = \int_\Gamma U^*_{\theta\alpha}\left(\mathbf{X}', \mathbf{x}\right) t^r_\alpha(\mathbf{x})\, d\Gamma$$
$$+ \int_\Omega U^*_{\theta\alpha}\left(\mathbf{X}', \mathbf{X}\right) b^r_\alpha(\mathbf{X})\, d\Omega - \lambda_{ij} \int_\Omega U^*_{\theta\alpha}\left(\mathbf{X}', \mathbf{X}\right) \ddot{u}^r_\alpha(\mathbf{X})\, d\Omega$$
$$-\frac{\rho}{h} \int_\Omega W^*_{ij}\left(\mathbf{X}', \mathbf{X}\right) \ddot{w}^r_j(\mathbf{X})\, d\Omega \tag{11.4}$$

Discretizing the contour, we obtain the discretized form for the above equations, which are written in matrix form as

$$\mathbf{M}^r_s \ddot{\mathbf{u}}^r + \mathbf{H}^r_s \mathbf{u}^r = \mathbf{G}^r_s \mathbf{t}^r - \mathbf{B}^r_s \mathbf{b}^r \tag{11.5}$$

The displacements and traction in each plate are referenced to their coordinate system by means of the vectors: $\ddot{\mathbf{u}}^r = \{\ddot{u}^r_1, \ddot{u}^r_2\}^T$, $\mathbf{u}^r = \{u^r_1, u^r_2\}^T$, $\mathbf{t}^r = \{t^r_1, t^r_2\}^T$ and $\mathbf{b}^r = \{b^r_1, b^r_2\}^T$. Compactly, for a plate, the above equations can be written in matrix form, after discretization as a discretization as Dirgantara [2]:

$$\begin{bmatrix} \mathbf{M}^r_p & \mathbf{0} \\ \mathbf{0} & \mathbf{M}^r_p \end{bmatrix} \begin{Bmatrix} \ddot{\mathbf{u}}^r \\ \ddot{\mathbf{w}}^r \end{Bmatrix} + \begin{bmatrix} \mathbf{H}^r_p & \mathbf{0} \\ \mathbf{0} & \mathbf{H}^r_s \end{bmatrix} \begin{Bmatrix} \mathbf{u}^r \\ \mathbf{w}^r \end{Bmatrix} =$$
$$\begin{bmatrix} \mathbf{G}^r_p & \mathbf{0} \\ \mathbf{0} & \mathbf{G}^r_s \end{bmatrix} \begin{Bmatrix} \mathbf{t}^r \\ \mathbf{p}^r \end{Bmatrix} + \begin{bmatrix} \mathbf{B}^r_p & \mathbf{0} \\ \mathbf{0} & \mathbf{B}^r_s \end{bmatrix} \begin{Bmatrix} \mathbf{q}^r \\ \mathbf{b}^r \end{Bmatrix} \tag{11.6}$$

Compactly written:

$$\mathbf{H}^r \ddot{\mathbf{d}}^r + \mathbf{H}^r \mathbf{d}^r = \mathbf{G}^r \mathbf{f}^r + \mathbf{B}^r \mathbf{s}^r \tag{11.7}$$

where $\ddot{\mathbf{d}}^r = \{\ddot{\mathbf{u}}^r, \ddot{\mathbf{w}}^r\}$, $\mathbf{d}^r = \{\mathbf{u}^r, \mathbf{w}^r\}$, $\mathbf{f}^r = \{\mathbf{t}^r, \mathbf{p}^r\}$ y $\mathbf{s}^r = \{\mathbf{q}^r, \mathbf{b}^r\}$.

In spatially assembled structures, plates are joined along common boundaries (see Fig. 11.1), where compatibility equations for displacements and equilibrium equations for tractions apply. To illustrate this approach, consider $N(r = 1, \ldots, N)$ plates assembled at the common joint boundary with the direction x^N_2 aligned with the same global direction along a straight joint line. In this case, the compatibility equations along the joint line are defined as (Fig. 11.2):

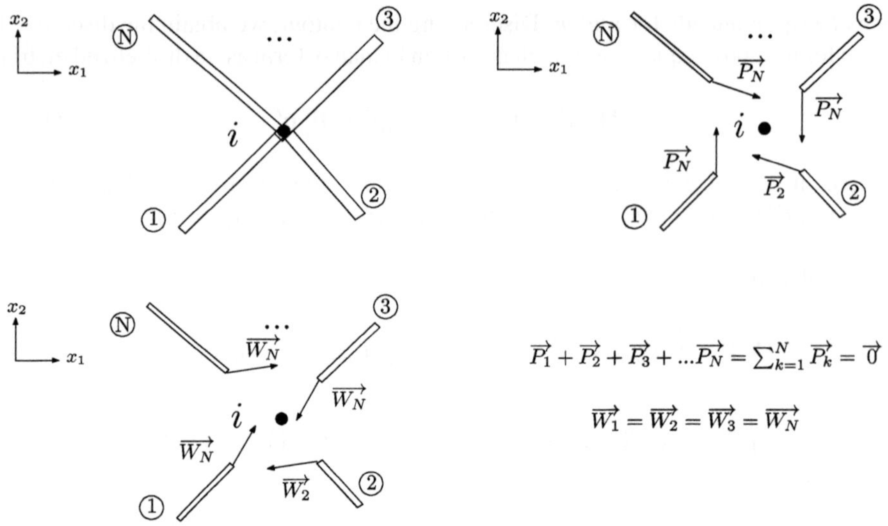

Fig. 11.2 Example

For x_1–displacements:

$$u_1^1 n_{11}^1 + w_3^1 n_{13}^1 = u_1^2 n_{11}^2 + w_3^2 n_{13}^2 = \cdots = u_1^N n_{11}^N + w_3^N n_{13}^N \tag{11.8}$$

For x_2-displacements:

$$u_2^1 = u_2^2 = \cdots = u_2^N \tag{11.9}$$

For x_3-displacements:

$$u_1^1 n_{31}^1 + w_3^1 n_{33}^1 = u_1^2 n_{31}^2 + w_3^2 n_{33}^2 = \cdots = u_1^N n_{31}^N + w_3^N n_{33}^N \tag{11.10}$$

For x_1-rotations:

$$w_1^1 = w_1^2 = \cdots = w_1^N = 0 \tag{11.11}$$

For x_2-rotations:

$$w_2^1 = w_2^2 = \cdots = w_2^N \tag{11.12}$$

where n_{ik}^N are the components of the rotation matrix, $\mathbf{T}^N$, of the plate N from local coordinates to global coordinates. Similarly, the compatibility equations for accelerations are expressed as: For acceleration along x_1-axis:

$$\ddot{u}_1^1 n_{11}^1 + \ddot{w}_3^1 n_{13}^1 = \ddot{u}_1^2 n_{11}^2 + \ddot{w}_3^2 n_{13}^2 = \cdots = \ddot{u}_1^N n_{11}^N + \ddot{w}_3^N n_{13}^N \tag{11.13}$$

For acceleration along x_2-axis:

$$\ddot{u}_2^1 = \ddot{u}_2^2 = \cdots = \ddot{u}_2^N \tag{11.14}$$

For acceleration along x_3-axis:

$$\ddot{u}_1^1 n_{31}^1 + \ddot{w}_3^1 n_{33}^1 = \ddot{u}_1^2 n_{31}^2 + \ddot{w}_3^2 n_{33}^2 = \cdots = \ddot{u}_1^N n_{31}^N + \ddot{w}_3^N n_{33}^N \tag{11.15}$$

For rotational acceleration over x_1-axis:

$$\ddot{w}_1^1 = \ddot{w}_1^2 = \cdots = \ddot{w}_1^N = 0 \tag{11.16}$$

For rotational acceleration over x_2-axis:

$$\ddot{w}_2^1 = \ddot{w}_2^2 = \cdots = \ddot{w}_2^N \tag{11.17}$$

Compatibility equations are supplemented by the equilibrium conditions along the joint-line: Equilibrium equation for x_1-direction:

$$\sum_{r=1}^{N} (t_1^r n_{11}^r + q_3^r n_{13}^r) = 0 \tag{11.18}$$

Equilibrium equation for x_2-direction:

$$\sum_{r=1}^{N} t_2^r = 0 \tag{11.19}$$

Equilibrium equation for x_3-direction:

$$\sum_{r=1}^{N} (t_1^r n_{31}^r + q_3^r n_{33}^r) = 0 \tag{11.20}$$

Equilibrium of momentum along for x_2-axis:

$$\sum_{r=1}^{N} q_2^r = 0 \tag{11.21}$$

As example consider the assembled plate structure. The boundary element system of equations is given by

$$\begin{bmatrix} \mathbf{M}^1 & \mathbf{0} & \mathbf{0} \\ \mathbf{0} & \mathbf{M}^2 & \mathbf{0} \\ \mathbf{0} & \mathbf{0} & \mathbf{M}^3 \end{bmatrix} \begin{Bmatrix} \ddot{\mathbf{d}}^1 \\ \ddot{\mathbf{d}}^2 \\ \ddot{\mathbf{d}}^3 \end{Bmatrix} + \begin{bmatrix} \mathbf{H}^1 & \mathbf{0} & \mathbf{0} \\ \mathbf{0} & \mathbf{H}^2 & \mathbf{0} \\ \mathbf{0} & \mathbf{0} & \mathbf{H}^3 \end{bmatrix} \begin{Bmatrix} \mathbf{d}^1 \\ \mathbf{d}^2 \\ \mathbf{d}^3 \end{Bmatrix} = \begin{bmatrix} \mathbf{G}^1 & \mathbf{0} & \mathbf{0} \\ \mathbf{0} & \mathbf{G}^2 & \mathbf{0} \\ \mathbf{0} & \mathbf{0} & \mathbf{G}^3 \end{bmatrix} \begin{Bmatrix} \mathbf{f}^1 \\ \mathbf{f}^2 \\ \mathbf{f}^3 \end{Bmatrix} + \begin{bmatrix} \mathbf{B}^1 & \mathbf{0} & \mathbf{0} \\ \mathbf{0} & \mathbf{B}^2 & \mathbf{0} \\ \mathbf{0} & \mathbf{0} & \mathbf{B}^3 \end{bmatrix} \begin{Bmatrix} \mathbf{s}^1 \\ \mathbf{s}^2 \\ \mathbf{s}^3 \end{Bmatrix} \tag{11.22}$$

Now, each plate's equations can be partitioned into two parts. One contains the nodes belonging to the standard contour c and another part with the nodes that do not belong d. Thus, the above system is written as

$$\begin{bmatrix} \mathbf{M}_d^1 & \mathbf{M}_c^1 & \mathbf{0} & \mathbf{0} & \mathbf{0} & \mathbf{0} \\ \mathbf{0} & \mathbf{0} & \mathbf{M}_d^2 & \mathbf{M}_c^2 & \mathbf{0} & \mathbf{0} \\ \mathbf{0} & \mathbf{0} & \mathbf{0} & \mathbf{0} & \mathbf{M}_d^3 & \mathbf{M}_c^3 \end{bmatrix} \begin{Bmatrix} \ddot{\mathbf{d}}_d^1 \\ \ddot{\mathbf{d}}_c^1 \\ \ddot{\mathbf{d}}_d^2 \\ \ddot{\mathbf{d}}_c^2 \\ \ddot{\mathbf{d}}_d^3 \\ \ddot{\mathbf{d}}_c^3 \end{Bmatrix} + \begin{bmatrix} \mathbf{H}_d^1 & \mathbf{H}_c^1 & \mathbf{0} & \mathbf{0} & \mathbf{0} & \mathbf{0} \\ \mathbf{0} & \mathbf{0} & \mathbf{H}_d^2 & \mathbf{H}_c^2 & \mathbf{0} & \mathbf{0} \\ \mathbf{0} & \mathbf{0} & \mathbf{0} & \mathbf{0} & \mathbf{H}_d^3 & \mathbf{H}_c^3 \end{bmatrix} \begin{Bmatrix} \mathbf{d}_d^1 \\ \mathbf{d}_c^1 \\ \mathbf{d}_d^2 \\ \mathbf{d}_c^2 \\ \mathbf{d}_d^3 \\ \mathbf{d}_c^3 \end{Bmatrix} =$$

$$\begin{bmatrix} \mathbf{G}_d^1 & \mathbf{G}_c^1 & \mathbf{0} & \mathbf{0} & \mathbf{0} & \mathbf{0} \\ \mathbf{0} & \mathbf{0} & \mathbf{G}_d^2 & \mathbf{G}_c^2 & \mathbf{0} & \mathbf{0} \\ \mathbf{0} & \mathbf{0} & \mathbf{0} & \mathbf{0} & \mathbf{G}_d^3 & \mathbf{G}_c^3 \end{bmatrix} \begin{Bmatrix} \mathbf{f}_d^1 \\ \mathbf{f}_c^1 \\ \mathbf{f}_d^2 \\ \mathbf{f}_c^2 \\ \mathbf{f}_d^3 \\ \mathbf{f}_c^3 \end{Bmatrix} + \begin{bmatrix} \mathbf{B}_d^1 & \mathbf{B}_c^1 & \mathbf{0} & \mathbf{0} & \mathbf{0} & \mathbf{0} \\ \mathbf{0} & \mathbf{0} & \mathbf{B}_d^2 & \mathbf{B}_c^2 & \mathbf{0} & \mathbf{0} \\ \mathbf{0} & \mathbf{0} & \mathbf{0} & \mathbf{0} & \mathbf{B}_d^3 & \mathbf{B}_c^3 \end{bmatrix} \begin{Bmatrix} \mathbf{s}_d^1 \\ \mathbf{s}_c^1 \\ \mathbf{s}_d^2 \\ \mathbf{s}_c^2 \\ \mathbf{s}_d^3 \\ \mathbf{s}_c^3 \end{Bmatrix} \tag{11.23}$$

Now, the compatibility equations of displacement and accelerations between each of the plates along the common edge c can be written as

$$\begin{aligned}\mathbf{T}^1\mathbf{d}_c^1 - \mathbf{T}^2\mathbf{d}_c^2 = \mathbf{0}, \qquad \mathbf{T}^1\ddot{\mathbf{d}}_c^1 - \mathbf{T}^2\ddot{\mathbf{d}}_c^2 = \mathbf{0}\\ \mathbf{T}^2\mathbf{d}_c^2 - \mathbf{T}^3\mathbf{d}_c^3 = \mathbf{0}, \qquad \mathbf{T}^2\ddot{\mathbf{d}}_c^2 - \mathbf{T}^3\ddot{\mathbf{d}}_c^3 = \mathbf{0}\end{aligned} \tag{11.24}$$

In compact form:

$$\begin{bmatrix} \mathbf{0} & \mathbf{T}^1 & \mathbf{0} & -\mathbf{T}^2 & \mathbf{0} & \mathbf{0} \\ \mathbf{0} & \mathbf{0} & \mathbf{0} & \mathbf{T}^2 & \mathbf{0} & -\mathbf{T}^3 \end{bmatrix} \begin{Bmatrix} \mathbf{d}_d^1 \\ \mathbf{d}_c^1 \\ \mathbf{d}_d^2 \\ \mathbf{d}_c^2 \\ \mathbf{d}_d^3 \\ \mathbf{d}_c^3 \end{Bmatrix} = \mathbf{C}_c\mathbf{d}_c \tag{11.25}$$

and

$$\begin{bmatrix} \mathbf{0} & \mathbf{T}^1 & \mathbf{0} & -\mathbf{T}^2 & \mathbf{0} & \mathbf{0} \\ \mathbf{0} & \mathbf{0} & \mathbf{0} & \mathbf{T}^2 & \mathbf{0} & -\mathbf{T}^3 \end{bmatrix} \begin{Bmatrix} \ddot{\mathbf{d}}_d^1 \\ \ddot{\mathbf{d}}_c^1 \\ \ddot{\mathbf{d}}_d^2 \\ \ddot{\mathbf{d}}_c^2 \\ \ddot{\mathbf{d}}_d^3 \\ \ddot{\mathbf{d}}_c^3 \end{Bmatrix} = \mathbf{A}_c\ddot{\mathbf{d}}_c \tag{11.26}$$

Now, the compatibility equations of displacement and accelerations between each of the plates along the common edge c can be written as

$$\mathbf{T}^1\mathbf{f}_c^1 + \mathbf{T}^2\mathbf{f}_c^2 + \mathbf{T}^2\mathbf{f}_c^3 = \mathbf{0} \tag{11.27}$$

In compact form:

$$\begin{bmatrix} \mathbf{0} & \mathbf{T}^1 & \mathbf{0} & \mathbf{T}^2 & \mathbf{0} & \mathbf{T}^3 \end{bmatrix} \begin{Bmatrix} \mathbf{f}_d^1 \\ \mathbf{f}_c^1 \\ \mathbf{f}_d^2 \\ \mathbf{f}_c^2 \\ \mathbf{f}_d^3 \\ \mathbf{f}_c^3 \end{Bmatrix} = \mathbf{E}_c\mathbf{f}_c \tag{11.28}$$

Replacing the matrices $\mathbf{C}_c$, $\mathbf{A}_c$ y $\mathbf{E}_c$ in Eq. (11.22) gives the system of equations of the structure:

$$\begin{bmatrix} \mathbf{M}_d^1 & \mathbf{M}_c^1 & \mathbf{0} & \mathbf{0} & \mathbf{0} & \mathbf{0} \\ \mathbf{0} & \mathbf{0} & \mathbf{M}_d^2 & \mathbf{M}_c^2 & \mathbf{0} & \mathbf{0} \\ \mathbf{0} & \mathbf{0} & \mathbf{0} & \mathbf{0} & \mathbf{M}_d^3 & \mathbf{M}_c^3 \\ \mathbf{0} & \mathbf{T}^1 & \mathbf{0} & -\mathbf{T}^2 & \mathbf{0} & \mathbf{0} \\ \mathbf{0} & \mathbf{0} & \mathbf{0} & \mathbf{T}^2 & \mathbf{0} & -\mathbf{T}^3 \end{bmatrix} \begin{Bmatrix} \ddot{\mathbf{d}}_d^1 \\ \ddot{\mathbf{d}}_c^1 \\ \ddot{\mathbf{d}}_d^2 \\ \ddot{\mathbf{d}}_c^2 \\ \ddot{\mathbf{d}}_d^3 \\ \ddot{\mathbf{d}}_c^3 \end{Bmatrix} +$$

$$\begin{bmatrix} \mathbf{H}_d^1 & \mathbf{H}_c^1 & \mathbf{0} & \mathbf{0} & \mathbf{0} & \mathbf{0} \\ \mathbf{0} & \mathbf{0} & \mathbf{H}_d^2 & \mathbf{H}_c^2 & \mathbf{0} & \mathbf{0} \\ \mathbf{0} & \mathbf{0} & \mathbf{0} & \mathbf{0} & \mathbf{H}_d^3 & \mathbf{H}_c^3 \\ \mathbf{0} & \mathbf{T}^1 & \mathbf{0} & -\mathbf{T}^2 & \mathbf{0} & \mathbf{0} \\ \mathbf{0} & \mathbf{0} & \mathbf{0} & \mathbf{T}^2 & \mathbf{0} & -\mathbf{T}^3 \end{bmatrix} \begin{Bmatrix} \mathbf{d}_d^1 \\ \mathbf{d}_c^1 \\ \mathbf{d}_d^2 \\ \mathbf{d}_c^2 \\ \mathbf{d}_d^3 \\ \mathbf{d}_c^3 \end{Bmatrix} =$$

$$\begin{bmatrix} \mathbf{G}_d^1 & \mathbf{G}_c^1 & \mathbf{0} & \mathbf{0} & \mathbf{0} & \mathbf{0} \\ \mathbf{0} & \mathbf{0} & \mathbf{G}_d^2 & \mathbf{G}_c^2 & \mathbf{0} & \mathbf{0} \\ \mathbf{0} & \mathbf{0} & \mathbf{0} & \mathbf{0} & \mathbf{G}_d^3 & \mathbf{G}_c^3 \\ \mathbf{0} & \mathbf{T}^1 & \mathbf{0} & \mathbf{T}^2 & \mathbf{0} & \mathbf{T}^3 \end{bmatrix} \begin{Bmatrix} \mathbf{f}_d^1 \\ \mathbf{f}_c^1 \\ \mathbf{f}_d^2 \\ \mathbf{f}_c^2 \\ \mathbf{f}_d^3 \\ \mathbf{f}_c^3 \end{Bmatrix} +$$

$$\begin{bmatrix} \mathbf{B}_d^1 & \mathbf{B}_c^1 & \mathbf{0} & \mathbf{0} & \mathbf{0} & \mathbf{0} \\ \mathbf{0} & \mathbf{0} & \mathbf{B}_d^2 & \mathbf{B}_c^2 & \mathbf{0} & \mathbf{0} \\ \mathbf{0} & \mathbf{0} & \mathbf{0} & \mathbf{0} & \mathbf{B}_d^3 & \mathbf{B}_c^3 \end{bmatrix} \begin{Bmatrix} \mathbf{s}_d^1 \\ \mathbf{s}_c^1 \\ \mathbf{s}_d^2 \\ \mathbf{s}_c^2 \\ \mathbf{s}_d^3 \\ \mathbf{s}_c^3 \end{Bmatrix} \quad (11.29)$$

By moving the unknown forces to the left side, one finally obtains

$$\begin{bmatrix} \mathbf{M}_d^1 & \mathbf{M}_c^1 & \mathbf{0} & \mathbf{0} & \mathbf{0} & \mathbf{0} \\ \mathbf{0} & \mathbf{0} & \mathbf{M}_d^2 & \mathbf{M}_c^2 & \mathbf{0} & \mathbf{0} \\ \mathbf{0} & \mathbf{0} & \mathbf{0} & \mathbf{0} & \mathbf{M}_d^3 & \mathbf{M}_c^3 \\ \mathbf{0} & \mathbf{T}^1 & \mathbf{0} & -\mathbf{T}^2 & \mathbf{0} & \mathbf{0} \\ \mathbf{0} & \mathbf{0} & \mathbf{0} & \mathbf{T}^2 & \mathbf{0} & -\mathbf{T}^3 \end{bmatrix} \begin{Bmatrix} \ddot{\mathbf{d}}_d^1 \\ \ddot{\mathbf{d}}_c^1 \\ \ddot{\mathbf{d}}_d^2 \\ \ddot{\mathbf{d}}_c^2 \\ \ddot{\mathbf{d}}_d^3 \\ \ddot{\mathbf{d}}_c^3 \end{Bmatrix} +$$

$$\begin{bmatrix} \mathbf{H}_d^1 & \mathbf{H}_c^1 & \mathbf{0} & \mathbf{0} & \mathbf{0} & \mathbf{0} & -\mathbf{G}_c^1 & \mathbf{0} & \mathbf{0} \\ \mathbf{0} & \mathbf{0} & \mathbf{H}_d^2 & \mathbf{H}_c^2 & \mathbf{0} & \mathbf{0} & \mathbf{0} & -\mathbf{G}_c^2 & \mathbf{0} \\ \mathbf{0} & \mathbf{0} & \mathbf{0} & \mathbf{0} & \mathbf{H}_d^3 & \mathbf{H}_c^3 & \mathbf{0} & \mathbf{0} & \mathbf{G}_c^3 \\ \mathbf{0} & \mathbf{T}^1 & \mathbf{0} & -\mathbf{T}^2 & \mathbf{0} & \mathbf{0} & \mathbf{0} & \mathbf{0} & \mathbf{0} \\ \mathbf{0} & \mathbf{0} & \mathbf{0} & \mathbf{T}^2 & \mathbf{0} & -\mathbf{T}^3 & \mathbf{0} & \mathbf{0} & \mathbf{0} \\ \mathbf{0} & \mathbf{0} & \mathbf{0} & \mathbf{0} & \mathbf{0} & \mathbf{0} & \mathbf{T}^1 & \mathbf{T}^2 & \mathbf{T}^3 \end{bmatrix} \begin{Bmatrix} \mathbf{d}_d^1 \\ \mathbf{d}_c^1 \\ \mathbf{d}_d^2 \\ \mathbf{d}_c^2 \\ \mathbf{d}_d^3 \\ \mathbf{d}_c^3 \\ \mathbf{f}_c^1 \\ \mathbf{f}_c^2 \\ \mathbf{f}_c^3 \end{Bmatrix} =$$

$$\begin{bmatrix} \mathbf{B}_d^1 & \mathbf{B}_c^1 & \mathbf{0} & \mathbf{0} & \mathbf{0} & \mathbf{0} \\ \mathbf{0} & \mathbf{0} & \mathbf{B}_d^2 & \mathbf{B}_c^2 & \mathbf{0} & \mathbf{0} \\ \mathbf{0} & \mathbf{0} & \mathbf{0} & \mathbf{0} & \mathbf{B}_d^3 & \mathbf{B}_c^3 \end{bmatrix} \begin{Bmatrix} \mathbf{s}_d^1 \\ \mathbf{s}_c^1 \\ \mathbf{s}_d^2 \\ \mathbf{s}_c^2 \\ \mathbf{s}_d^3 \\ \mathbf{s}_c^3 \\ \mathbf{0} \\ \mathbf{0} \\ \mathbf{0} \end{Bmatrix} \qquad (11.30)$$

11.3 Numerical Examples

Thin-walled beam with axial displacement

Considere una thin-walled beam empotrada en su base con dimensiones $length =$ 6 m, with section 1 $m \times 0.5$ m and thickness $h = 0.1$ m, tal como se presenta en la Fig. 11.3. La viga tiene propiedades elásticas $E = 210$ GPa y $\nu = 0.3$. El desplazamiento aplicado a su extremo libre es de $w_0 = 0.01$ metros.

El modelo de elementos finitos consiste en 4 placas para cada una de las cuales su contorno ha sido discretizado con 4 elementos cuadráticos discontinuos. Se emplean 9 puntos de colocación internos uniformemente espaciados en cada una de ellas con el fin de calcular los desplazamientos en el contorno. En la parte inferior de la Fig. 11.3 se muestra la distribución de desplazamiento encontrada.

11.4 Cracked Plates Repaired with Composite Patch

A contour integral formulation describes the mechanical behavior of isotropic cracked thick plates repaired with a bonded composite patch using Reissner's theory for isotropic and laminated plates Useche et al. [6, 7]. To model the plate with isotropic cracks, the boundary element formulation proposed by Vander Weeën [8]

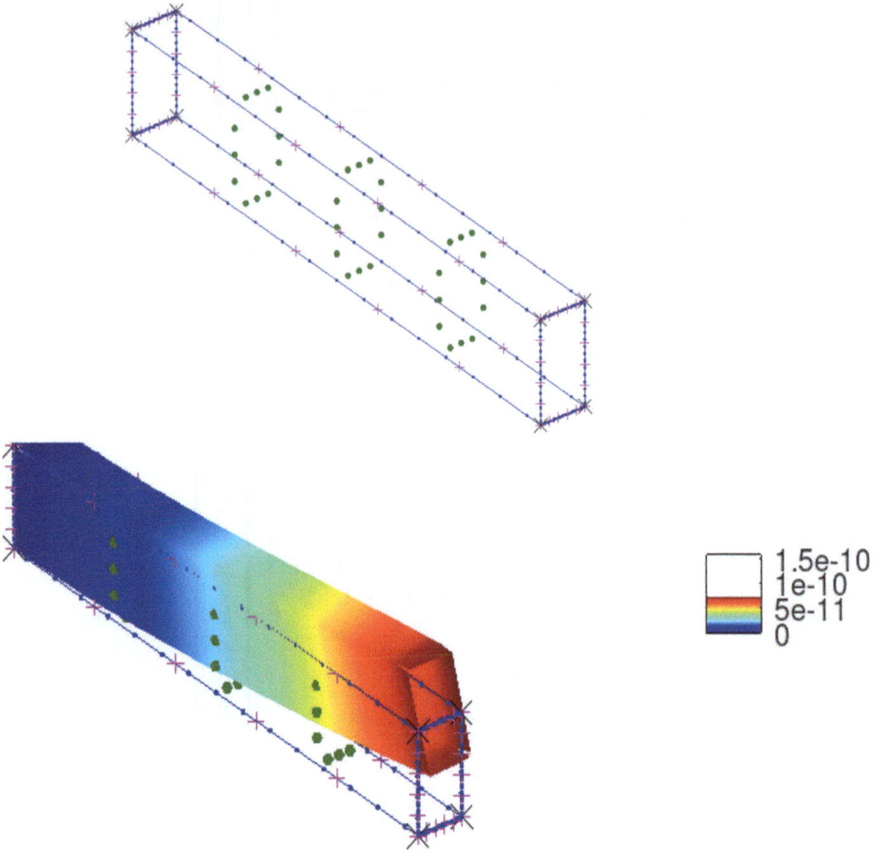

Fig. 11.3 Thin-walled beam. Left: BEM model. Right: Total displacements for thin-walled beam

and the dual boundary element formulation proposed by Dirgantara [3] are used for crack treatment. A three-parameter contour integral formulation for anisotropic plates is proposed to model the mechanical response of the composite repair. Integral equations are established for the problem at hand. Coupling equations are established based on the kinematic compatibility of the displacements and rotations of the isotropic plate, the anisotropic repair, and the equilibrium of forces and moments acting on the adhesive. Therefore, a general system of equations for the problem is developed. Numerical examples and preliminary conclusions are presented.

11.5 Boundary Integral Formulation for Isotropic Plates

11.5.1 Displacement Integral Formulation for Plane Stress

Figure 11.4 shows a schematic representation of the problem. A cracked isotropic plate with arbitrary geometry (but represented here as rectangular) is subjected to

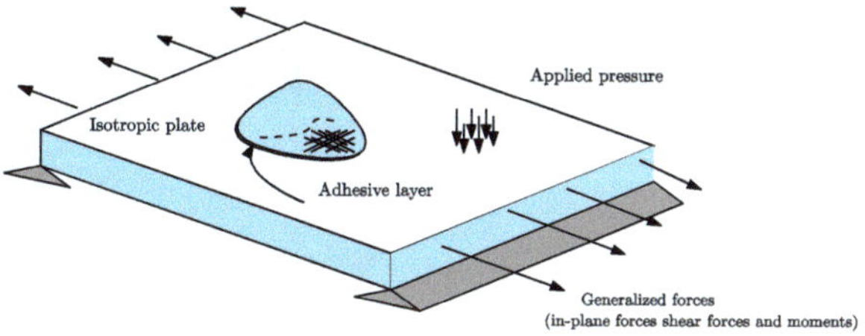

Fig. 11.4 Schematic representation of the cracked plate repaired problem

in-plane forces (membrane forces), bending moments, and distributed pressure. The plate is repaired using a bonded composite patch. This work does not consider any forces acting on this patch. The two-dimensional boundary integral equation for displacements at the boundary point $\mathbf{x}' \in \Gamma$ describing the membrane effects can be written as shown in Chap. 4:

$$c_{ij}^{P}\left(\mathbf{x}'\right) u_{\beta}\left(\mathbf{x}'\right) = \int_{\Gamma} U_{\alpha\beta}^{P}\left(\mathbf{x}', \mathbf{x}\right) t_{\beta} d\Gamma - \int_{\Gamma} T_{\alpha\beta}^{P}\left(\mathbf{x}', \mathbf{x}\right) u_{\beta} d\Gamma + \frac{1}{h_p}\int_{A} U_{\alpha\beta}^{P}\left(\mathbf{x}', \mathbf{x}\right) f_{\beta} dA \tag{11.31}$$

where $\alpha, \beta = 1, 2$ and $c_{ij}^{P}(\mathbf{x}')$ is a function of the geometry at the placement points that can be determined by considering rigid body motions. The limiting displacements and tractions for the shell are denoted by u_α and t_α, respectively; the fundamental displacement and traction solutions for the plane stress condition are $U_{\alpha\beta}^{P}(\mathbf{x}', \mathbf{x})$ and $T_{\alpha\beta}^{P}(\mathbf{x}', \mathbf{x})$ respectively, $f_\beta(\mathbf{x})$ denotes two-dimensional body forces per unit area over a region A of the patch and h_p is the plate thickness. In this work, other body forces in the plane will be disregarded.

11.5.2 Two-Dimensional Traction Integral Formulation

To model cracked plates, the Dual Boundary Element Method (DBEM) will be used. In this method, the displacement integral equation is formulated for the source points on one surface of the crack, and the tensile integral equation is formulated for the other surface. Subsequently, using the stress-strain relationship for plane stress, the tensile integral equation can be derived for two-dimensional problems on a smooth boundary, as presented in Chap. 7:

$$\frac{1}{2} t_{\alpha}\left(\mathbf{x}'\right) = n_{\beta}\left(\mathbf{x}'\right) \int_{\Gamma} U_{\alpha\beta\gamma}^{P}\left(\mathbf{x}', \mathbf{x}\right) t_{\gamma} d\Gamma - n_{\beta}\left(\mathbf{x}'\right) \int_{\Gamma} T_{\alpha\beta\gamma}^{P}\left(\mathbf{x}', \mathbf{x}\right) u_{\gamma} d\Gamma + n_{\beta}\left(\mathbf{x}'\right) \frac{1}{h_p} \int_{A} U_{\alpha\beta\gamma}^{P}\left(\mathbf{x}', \mathbf{x}\right) f_{\beta} dA \tag{11.32}$$

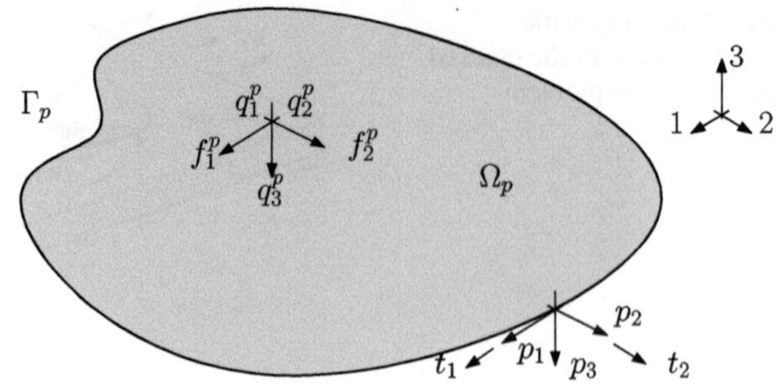

Fig. 11.5 Definition of boundary and distributed body forces and moments at isotropic plate

11.5.3 Displacement Integral Equation for Plate Bending

If w_α are defined as rotations in the x_α direction, w_3 is the deflection of the plate along x_3 as shown in Fig. 11.5, q_α^P and q_3^P are the distribution of moment body forces and out-of-plane body force per unit area, respectively, in the patch area A and p_o is the applied pressure force in the plate domain Ω (considered constant in this work); the boundary integrals for the plate bending problem can be obtained as (see Chap. 5):

$$c_{ik}^P\left(\mathbf{x}'\right) w_k\left(\mathbf{x}'\right) = \int_\Gamma W_{ik}^P\left(\mathbf{x}', \mathbf{x}\right) p_k d\Gamma - \int_\Gamma P_{ik}^P\left(\mathbf{x}', \mathbf{x}\right) w_k d\Gamma + p_o \int_\Omega W_{i3}^P\left(\mathbf{x}', \mathbf{x}\right) d\Omega + \int_A W_{ik}^P\left(\mathbf{x}', \mathbf{x}\right) q_k^P dA \tag{11.33}$$

11.5.4 Traction Integral Equation for Plate Bending

Similarly, fracture mechanics problems involving plate bending can be modeled using DBEM. In this case, the tensile equation can be written as

$$\frac{1}{2} p_i\left(\mathbf{x}'\right) = n_\beta\left(\mathbf{x}'\right) \int_\Gamma W_{i\beta k}^P\left(\mathbf{x}', \mathbf{x}\right) p_k d\Gamma - n_\beta\left(\mathbf{x}'\right) \int_\Gamma P_{i\beta k}^P\left(\mathbf{x}', \mathbf{x}\right) w_k d\Gamma + n_\beta\left(\mathbf{x}'\right) p_o \int_\Omega W_{i\beta 3}^P\left(\mathbf{x}', \mathbf{x}\right) d\Omega + n_\beta\left(\mathbf{x}'\right) \int_A W_{i\beta k}^P\left(\mathbf{x}', \mathbf{x}\right) q_k^P dA \tag{11.34}$$

where $W_{i\beta\gamma}^P(\mathbf{x}', \mathbf{x})$ and $P_{i\beta\gamma}^P(\mathbf{x}', \mathbf{x})$ are the traction fundamental solution for Reissner's plate, as showed in Chap. 7.

11.6 Boundary Integral Equation for the Anisotropic Repair

11.6.1 Displacement Integral Formulation for Plane Stress

Similar to the isotropic case, the in-plane displacements of a point $\mathbf{x}'$ in the anisotropic patch are given by (see Chap. 3):

$$c_{\alpha\beta}^{R}\left(\mathbf{x}'\right)u_{\beta}^{R}+\int_{\Gamma_R}T_{\alpha\beta}^{R}\left(\mathbf{x}',\mathbf{x}\right)u_{\beta}^{R}d\Gamma=\frac{1}{h_R}\int_{A}U_{\alpha\beta}^{R}\left(\mathbf{x}',\mathbf{x}\right)f_{\beta}^{R}dA \tag{11.35}$$

The plate bending response of the repair is modeled using the thick composite plate theory as presented in Chap. 8.

11.6.2 Coupling Equations

The isotropic plate equations have fifteen unknowns: five displacements (or tractions) at the boundary, five unknown displacements, and five body interaction forces at any point in the repair region. In addition, ten unknowns appear in the repair: five displacements (at the boundary and in the domain) and five body interaction forces (in the domain). Thus, the problem has twenty-five unknowns.

The boundary integral equations for the isotropic plate and the anisotropic repair presented represent fifteen equations and twenty-five unknowns. Ten additional equations must be provided. Additional equations can be written if displacement compatibility between the plate and the repair and equilibrium conditions in the adhesive layer are considered. Thus, a total of twenty-five equations could be written.

11.6.3 Equilibrium Conditions at Adhesive Layer

The equilibrium of the adhesive layer in the $x_3 - x_1$ plane can be written as (see Fig. 11.6) Useche et al. [6, 7]:

$$\begin{aligned}\sum F_{x_1} &= f_1^P + f_1^R = 0\\ \sum F_{x_3} &= q_3^P + q_3^R = 0\\ \sum M_{x_2} &= q_1^P + q_1^R + f_1^R\left(h_A+\frac{h_P+h_R}{2}\right) = 0\end{aligned} \tag{11.36}$$

Fig. 11.6 Forces and moments at $x_1 - z$ plane for equilibrium equation

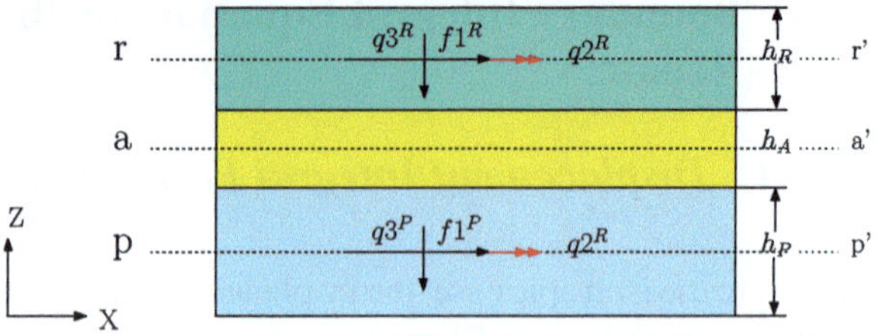

where h_A represents the thickness of the adhesive, similarly, equilibrium equations can be written for the $x_1 - x_2$ and $x_2 - x_3$ planes. In vector form, the above equations can be expressed ([7]) as

$$\begin{aligned} f_\alpha^P + f_\alpha^R &= 0 \\ q_3^P + q_3^R &= 0 \\ q_\alpha^P + q_\alpha^R + f_\alpha^R \left(h_A + \frac{h_P + h_R}{2} \right) &= 0 \end{aligned} \tag{11.37}$$

11.6.4 Displacement Compatibility Equations

The sheer force $\tau_{3\alpha}^A$, acting inside the adhesive layer, can be written as

$$\tau_{3\alpha}^A = f_\alpha^R = \mu_A \gamma_{3\alpha} \tag{11.38}$$

where μ_A is the shear modulus of and $\gamma_{3\alpha}$ is the shear strain in the α direction which can be written as (see Fig. 11.7):

$$\gamma_{3\alpha} = \frac{U_\alpha^R - U_\alpha^P}{h_A} \cong \frac{1}{h_A} \left\{ \left(u_\alpha^R - \frac{h_R}{2} w_\alpha^R \right) - \left(u_\alpha^P + \frac{h_P}{2} w_\alpha^P \right) \right\} \tag{11.39}$$

Substituting in Eq. (11.38) we have

$$\frac{h_A}{G_A} f_\alpha^R = \left(u_\alpha^R - \frac{h_R}{2} w_\alpha^R \right) - \left(u_\alpha^R + \frac{h_R}{2} w_\alpha^R \right) \tag{11.40}$$

Fig. 11.7 Displacement components at adhesive interfaces for shear stress definition

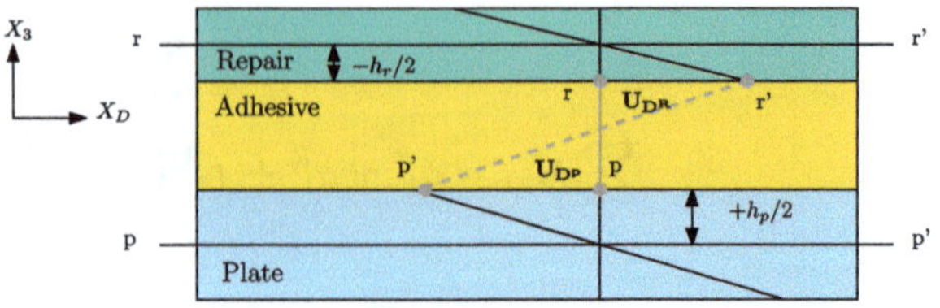

Finally, we can consider that the angles of deflection and rotation at coincident points on the plate and the repair are related by

$$w_3^P = w_3^R$$
$$q_\alpha^P = C\left(w_{,\alpha}^R - w_\alpha^P\right) \tag{11.41}$$

where $C = D(1-\nu)\lambda^2/2$, for the isotropic plate. Therefore, the Eqs. (11.37)–(11.41) represent ten additional equations obtained by considering the equilibrium and kinematic compatibility conditions in the adhesive layer.

11.6.5 Boundary Element Equations for the Plate

11.6.6 In-Plane Equations

Using the boundary element method, the discretized boundary equations for the plane elasticity (upper index m) for the isotropic plate p can be written in matrix form as

$$\begin{bmatrix} {}^p\mathbf{H}_c^m & \mathbf{0} \\ {}^p\mathbf{H}_d^m & \mathbf{I} \end{bmatrix} \begin{Bmatrix} \mathbf{u}_c^p \\ \mathbf{u}_d^p \end{Bmatrix} = \begin{bmatrix} {}^p\mathbf{G}_c^m \\ {}^p\mathbf{G}_d^m \end{bmatrix} \{\mathbf{t}^p\} + \begin{bmatrix} {}^p\mathbf{B}_c^m \\ {}^p\mathbf{B}_d^m \end{bmatrix} \{\mathbf{t}^p\} + \begin{Bmatrix} {}^p\mathbf{f}_c^0 \\ {}^p\mathbf{f}_d^0 \end{Bmatrix} \tag{11.42}$$

where the subindices c and d refer to the boundary and domain collocation points, respectively. Equation (11.42) can be written in compact form as (ignoring the body forces $\mathbf{f}^0$):

$${}^P\mathbf{H}^M\mathbf{u}^P = {}^P\mathbf{G}^M\mathbf{t}^P + {}^P\mathbf{B}^M\mathbf{f}^P \tag{11.43}$$

Using the cell method, the matrix ${}^P\mathbf{B}^M$ is given by

$$\int_\Omega U_{ij}^* f_i^P d\Omega = \left[\sum_{i=1}^{Ncell} \int_{\Omega cell} U_{ij}^* N_{kj} d\Omega_{cell}\right] f_i^P = {}^P\mathbf{B}^M\mathbf{f}^P \quad k = 1\ldots ND \tag{11.44}$$

where ND and NC are the number of collocation points in the domain and the isotropic plate boundary, respectively.

11.6.7 Bending Equations

The bending equations (identified by b) for the isotropic plate are given by

$$\begin{bmatrix} {}^p\mathbf{H}_c^b & \mathbf{0} \\ {}^p\mathbf{H}_d^b & \mathbf{I} \end{bmatrix} \begin{Bmatrix} \mathbf{w}_c^p \\ \mathbf{w}_d^p \end{Bmatrix} = \begin{bmatrix} {}^p\mathbf{G}_c^b \\ {}^p\mathbf{G}_d^b \end{bmatrix} \{\mathbf{p}^p\} + \begin{bmatrix} {}^p\mathbf{B}_c^b \\ {}^p\mathbf{B}_d^b \end{bmatrix} \{\mathbf{q}^p\} + \begin{Bmatrix} {}^p\mathbf{q}_c^0 \\ {}^p\mathbf{q}_d^0 \end{Bmatrix} \tag{11.45}$$

In compact form, this equation can be written as

$$ {}^{P}\mathbf{H}^{B}\mathbf{w}^{P} = {}^{P}\mathbf{G}^{B}\mathbf{p}^{P} + {}^{P}\mathbf{B}^{B}\mathbf{q}^{P} + \mathbf{q}^{\mathbf{0}} \tag{11.46} $$

where the vector $\mathbf{q}^0$ represents the pressure forces acting on the plate.

11.6.8 Isotropic Plate Equations

The complete boundary element equations for the isotropic plate can be written in matrix form, considering Eqs. (11.42) and (11.45) as

$$ \begin{bmatrix} {}^{P}\mathbf{H}_{c}^{m} & \mathbf{0} & \mathbf{0} & \mathbf{0} \\ {}^{P}\mathbf{H}_{d}^{m} & \mathbf{I} & \mathbf{0} & \mathbf{0} \\ \mathbf{0} & \mathbf{0} & {}^{P}\mathbf{H}_{c}^{b} & \mathbf{0} \\ \mathbf{0} & \mathbf{0} & {}^{P}\mathbf{H}_{d}^{b} & \mathbf{I} \end{bmatrix} \begin{Bmatrix} \mathbf{u}_{c}^{P} \\ \mathbf{u}_{d}^{P} \\ \mathbf{w}_{c}^{P} \\ \mathbf{w}_{d}^{P} \end{Bmatrix} = \begin{bmatrix} {}^{P}\mathbf{G}_{c}^{m} & \mathbf{0} \\ {}^{P}\mathbf{G}_{d}^{m} & \mathbf{0} \\ \mathbf{0} & {}^{P}\mathbf{G}_{c}^{b} \\ \mathbf{0} & {}^{P}\mathbf{G}_{d}^{b} \end{bmatrix} \begin{Bmatrix} \mathbf{t}^{P} \\ \mathbf{p}^{P} \end{Bmatrix} + \begin{bmatrix} {}^{P}\mathbf{B}_{c}^{m} & \mathbf{0} \\ {}^{P}\mathbf{B}_{d}^{m} & \mathbf{0} \\ \mathbf{0} & {}^{P}\mathbf{B}_{c}^{b} \\ \mathbf{0} & {}^{P}\mathbf{B}_{d}^{b} \end{bmatrix} \begin{Bmatrix} \mathbf{f}^{P} \\ \mathbf{q}^{P} \end{Bmatrix} + \begin{Bmatrix} {}^{P}\mathbf{f}_{c}^{0} \\ {}^{P}\mathbf{f}_{d}^{0} \\ {}^{P}\mathbf{q}_{c}^{0} \\ {}^{P}\mathbf{q}_{d}^{0} \end{Bmatrix} \tag{11.47} $$

In compact form, this equation can be written as

$$ {}^{P}\mathbf{H}\mathbf{x}^{P} = {}^{P}\mathbf{G}\mathbf{y}^{P} + {}^{P}\mathbf{B}\mathbf{z}^{P} + \mathbf{q}^{0P} \tag{11.48} $$

Considering that vector $\mathbf{z}^P$ is a vector of unknowns, Eq. (11.48) can be re-written as

$$ \begin{bmatrix} {}^{P}\mathbf{H} & -{}^{P}\mathbf{B} \end{bmatrix} \begin{Bmatrix} \mathbf{x}^{P} \\ \mathbf{z}^{P} \end{Bmatrix} = {}^{P}\mathbf{G}\mathbf{y}^{P} + \mathbf{q}^{0P} \tag{11.49} $$

Expanding, we have:

$$ \begin{bmatrix} {}^{P}\mathbf{H}_{c}^{m} & \mathbf{0} & \mathbf{0} & \mathbf{0} & -{}^{P}\mathbf{B}_{c}^{m} & \mathbf{0} \\ {}^{P}\mathbf{H}_{d}^{m} & \mathbf{I} & \mathbf{0} & \mathbf{0} & -{}^{P}\mathbf{B}_{d}^{m} & \mathbf{0} \\ \mathbf{0} & \mathbf{0} & {}^{P}\mathbf{H}_{c}^{b} & \mathbf{0} & \mathbf{0} & -{}^{P}\mathbf{B}_{c}^{b} \\ \mathbf{0} & \mathbf{0} & {}^{P}\mathbf{H}_{d}^{b} & \mathbf{I} & \mathbf{0} & -{}^{P}\mathbf{B}_{d}^{b} \end{bmatrix} \begin{Bmatrix} \mathbf{u}_{c}^{P} \\ \mathbf{u}_{d}^{P} \\ \mathbf{w}_{c}^{P} \\ \mathbf{w}_{d}^{P} \\ \mathbf{f}^{P} \\ \mathbf{q}^{P} \end{Bmatrix} = \begin{bmatrix} {}^{P}\mathbf{G}_{c}^{m} & \mathbf{0} \\ {}^{P}\mathbf{G}_{d}^{m} & \mathbf{0} \\ \mathbf{0} & {}^{P}\mathbf{G}_{c}^{b} \\ \mathbf{0} & {}^{P}\mathbf{G}_{d}^{b} \end{bmatrix} \begin{Bmatrix} \mathbf{t}^{P} \\ \mathbf{p}^{P} \end{Bmatrix} + \begin{Bmatrix} {}^{P}\mathbf{f}_{c}^{0} \\ {}^{P}\mathbf{f}_{d}^{0} \\ {}^{P}\mathbf{q}_{c}^{0} \\ {}^{P}\mathbf{q}_{d}^{0} \end{Bmatrix} \tag{11.50} $$

11.7 Boundary Element Equations for the Repair

In a similar way, as the case of isotropic plate, boundary element equations for the repair are

$$\begin{bmatrix} {}^R\mathbf{H}_c^m & \mathbf{0} & \mathbf{0} & \mathbf{0} & -{}^R\mathbf{B}_c^m & \mathbf{0} \\ {}^R\mathbf{H}_d^m & \mathbf{I} & \mathbf{0} & \mathbf{0} & -{}^R\mathbf{B}_d^m & \mathbf{0} \\ \mathbf{0} & \mathbf{0} & {}^R\mathbf{H}_c^b & \mathbf{0} & \mathbf{0} & -{}^R\mathbf{B}_c^b \\ \mathbf{0} & \mathbf{0} & {}^R\mathbf{H}_d^b & \mathbf{I} & \mathbf{0} & -{}^R\mathbf{B}_d^b \end{bmatrix} \begin{Bmatrix} \mathbf{u}_c^R \\ \mathbf{u}_d^R \\ \mathbf{w}_c^R \\ \mathbf{w}_d^R \\ \mathbf{f}^R \\ \mathbf{q}^R \end{Bmatrix} = \begin{Bmatrix} {}^R\mathbf{f}_c^0 \\ {}^R\mathbf{f}_d^0 \\ {}^R\mathbf{q}_c^0 \\ {}^R\mathbf{q}_d^0 \end{Bmatrix} \tag{11.51}$$

Considering that all collocation points of the repair are coupled with the domain collocation points of the plate, the above equation can be written as

$$\begin{bmatrix} {}^R\mathbf{H}^m & \mathbf{0} & -{}^R\mathbf{B}^m & \mathbf{0} \\ \mathbf{0} & {}^R\mathbf{H}^b & \mathbf{0} & -{}^R\mathbf{B}^b \end{bmatrix} \begin{Bmatrix} \mathbf{u}^R \\ \mathbf{w}^R \\ \mathbf{f}^R \\ \mathbf{q}^R \end{Bmatrix} = \begin{Bmatrix} {}^R\mathbf{f}^0 \\ {}^R\mathbf{q}^0 \end{Bmatrix} \tag{11.52}$$

where

$${}^R\mathbf{H}^m = \begin{bmatrix} {}^R\mathbf{H}_c^m & \mathbf{0} \\ {}^R\mathbf{H}_d^m & \mathbf{I} \end{bmatrix};\ {}^R\mathbf{H}^b = \begin{bmatrix} {}^R\mathbf{H}_c^b & \mathbf{0} \\ {}^R\mathbf{H}_d^b & \mathbf{I} \end{bmatrix}. \tag{11.53}$$

11.8 Equilibrium Equations for the Adhesive Layer

First equation in (11.37) can be written in matrix form as

$$\begin{Bmatrix} \mathbf{f}_1^P \\ \mathbf{f}_2^P \\ \vdots \\ \mathbf{f}_{ND}^P \end{Bmatrix} + \begin{Bmatrix} \mathbf{f}_1^R \\ \mathbf{f}_2^R \\ \vdots \\ \mathbf{f}_{ND}^R \end{Bmatrix} = \begin{Bmatrix} \mathbf{0} \\ \mathbf{0} \\ \vdots \\ \mathbf{0} \end{Bmatrix} \Rightarrow \mathbf{I}\mathbf{f}^P + \mathbf{I}\mathbf{f}^R = \mathbf{0} \tag{11.54}$$

Second and third equations in (11.37) can be written in compact form as

$$\begin{Bmatrix} \mathbf{q}_1^P \\ \mathbf{q}_2^P \\ \vdots \\ \mathbf{q}_{ND}^P \end{Bmatrix} + \begin{Bmatrix} \mathbf{q}_1^R \\ \mathbf{q}_2^R \\ \vdots \\ \mathbf{q}_{ND}^R \end{Bmatrix} + \mathbf{C}_1 \begin{Bmatrix} \mathbf{f}_1^R \\ \mathbf{f}_2^R \\ \vdots \\ \mathbf{f}_{ND}^R \end{Bmatrix} = \begin{Bmatrix} \mathbf{0} \\ \mathbf{0} \\ \vdots \\ \mathbf{0} \end{Bmatrix} \Rightarrow \mathbf{I}\mathbf{q}^P + \mathbf{I}\mathbf{q}^R + \mathbf{C}_1\mathbf{f}^R = \mathbf{0} \tag{11.55}$$

where

$$\mathbf{C}_1 = \left(h_A + \frac{h_P + h_R}{2}\right) \begin{bmatrix} 1 & & & & & & \\ & 1 & & & & & \\ & & 0 & & & & \\ & & & 1 & & & \\ & & & & 1 & & \\ & & & & & 0 & \\ & & & & & & \ddots \end{bmatrix}. \tag{11.56}$$

11.9 Displacement Compatibility Equations

Equation (11.40) can be written in matrix form as

$$\begin{Bmatrix} \mathbf{u}_1^R \\ \mathbf{u}_2^R \\ \vdots \\ \mathbf{u}_{ND}^R \end{Bmatrix} - \begin{Bmatrix} \mathbf{u}_{1d}^P \\ \mathbf{u}_{2d}^P \\ \vdots \\ \mathbf{u}_{(ND)d}^P \end{Bmatrix} - \frac{h_R}{2}\mathbf{C}_1^T \begin{Bmatrix} \mathbf{w}_1^R \\ \mathbf{w}_2^R \\ \vdots \\ \mathbf{w}_{ND}^R \end{Bmatrix} - \frac{h_P}{2}\mathbf{C}_1^T \begin{Bmatrix} \mathbf{w}_{1d}^P \\ \mathbf{w}_{2d}^P \\ \vdots \\ \mathbf{w}_{(ND)d}^P \end{Bmatrix} = \frac{h_A}{G_A} \begin{Bmatrix} \mathbf{f}_1^R \\ \mathbf{f}_2^R \\ \vdots \\ \mathbf{f}_{ND}^R \end{Bmatrix} \tag{11.57}$$

or,

$$\mathbf{I}\mathbf{u}^R - \mathbf{I}\mathbf{u}_d^P - \frac{h_R}{2}\mathbf{C}_1^T\mathbf{w}^R - \frac{h_R}{2}\mathbf{C}_1^T\mathbf{w}_d^P - \frac{h_A}{G_A}\mathbf{f}^R = \mathbf{0} \tag{11.58}$$

Finally, Eqs. (11.41) can be written as

$$\mathbf{C}\left(\mathbf{w^R} - \mathbf{w_d^P}\right) = \mathbf{q^P} \tag{11.59}$$

where **C** is an incidence matrix.

11.10 Plate-Repair Coupling Equations

Substituting Eq. (11.54) into (11.55), we have

$$\mathbf{I}\mathbf{q}^P + \mathbf{I}\mathbf{q}^R - \mathbf{C}^1\mathbf{f}^P = \mathbf{0} \tag{11.60}$$

In similar way, substituting Eqs. (11.54) and (11.59) into Eq. (11.58), we obtain

$$\mathbf{I}\mathbf{u}^R - \mathbf{I}\mathbf{u}_d^P - \mathbf{C}_2\mathbf{q}^P - \mathbf{C}_1^T\mathbf{w}_d^P + \mathbf{A}\mathbf{f}^P = \mathbf{0} \tag{11.61}$$

where the constant $1/2(h_R + h_P)$ was included into matrix $\mathbf{C}_1^T$, $\mathbf{C}_2 = (h_R/2)\mathbf{C}_1{}^T\mathbf{C}^{-1}$ and $\mathbf{A} = h_A/G_A\mathbf{I}$. Replacing Eqs. (11.54) and (11.59) into Eq. (11.52), we obtain:

$$\begin{bmatrix} {}^R\mathbf{H}^m & \mathbf{0} & {}^R\mathbf{B}^m & \mathbf{0} \\ \mathbf{0} & {}^R\mathbf{H}^b & \mathbf{0} & -{}^R\mathbf{B}^b \end{bmatrix} \begin{Bmatrix} \mathbf{u}^R \\ \mathbf{w}_d^P \\ \mathbf{f}^P \\ \mathbf{q}^R \end{Bmatrix} = \begin{Bmatrix} {}^R\mathbf{f}^0 \\ {}^R\mathbf{q}^0 \end{Bmatrix} \tag{11.62}$$

Finally, considering Eqs. (11.50) and (11.60) through (11.62), we obtain the complete equations system for the problem:

$$\begin{bmatrix} {}^p\mathbf{H}_c^m & \mathbf{0} & \mathbf{0} & \mathbf{0} & -{}^p\mathbf{B}_c^m & \mathbf{0} & \mathbf{0} & \mathbf{0} \\ {}^p\mathbf{H}_d^m & \mathbf{I} & \mathbf{0} & \mathbf{0} & -{}^p\mathbf{B}_d^m & \mathbf{0} & \mathbf{0} & \mathbf{0} \\ \mathbf{0} & \mathbf{0} & {}^p\mathbf{H}_c^b & \mathbf{0} & \mathbf{0} & -{}^p\mathbf{B}_c^b & \mathbf{0} & \mathbf{0} \\ \mathbf{0} & \mathbf{0} & {}^p\mathbf{H}_d^b & \mathbf{I} & \mathbf{0} & -{}^p\mathbf{B}_d^b & \mathbf{0} & \mathbf{0} \\ \mathbf{0} & \mathbf{0} & \mathbf{0} & \mathbf{0} & {}^R\mathbf{B}^m & \mathbf{0} & {}^R\mathbf{H}^m & \mathbf{0} \\ \mathbf{0} & \mathbf{0} & \mathbf{0} & {}^R\mathbf{H}^b & \mathbf{0} & \mathbf{0} & \mathbf{0} & -{}^R\mathbf{B}^b \\ \mathbf{0} & \mathbf{0} & \mathbf{0} & \mathbf{0} & -\mathbf{C}_1 & \mathbf{I} & \mathbf{0} & \mathbf{I} \\ \mathbf{0} & -\mathbf{I} & \mathbf{C}_2 & -\mathbf{C}_1^T & \mathbf{A} & \mathbf{0} & \mathbf{I} & \mathbf{0} \end{bmatrix} \begin{Bmatrix} \mathbf{u}_c^p \\ \mathbf{u}_d^p \\ \mathbf{w}_c^p \\ \mathbf{w}_d^p \\ \mathbf{f}^p \\ \mathbf{q}^p \\ \mathbf{u}^R \\ \mathbf{q}^R \end{Bmatrix} = \tag{11.63}$$

$$\begin{bmatrix} {}^p\mathbf{G}_c^m & \mathbf{0} \\ {}^p\mathbf{G}_d^m & \mathbf{0} \\ \mathbf{0} & {}^p\mathbf{G}_c^b \\ \mathbf{0} & {}^p\mathbf{G}_c^b \\ \mathbf{0} & \mathbf{0} \\ \mathbf{0} & \mathbf{0} \\ \mathbf{0} & \mathbf{0} \\ \mathbf{0} & \mathbf{0} \end{bmatrix} \begin{Bmatrix} \mathbf{t}^p \\ \mathbf{p}^p \end{Bmatrix} + \begin{Bmatrix} {}^p\mathbf{f}_c^0 \\ {}^p\mathbf{f}_d^0 \\ {}^p\mathbf{q}_c^0 \\ {}^p\mathbf{q}_d^0 \\ {}^R\mathbf{f}^0 \\ {}^R\mathbf{q}^0 \\ \mathbf{0} \\ \mathbf{0} \end{Bmatrix} \tag{11.64}$$

11.11 Numerical Examples

11.11.1 Plate with Adhesively Bonded Isotropic Circular Patch

To test the formulation developed here, a square isotropic plate with adhesively bonded isotropic circular patch modeled using the Reissner plate theory will be analyzed and shear stress distribution in the repair zone will be compared with the theoretical solution given by Rose [4] for isotropic sheet repaired with isotropic repair.

The width of the plate is 90 mm, thickness 1.5 mm and it is subject to an in-plane load σ_0. The material constants are chosen as $E = 70$ GPa, $\nu = 0.3$. A circular isotropic patch of radius R = 30 mm is bonded to the sheet over the region $R^2 = x^2 + y^2$. The patch has the same material as the plate with thickness h_R = 1.5 mm. The adhesive layer has thickness $h_a = 0.15$ mm and shear modulus $G_a = 0.6$ GPa. The boundary of the plate is subdivided into 25 quadratic discontinuous elements and 24 elements at boundary patch (see Fig. 11.8). 56 continuous and constant bi-quadratic cells have been used. Simply supported conditions are applied to the plate. Figure 11.8 shows the shear stress distribution in the adhesive layer and Fig. 11.9 presents the normalized shear stress in the adhesive along y-axis compared with analytic solution given by Rose [4] for the two-dimensional case.

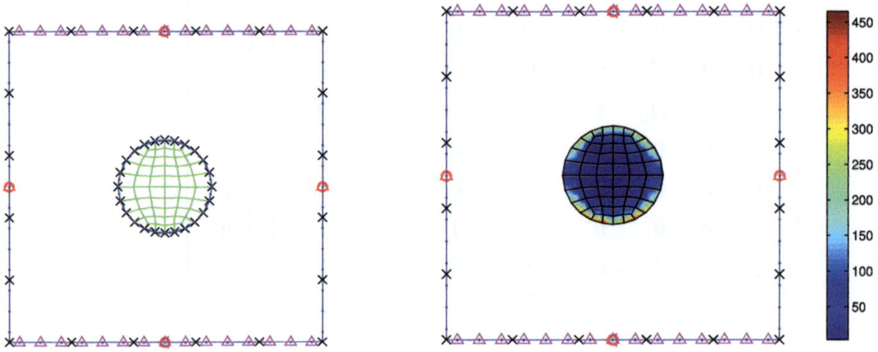

Fig. 11.8 Left: BEM for plate with adhesive isotropic patch. Right: Shear stress at adhesive layer

Fig. 11.9 Shear stress distribution in the adhesive layer along y-axis for isotropic patch

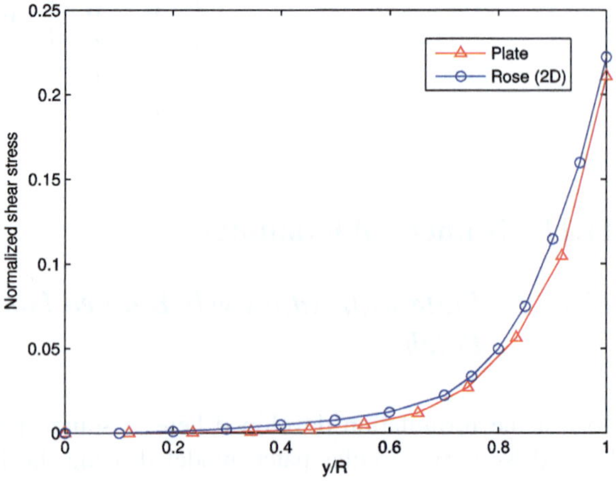

11.11.2 Plate with Adhesively Bonded Anisotropic Circular Patch

The same problem is analyzed but now considering an anisotropic patch. Mechanical properties considered are $E_1 = 25$ GPa, $E_2 = 208$ GPa, $G_{12} = 72.4$ GPa and $\nu = 0.02$. Twelve boundary element was used to discretized the plate border and 24 quadratic discontinuous boundary elements in the repair boundary (see Fig. 11.10). Similar boundary conditions and loads were applied.

Figure 11.10(right) shows the normalized shear stress distribution in the adhesive layer. A similar distribution with that founded for the isotropic repair is observed. Figure 11.10(center) shows the normalized shear stress in the adhesive along y-axis compared with analytic solution given by Rose [4] for two-dimensional isotropic repair.

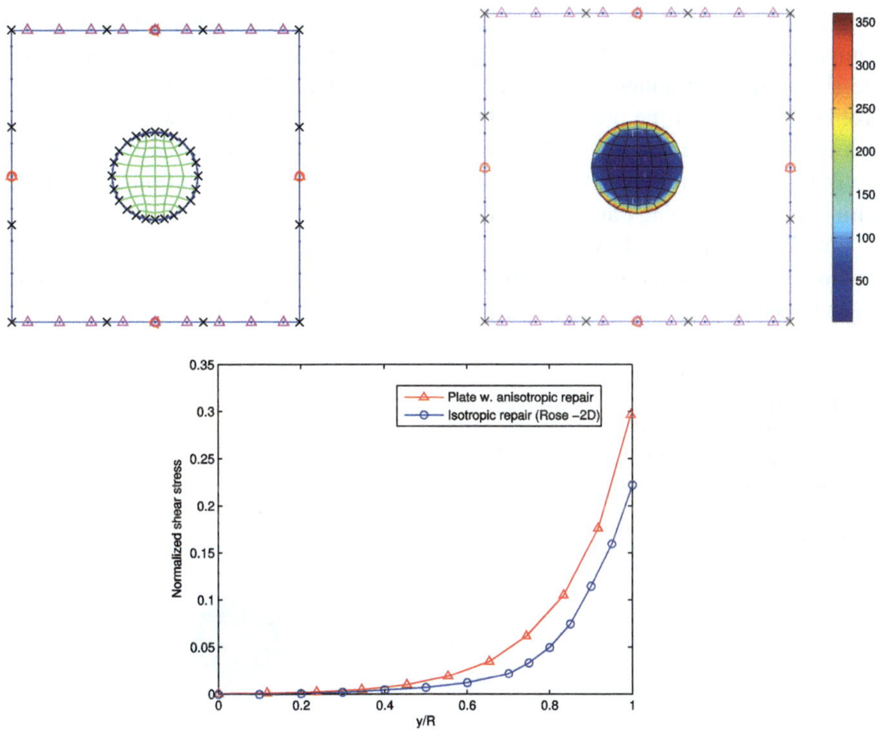

Fig. 11.10 Left: DBEM model for plate with anisotropic patch. Right: Normalized shear stress map in the adhesive Center: Normalized Shear stress in the adhesive layer along y-axis

11.11.3 Cracked Plate with Bonded Anisotropic Circular Patch

A Square isotropic cracked plate with adhesively bonded anisotropic circular patch is analyzed. The length of the crack is $2a = 15$ mm. The wide of the plate is 100 mm, thickness 1.5 mm and it is subject to in-plane load σ_0, along of $y = \pm 50$ mm. The material constants are chosen as $E = 70$ GPa, $\nu = 0.3$. A circular orthotropic patch with radius $R = 30$ mm is bonded to the plate. The material properties are $E_1 = 25$ GPa, $E_2 = 208$ GPa, $G_{12} = 72.4$ GPa and $\nu = 0.02$, with thickness of 1.5 mm. The adhesive layer has thickness $h_a = 0.15$ mm and shear modulus $G_a = 0.6$ GPa.

A similar BEM model used in the example 8.5.2. Seven boundary element was used to discretized the plate border and 24 quadratic discontinuous boundary elements in the repair boundary (see Fig. 11.11). Plate's border at $y = \pm 50$ mm is simply supported. Figure 11.10(right) shows the shear stress distribution in the adhesive layer. A similar distribution with that encountered for the isotropic repair is observed. Figure 11.11(center) shows the normalized shear stress in the adhesive along y-axis compared with solution obtained for the two-dimensional problem. As can be seen the shear stress in adhesive is lower than those obtained for the two-dimensional problem because of the presence of bending effect in the plate. Additional verification is presented in this figure. The anisotropic formulation proposed in this chapter is verified using an quasi-isotropic model for the repair and solution is compared with the fully isotropic patch. Good correlation is obtained.

Finally, Fig. 11.12 shows the K_{RMS} as function of crack length. As in two-dimensional problems an asymptotic behavior is presented. The analytical value for the asymptotic value predicted by equation is 355.6 MPa.m$^{1/2}$.

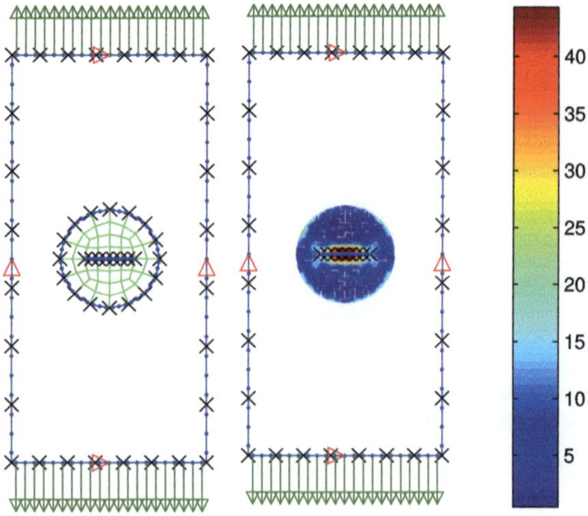

Fig. 11.11 Boundary element model for cracked isotropic plate repairs with composite circular patch and shear stress distribution in the adhesive

Fig. 11.12 K_{RMS} variation for different crack lengths

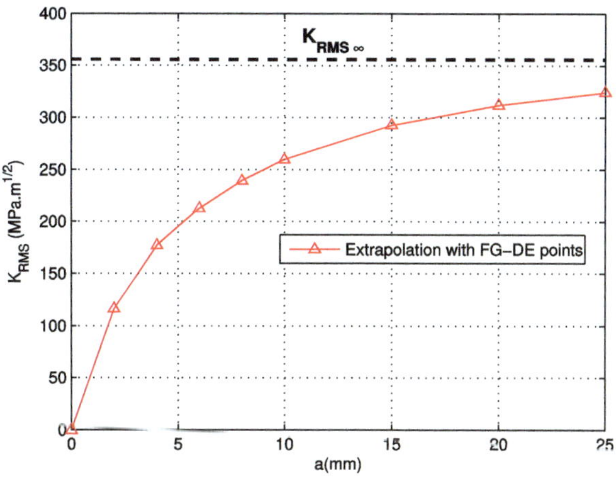

11.12 Summary

This chapter presents a methodology for boundary element analysis of thin-walled spatial structures by assembling plates in space. The system of equations was obtained by mathematically coupling the equations of the plates from conditions of continuity of displacements and rotations and equilibrium of forces and moments in contours and coupling surface between them. By means of application examples, the application of the methodology to the analysis of spatial structures and to the analysis of fractured plates by means of composite patches was presented.

References

1. Baiz PM, Aliabadi MH (2009) Local buckling of thin-walled structures by the boundary element method. Eng Anal Bound Elem 33(3):302–313
2. Dirgantara T (2002) Boundary element analysis of cracks in shear deformable plates and shells. WIT Press, Southampton
3. Dirgantara T, Aliabadi MH (2001) Dual boundary element formulation for fracture mechanics analysis of shear deformable shells. Int J Solids Struct 38:7769–7800
4. Rose LRF, Wang CH (2002) Analytical methods for designing composite repairs. In: Advances in the bonded composite repair of metallic aircraft structure, vol 1
5. Di Pisa C, Aliabadi MH, Young A (2010) Boundary element method analysis of assembled plate structures undergoing large deflection. J Strain Anal Eng Des 45
6. Useche J, Sollero P, Albuquerque EL (2008) Boundary element analysis of cracked sheets repaired with bonded anisotropic patches. Key Eng Mater J 383
7. Useche J, Sollero P, Albuquerque EL, Palermo L (2008) Thick plates repaired with adhesively bonded composite patches. Struct Durab Health Monit J 4:107–115
8. Vander Weeën F (1982) Application of the direct boundary integral equation method to reissner's plate model. Int J Num Meth Eng 67, 1–10

Chapter 12
Boundary Element Analysis of Buckling Analysis of Plates and Shells

Abstract This chapter presents a methodology based on the Boundary Element Method for buckling analysis of shear deformables plates and shells. This formulation is based on the plate and shell equations presented in previous chapters. The buckling equations are established from the initial deflection calculation, and the generalized forces acting on the plate are calculated in the linear elastic problem. Thus, the buckling problem is established as a standard eigenvalue and eigenvector problem by reorganizing the plate equations.

12.1 Introduction

The application of the boundary element method to elastic plate buckling dates back to the work of Stern [16], where singular integral equations for displacement, normal rotation, bending moment, and shear on the thin plate boundary are presented. Later in Costa and Brebbia [7] presented a boundary element formulation for buckling analysis of thin plates. Tanaka and Miyazaki [19] presents a boundary element formulation for the buckling analysis of assembled thin walled structures. In Tanaka and Miyazaki [18] an integral equation formulation for the buckling problem using the biharmonic fundamental solution is presented. In 1990, Shi [14] developed the boundary element method analysis of buckling of elastic (isotropic and orthotropic) plates, according to Kirchhoff's theory. Syngellakis and Elzein [17] present the plate buckling problem combinations of loading and boundary conditions. In the work of Chang Jun and Wang [6], boundary integral equations are established to solve the problem of plane stresses and critical loads. Post-buckling for perforated thin plates is based on general bifurcation theory and a mathematical model for stability analysis of perforated thin plates. Later in 1999, Lin et al. [8] presents a boundary element formulation for buckling of plate using a on-boundary discretization. Purbasukano [10] proposed a boundary element analysis of buckling and postbuckling of cracked thick plates. Purbasukano and Aliabadi [11] present the derivation and numerical implementation of boundary integral equations for the buckling analysis of shear-deformable plates. The plate buckling equations are derived as a standard eigenvalue problem. The formulation is formed by coupling two-dimensional plane

J. Useche Vivero, *The Boundary Element Method for Engineering Analysis of Plates and Shells*, Solid Mechanics and Its Applications 281,
https://doi.org/10.1007/978-3-031-93050-8_12

stress elasticity and shear-deformable plate boundary element formulations. Baiz and Aliabadi [4] extend this formulation to analyzing thick shells. The buckling analysis of fractured thick plates is presented in Baiz and Aliabadi [2]. The analysis of thin-walled spatial structures is presented by Baiz and Aliabadi [1], where the spatial assembly methodology of plates and shells is used (see previous chapter).

This chapter presents a methodology based on the Boundary Element Method for buckling analysis of shear deformables plates and shells. This formulation is based on the plate and shell equations presented in previous chapters for isotropic and composite materials. The buckling equations are established from the initial deflection calculation, and the generalized forces acting on the plate are calculated in the linear elastic problem. Thus, the buckling problem is established as a standard eigenvalue and eigenvector problem by reorganizing the plate equations.

12.2 Buckling Analysis of Thick Plates

12.2.1 Gobering Equations

Consider an isotropic linear elastic shallow shell as presented in Fig. 12.1. The shell has a uniform thickness h, an elastic modulus E, a Poisson modulus ν, and a shear modulus G. The midsurface of the shell is defined by a quadratic surface, as presented in Chap. 6.

The equilibrium equations for plate bending are obtained by balancing the forces and moments acting on this element without considering body forces (see Reddy [12], Baiz and Aliabadi [1, 4]):

$$\begin{aligned} M_{\alpha\beta,\beta} - Q_\alpha + q_\alpha &= 0 \\ Q_{\alpha,\alpha} + (N_{\alpha\beta} w_{3,\beta})_{,\alpha} + q_3 &= 0 \\ N_{\alpha\beta,\beta} &= 0 \end{aligned} \tag{12.1}$$

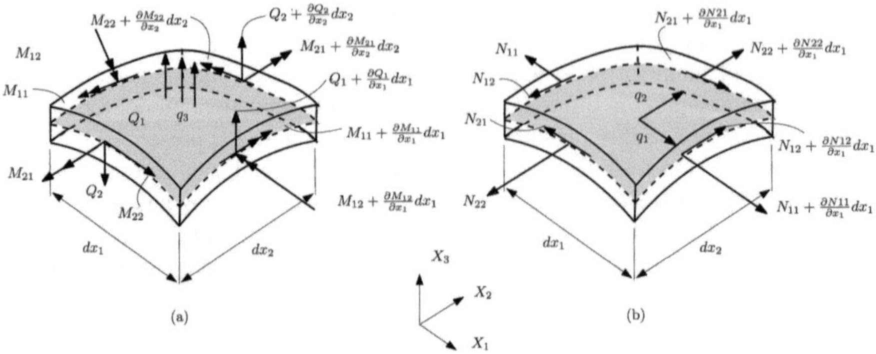

Fig. 12.1 Stress resultant equilibrium of a shell element

It is observed that there is a coupling between the resultant $N_{\alpha\beta}$ and the transverse forces Q_α through the deflection of the plate w_3. The constitutive equations based on Reissner's theory, as Reissner [13]:

$$N_{\alpha\beta} = \frac{1-\nu}{2} B \left(u_{\alpha,\beta} + u_{\beta,\alpha} + \frac{2\nu}{1-\nu} u_{\gamma,\gamma} \delta_{\alpha\beta} \right) \quad (12.2)$$

$$M_{\alpha\beta} = \frac{1-\nu}{2} D \left(w_{\alpha,\beta} + w_{\beta,\alpha} + \frac{2\nu}{1-\nu} w_{\gamma,\gamma} \delta_{\alpha\beta} \right) + \frac{\nu}{(1-\nu^2)\lambda^2} q_3 \delta_{\alpha\beta}$$

$$Q_\alpha = C \left(w_\alpha + w_{3,\alpha} \right) \quad (12.3)$$

Replacing the equilibrium equations of thick plates as presented in Chap. 5 the equilibrium equations in terms of displacements are obtained:

$$D\nabla^2 w_1 + \frac{D}{2}(1+\nu)\frac{\partial}{\partial x_2}(-\frac{\partial w_1}{\partial x_2} + \frac{\partial w_2}{\partial x_1}) - C w_1 - C\frac{\partial w_3}{\partial x_1} = 0 \quad (12.4)$$

$$\frac{D}{2}(1+\nu)\frac{\partial}{\partial x_1}(-\frac{\partial w_1}{\partial x_2} + \frac{\partial w_2}{\partial x_1}) + D\nabla^2 w_2 - C w_2 - C\frac{\partial w_3}{\partial x_2} = 0 \quad (12.5)$$

$$C\nabla^2 w_3 + C\frac{\partial w_1}{\partial x_1} + C\frac{\partial w_2}{\partial x_2} + (N_{\alpha\beta} w_{3,\beta})_{,\alpha} + q_3 = 0 \quad (12.6)$$

12.2.2 Boundary Integral Buckling Equations

The linearized buckling problem initially involves calculating the plane's internal force generated by acting forces and then evaluating the bending problem. To reduce the linear buckling problem to an eigenvalue formulation, the distribution of membrane stresses $N_{\alpha\beta}$ when the structure is subjected to the applied load must be known. In the absence of body forces, the contour integral equation for the plane stress problem is given by

$$c_{ij} u_\theta (\mathbf{x}') + \int_\Gamma T^*_{\theta\alpha} (\mathbf{x}', \mathbf{x}) \, u_\alpha (\mathbf{x}) \, d\Gamma = \int_\Gamma U^*_{\theta\alpha} (\mathbf{x}', \mathbf{x}) \, t_\alpha (\mathbf{x}) \, d\Gamma \quad (12.7)$$

The contour integral equation in the domain is given by ($c_{ij} = 1$):

$$u_\theta (\mathbf{x}') + \int_\Gamma T^*_{\theta\alpha} (\mathbf{X}', \mathbf{x}) \, u_\alpha (\mathbf{x}) \, d\Gamma = \int_\Gamma U^*_{\theta\alpha} (\mathbf{X}', \mathbf{x}) \, t_\alpha (\mathbf{x}) \, d\Gamma \quad (12.8)$$

The resultant dominance force $N_{\alpha\beta}$ can be calculated as

$$N^l_{\alpha\beta}(\mathbf{X}') = \int_\Gamma U^*_{\gamma\theta\alpha} (\mathbf{X}', \mathbf{x}) \, t_\gamma (\mathbf{x}) \, d\Gamma - \int_\Gamma T^*_{\gamma\theta\alpha} (\mathbf{X}', \mathbf{x}) \, u_\gamma (\mathbf{x}) \, d\Gamma \quad (12.9)$$

The buckling equations are obtained by introducing the load factor λ into the plate bending equations. The critical buckling load is thus assumed to be λ-times the internal force distribution $N_{\alpha\beta}$ of the membrane. Therefore, the bending boundary equation for the equations is given by Purbasukano and Aliabadi [11], Baiz and Aliabadi [2–4]:

$$c_{ij}\left(\mathbf{x}'\right) w_j\left(\mathbf{x}'\right) + \int_\Gamma P_{ij}^*\left(\mathbf{x}', \mathbf{x}\right) w_j\left(\mathbf{x}'\right) d\Gamma = \int_\Gamma W_{ij}^*\left(\mathbf{x}', \mathbf{x}\right) p_j\left(\mathbf{x}'\right) d\Gamma$$

$$+\lambda \int_\Omega W_{i3}^*\left(\mathbf{x}', \mathbf{X}\right) q_3 d\Omega + \lambda \int_\Omega W_{i3}^*\left(\mathbf{x}', \mathbf{X}\right) \left(N_{\alpha\beta}^l w_{3,\beta}\right)_{,\alpha} (\mathbf{X}) d\Omega \quad (12.10)$$

Expanding the last term of this equation:

$$c_{ij}\left(\mathbf{x}'\right) w_j\left(\mathbf{x}'\right) + \int_\Gamma P_{ij}^*\left(\mathbf{x}', \mathbf{x}\right) w_j\left(\mathbf{x}'\right) d\Gamma =$$

$$\int_\Gamma W_{ij}^*\left(\mathbf{x}', \mathbf{x}\right) p_j\left(\mathbf{x}'\right) d\Gamma + \lambda \int_\Omega W_{i3}^*\left(\mathbf{x}', \mathbf{X}\right) q_3 d\Omega +$$

$$\lambda \int_\Omega W_{i3}^*\left(\mathbf{x}', \mathbf{X}\right) \left(N_{\alpha\beta,\alpha}^l w_{3,\beta} + N_{\alpha\beta}^l w_{3,\beta\alpha}\right) (\mathbf{X}) d\Omega \quad (12.11)$$

Some authors eliminate the term $N_{\alpha\beta,\alpha}$ from the above equation, arguing that it is implicitly satisfied by the force equilibrium condition in the plane. However, it is essential to highlight that the equilibrium of the resulting forces in the plane $N_{\alpha\beta}$ is not implicitly satisfied by the two-dimensional BEM formulation. Therefore, it is essential to retain the term $N_{\alpha\beta,\alpha}$ in the plate equation Purbasukano and Aliabadi [11]. The evaluation of the last integral on the right-hand side requires the evaluation of the deflection equation w_3 at the points of the domain $\mathbf{X}$, which is obtained from the above equation as

$$c_{ij}\left(\mathbf{x}'\right) w_i\left(\mathbf{x}'\right) + \int_\Gamma P_{ij}^*\left(\mathbf{x}', \mathbf{x}\right) w_j\left(\mathbf{x}\right) d\Gamma =$$

$$\int_\Gamma W_{ij}^*\left(\mathbf{x}', \mathbf{x}\right) p_j\left(\mathbf{x}\right) d\Gamma + \lambda \int_\Omega W_{i3}^*\left(\mathbf{x}', \mathbf{X}\right) q_3(\mathbf{X}) d\Omega +$$

$$\lambda \int_\Omega W_{i3}^*\left(\mathbf{x}', \mathbf{X}\right) \left(N_{\alpha\beta,\alpha}^l w_{3,\beta} + N_{\alpha\beta}^l w_{3,\beta\alpha}\right) (\mathbf{X}) d\Omega \quad (12.12)$$

The evaluation of the last integral on the right-hand side requires the evaluation of the deflection equation w_3 at the points of the domain $\mathbf{X}$, which is obtained from the previous equation as

$$w_3\left(\mathbf{X}'\right) = \int_{\Gamma} W_{3j}^*\left(\mathbf{X}', \mathbf{x}\right) p_j\left(\mathbf{x}\right) d\Gamma - \int_{\Gamma} P_{3j}^*\left(\mathbf{X}', \mathbf{x}\right) w_j\left(\mathbf{x}\right) d\Gamma$$
$$+\lambda \int_{\Omega} W_{33}^*\left(\mathbf{X}', \mathbf{X}\right) q_3(\mathbf{X}) d\Omega +$$
$$\lambda \int_{\Omega} W_{33}^*\left(\mathbf{X}', \mathbf{X}\right) \left(N_{\alpha\beta,\alpha}^l w_{3,\beta} + N_{\alpha\beta}^l w_{3,\beta\alpha}\right)(\mathbf{X}) d\Omega \tag{12.13}$$

The transformation of the above equations into an eigenvalue equation involves approximating the terms $w_{3,\beta}$ and $w_{3,\beta\alpha}$ in terms of w_3. In the Domain, w_3 is approximated by using radial basis functions $f(r)$ Soares et al. [15], Nunes-Rangel et al. [9]:

$$w_3(\mathbf{X}) = \sum_{m=1}^{M} f(r)^m \psi^m = \mathbf{f}\boldsymbol{\psi} \tag{12.14}$$

where M is the total number of approximation points used, the radial basis functions used are usually of the type $f(r) = 1 + r^n$, with $n = 2, 3 \cdots$ (the most common being $n = 1, 2$), which have demonstrated good performance in buckling problems. The function ψ^m is calculated by taking values of w_3 at the M approximation points, thus obtaining a system of equations with ψ^m as unknowns:

$$\mathbf{w}_3 = \mathbf{F}\boldsymbol{\psi} \tag{12.15}$$

where each row of the matrix $\mathbf{F}$ contains the values of the function $f(r)$ evaluated at the M points. Similarly, the vector $\boldsymbol{\psi}$ contains the values of ψ^m for each approximation point. Thus, $\boldsymbol{\psi}$ is obtained by inverting the above system of equations:

$$\boldsymbol{\psi} = \mathbf{F}^{-1}\mathbf{w}_3 \tag{12.16}$$

Similarly, the approximation for $w_{3,\beta}$ is given by

$$w_{3,\beta}(\mathbf{X}) = \sum_{m=1}^{M} f(r)_{,\beta}^m \psi^m = \mathbf{f}(r)_{,\beta}\boldsymbol{\psi} \tag{12.17}$$

and therefore replacing $\boldsymbol{\psi}$ in this expression, we obtain $w_{3,\beta}$ in terms of $\mathbf{w}_3$:

$$w_{3,\beta}(\mathbf{X}) = \mathbf{f}(r)_{,\beta}\mathbf{F}^{-1}\mathbf{w}_3 \tag{12.18}$$

Now, to find $w_{3,\alpha\beta}$, $w_{3,\beta}$ is approximated as

$$w_{3,\beta}(\mathbf{X}) = \mathbf{f}(r)\boldsymbol{\psi} \tag{12.19}$$

where $\boldsymbol{\psi} = \mathbf{F}^{-1}\mathbf{w}_{3,\beta}$.

From this expression, the derivative $w_{3,\alpha\beta}$ is given by

$$w_{3,\beta\alpha}(\mathbf{X}) = (w_{3,\beta}(\mathbf{X}))_{,\alpha} = \mathbf{f}(r)_{,\alpha}\psi \tag{12.20}$$

where $\psi = \mathbf{F}^{-1}\mathbf{w}_{3,\beta} = \mathbf{f}(r)_{,\beta}\mathbf{F}^{-1}\mathbf{w}_3$. Replacing this expression we finally obtain

$$w_{3,\beta\alpha}(\mathbf{X}) = \mathbf{f}(r)_{,\alpha}\mathbf{f}(r)_{,\beta}\mathbf{F}^{-1}\mathbf{w}_3 \tag{12.21}$$

Using the above procedure, the derivative of the internal resultant force is given by

$$N^l_{\alpha\beta,\alpha}(\mathbf{X}) = \mathbf{f}(r)_{,\alpha}\mathbf{F}^{-1}\mathbf{N}_{\alpha\beta} \tag{12.22}$$

where $\mathbf{N}^l_{\alpha\beta}$ is a vector with the $N^l_{\alpha\beta}$ values at the M approximation points, obtained using the integral equation (12.9).

In this way Eqs. (12.12) and (12.13) can be rewritten as

$$\begin{aligned} c_{ij}\left(\mathbf{x}'\right) w_i\left(\mathbf{x}'\right) + \int_\Gamma P^*_{ij}\left(\mathbf{x}',\mathbf{x}\right) w_j\left(\mathbf{x}\right) d\Gamma = \\ \int_\Gamma W^*_{ij}\left(\mathbf{x}',\mathbf{x}\right) p_j\left(\mathbf{x}\right) d\Gamma + \lambda \int_\Omega W^*_{i3}\left(\mathbf{x}',\mathbf{X}\right) q_3(\mathbf{X}) d\Omega + \\ \lambda \int_\Omega W^*_{i3}\left(\mathbf{x}',\mathbf{X}\right)\left(N_{\alpha\beta,\alpha}\mathbf{f}(r)_{,\beta}\mathbf{F}^{-1}\mathbf{w}_3\right)(\mathbf{X}) d\Omega + \\ \lambda \int_\Omega W^*_{i3}\left(\mathbf{x}',\mathbf{X}\right)\left(N_{\alpha\beta}\mathbf{f}(r)_{,\alpha}\mathbf{f}(r)_{,\beta}\mathbf{F}^{-1}\mathbf{w}_3\right)(\mathbf{X}) d\Omega \end{aligned} \tag{12.23}$$

and

$$\begin{aligned} w_3\left(\mathbf{X}'\right) = \int_\Gamma W^*_{3j}\left(\mathbf{X}',\mathbf{x}\right) p_j\left(\mathbf{x}\right) d\Gamma - \int_\Gamma P^*_{3j}\left(\mathbf{X}',\mathbf{x}\right) w_j\left(\mathbf{x}\right) d\Gamma \\ +\lambda \int_\Omega W^*_{33}\left(\mathbf{X}',\mathbf{X}\right) q_3(\mathbf{X}) d\Omega + \\ \lambda \int_\Omega W^*_{33}\left(\mathbf{X}',\mathbf{X}\right)\left(N_{\alpha\beta,\alpha}\mathbf{f}(r)_{,\beta}\mathbf{F}^{-1}\mathbf{w}_3\right)(\mathbf{X}) d\Omega + \\ \lambda \int_\Omega W^*_{33}\left(\mathbf{X}',\mathbf{X}\right)\left(N_{\alpha\beta}\mathbf{f}(r)_{,\alpha}\mathbf{f}(r)_{,\beta}\mathbf{F}^{-1}\mathbf{w}_3\right)(\mathbf{X}) d\Omega \end{aligned} \tag{12.24}$$

12.2.3 Domain Integral Transformations

The domain integrals in these equations are transformed to the boundary using the Dual Reciprocity Boundary Element Method. The transformation of the domain integral related to the body force q has been presented in previous chapters. The last two domain integrals related to the body force resultants can be condensed into a single domain integral by defining an equivalent body force vector as

$$q_b(\mathbf{X}) = N^l_{\alpha\beta,\alpha}\mathbf{f}(r)_{,\beta}\mathbf{F}^{-1}\mathbf{w}_3 + N^l_{\alpha\beta}\mathbf{f}(r)_{,\alpha}\mathbf{f}(r)_{,\beta}\mathbf{F}^{-1}\mathbf{w}_3 \tag{12.25}$$

So the equations are as follows:

$$c_{ij}\left(\mathbf{x}'\right) w_i\left(\mathbf{x}'\right) + \int_{\Gamma} P^*_{ij}\left(\mathbf{x}',\mathbf{x}\right) w_j\left(\mathbf{x}\right) d\Gamma = \int_{\Gamma} W^*_{ij}\left(\mathbf{x}',\mathbf{x}\right) p_j\left(\mathbf{x}\right) d\Gamma + \lambda \int_{\Omega} W^*_{i3}\left(\mathbf{x}',\mathbf{X}\right) q_3(\mathbf{X}) d\Omega + \lambda \int_{\Omega} W^*_{i3}\left(\mathbf{x}',\mathbf{X}\right) q_b(\mathbf{X}) d\Omega \tag{12.26}$$

and

$$w_3\left(\mathbf{X}'\right) = \int_{\Gamma} W^*_{3j}\left(\mathbf{X}',\mathbf{x}\right) p_j\left(\mathbf{x}\right) d\Gamma - \int_{\Gamma} P^*_{3j}\left(\mathbf{X}',\mathbf{x}\right) w_j\left(\mathbf{x}\right) d\Gamma + \lambda \int_{\Omega} W^*_{33}\left(\mathbf{X}',\mathbf{X}\right) q_3(\mathbf{X}) d\Omega + \lambda \int_{\Omega} W^*_{33}\left(\mathbf{X}',\mathbf{X}\right) q_b(\mathbf{X}) d\Omega \tag{12.27}$$

Using the Dual Reciprocity Boundary Element Method, the term q_b is approximated by the following radial basis function:

$$q_b(\mathbf{X}) = \sum_{m=1}^{M} f(r)^m \psi^m = \mathbf{f}\psi \tag{12.28}$$

In this way, in the previous equations, considering ($q_3 = 0$) they are transformed by DRBEM as

$$c_{ij}\left(\mathbf{x}'\right) w_i\left(\mathbf{x}'\right) + \int_{\Gamma} P^*_{ij}\left(\mathbf{x}',\mathbf{x}\right) w_j\left(\mathbf{x}\right) d\Gamma = \int_{\Gamma} W^*_{ij}\left(\mathbf{x}',\mathbf{x}\right) p_j\left(\mathbf{x}\right) d\Gamma + \lambda \sum_{m=1}^{m} \left[c_{ij}\hat{w}^3_{mk}(\mathbf{x}') + \int_{\Gamma} P^*_{ik}\left(\mathbf{x}',\mathbf{x}\right) \hat{w}^3_{mk}(\mathbf{x}) d\Gamma - \int_{\Omega} W^*_{ik}\left(\mathbf{x}',\mathbf{x}\right) \hat{p}^3_{mk}(\mathbf{x}) \right] \psi_m \, d\Gamma \tag{12.29}$$

and

$$
\begin{aligned}
w_3\left(\mathbf{X}'\right) = \int_\Gamma W^*_{3j}\left(\mathbf{X}', \mathbf{x}\right) p_j\left(\mathbf{x}\right) d\Gamma - \int_\Gamma P^*_{3j}\left(\mathbf{X}', \mathbf{x}\right) w_j\left(\mathbf{x}\right) d\Gamma \\
+\lambda \sum_{m=1}^{m} \left[\hat{w}^3_{m3}(\mathbf{x}') + \fint_\Gamma P^*_{33}\left(\mathbf{x}', \mathbf{x}\right) \hat{w}^3_{m3}(\mathbf{x}) d\Gamma \right. \\
\left. - \int_\Omega W^*_{33}\left(\mathbf{x}', \mathbf{x}\right) \hat{p}^3_{m3}(\mathbf{x}) \right] \psi_m \, d\Gamma
\end{aligned}
\tag{12.30}
$$

where $\psi_m = (\mathbf{F}^{-1}\mathbf{q}_b)_m$. The particular solutions for $f(r) = 1 + r$, used to approximate q_b.

12.3 Buckling Analysis of Thick Shells

Unlike plates, the plane stress and bending equations in plates involve a coupling through the curvature terms, deflections, and membrane displacements Baiz and Aliabadi [1, 4, 5]. Therefore, in the shell buckling problem, the displacements and tractions u_α, w_j, t_α, and p_j due to the applied loads on the shell must be solved initially as a prerequisite for solving the buckling problem.

The shell equations, seen in Chap. 6, are given by the integral equation:

$$
\begin{aligned}
c_{ij}(\mathbf{x}')w_j\left(\mathbf{x}'\right) + \fint_\Gamma P^*_{ij}\left(\mathbf{x}', \mathbf{x}\right) w_j\left(\mathbf{x}\right) d\Gamma = \int_\Gamma W^*_{ij}\left(\mathbf{x}', \mathbf{x}\right) p_j\left(\mathbf{x}\right) d\Gamma \\
- \int_\Omega \kappa_{\alpha\beta} B \left(\frac{1-\nu}{2}\right) \left[u_{\alpha,\beta}(\mathbf{X}) + u_{\beta,\alpha}(\mathbf{X}) + \frac{2\nu}{1-\nu} u_{\gamma,\gamma}(\mathbf{X}) \delta_{\alpha\beta} \right] W^*_{i3}\left(\mathbf{x}', \mathbf{X}\right) d\Gamma \\
- \int_\Gamma \kappa_{\alpha\beta} B[(1-\nu)\kappa_{\alpha\beta} + \nu\delta_{\alpha\beta}\kappa_{\gamma\gamma}] w_3\left(\mathbf{X}\right) W^*_{i3}\left(\mathbf{x}', \mathbf{X}\right) d\Omega \\
+\lambda \int_\Omega W^*_{i3}\left(\mathbf{x}', \mathbf{X}\right) q_3\left(\mathbf{X}\right) d\Omega
\end{aligned}
\tag{12.31}
$$

Similarly, for the membrane problem:

$$
c_{\theta\alpha}\left(\mathbf{x}'\right) u_\theta\left(\mathbf{x}'\right) + \fint_\Gamma T^{(i)*}_{\theta\alpha}\left(\mathbf{x}', \mathbf{x}\right) u_\alpha\left(\mathbf{x}\right) d\Gamma
$$

$$+\int_{\Gamma} U^*_{\theta\alpha,\beta}\left(\mathbf{x}',\mathbf{x}\right) B\left[(1-\nu)\kappa_{\alpha\beta}+\nu\delta_{\alpha\beta}\kappa_{\gamma\gamma}\right] w_3\left(\mathbf{x}\right) n_\beta(\mathbf{x}) d\Omega$$

$$-\int_{\Gamma} U^*_{\theta\alpha}\left(\mathbf{x}',\mathbf{X}\right) B\left[(1-\nu)\kappa_{\alpha\beta}+\nu\delta_{\alpha\beta}\kappa_{\gamma\gamma}\right] w_{3,\beta}\left(\mathbf{X}\right) d\Omega$$

$$=\int_{\Gamma} U^*_{\theta\alpha}\left(\mathbf{X}',\mathbf{x}\right) t_\alpha\left(\mathbf{x}\right) d\Gamma+\int_{\Omega} U^*_{\theta\alpha}\left(\mathbf{x}',\mathbf{X}\right) q_\alpha\left(\mathbf{X}\right) d\Omega \tag{12.32}$$

12.3.1 Boundary Integral Buckling Equations

The following expression gives the contour integral equation for bending:

$$c_{ij}(\mathbf{x}')w_j\left(\mathbf{x}'\right)+\fint_{\Gamma} P^*_{ij}\left(\mathbf{x}',\mathbf{x}\right) w_j\left(\mathbf{x}\right) d\Gamma=\int_{\Gamma} W^*_{ij}\left(\mathbf{x}',\mathbf{x}\right) p_j\left(\mathbf{x}\right) d\Gamma$$
$$-\int_{\Omega} \kappa_{\alpha\beta} B\left(\frac{1-\nu}{2}\right)\left[u_{\alpha,\beta}(\mathbf{X})+u_{\beta,\alpha}(\mathbf{X})+\frac{2\nu}{1-\nu}u_{\gamma,\gamma}(\mathbf{X})\delta_{\alpha\beta}\right] W^*_{i3}\left(\mathbf{x}',\mathbf{X}\right) d\Gamma$$
$$-\int_{\Omega} \kappa_{\alpha\beta} B[(1-\nu)\kappa_{\alpha\beta}+\nu\delta_{\alpha\beta}\kappa_{\gamma\gamma}]w_3\left(\mathbf{X}\right) W^*_{i3}\left(\mathbf{x}',\mathbf{X}\right) d\Omega$$
$$+\lambda\int_{\Omega} W^*_{i3}\left(\mathbf{x}',\mathbf{X}\right) q_3\left(\mathbf{X}\right) d\Omega+\lambda\int_{\Omega} W^*_{i3}\left(\mathbf{x}',\mathbf{X}\right)\left(N_{\alpha\beta}w_{3,\beta}\right)_\alpha\left(\mathbf{X}\right) d\Omega \tag{12.33}$$

Similarly, the deflection equation w_3 in the domain is

$$w_3\left(\mathbf{X}'\right)=\int_{\Gamma} W^*_{3j}\left(\mathbf{X}',\mathbf{x}\right) p_j\left(\mathbf{x}\right) d\Gamma-\fint_{\Gamma} P^*_{3j}\left(\mathbf{X}',\mathbf{x}\right) w_j\left(\mathbf{x}\right) d\Gamma$$
$$-\int_{\Omega} \kappa_{\alpha\beta} B\left(\frac{1-\nu}{2}\right)\left[u_{\alpha,\beta}(\mathbf{X})+u_{\beta,\alpha}(\mathbf{X})+\frac{2\nu}{1-\nu}u_{\gamma,\gamma}(\mathbf{X})\delta_{\alpha\beta}\right] W^*_{33}\left(\mathbf{X}',\mathbf{X}\right) d\Gamma$$
$$-\int_{\Gamma} \kappa_{\alpha\beta} B[(1-\nu)\kappa_{\alpha\beta}+\nu\delta_{\alpha\beta}\kappa_{\gamma\gamma}]w_3\left(\mathbf{X}\right) W^*_{i3}\left(\mathbf{X}',\mathbf{X}\right) d\Omega$$
$$+\lambda\int_{\Omega} W^*_{i3}\left(\mathbf{X}',\mathbf{X}\right) q_3\left(\mathbf{X}\right) d\Omega+\lambda\int_{\Omega} W^*_{i3}\left(\mathbf{X}',\mathbf{X}\right)\left(N_{\alpha\beta}w_{3,\beta}\right)_\alpha\left(\mathbf{X}\right) d\Omega \tag{12.34}$$

The integral equation related to the plane stress equations is given by

$$c_{\theta\alpha}\left(\mathbf{x}'\right) u_\theta\left(\mathbf{x}'\right) + \int\!\!\!\!\!\!-_{\Gamma} T_{\theta\alpha}^{(i)*}\left(\mathbf{x}',\mathbf{x}\right) u_\alpha\left(\mathbf{x}\right) d\Gamma$$

$$+ \int_\Gamma U_{\theta\alpha,\beta}^{*}\left(\mathbf{x}',\mathbf{x}\right) B\left[(1-\nu)\kappa_{\alpha\beta} + \nu\delta_{\alpha\beta}\kappa_{\gamma\gamma}\right] w_3\left(\mathbf{x}\right) n_\beta(\mathbf{x}) d\Omega$$

$$- \int_\Gamma U_{\theta\alpha}^{*}\left(\mathbf{x}',\mathbf{X}\right) B\left[(1-\nu)\kappa_{\alpha\beta} + \nu\delta_{\alpha\beta}\kappa_{\gamma\gamma}\right] w_{3,\beta}\left(\mathbf{X}\right) d\Omega$$

$$= \int_\Gamma U_{\theta\alpha}^{*}\left(\mathbf{X}',\mathbf{x}\right) t_\alpha\left(\mathbf{x}\right) d\Gamma + \int_\Omega U_{\theta\alpha}^{*}\left(\mathbf{x}',\mathbf{X}\right) q_\alpha\left(\mathbf{X}\right) d\Omega \tag{12.35}$$

After solving Eqs. (12.33), (12.34) and (12.35) the values of u_α, w_j, t_α, p_j and w_3 in the domain, the membrane tension $N_{\alpha\beta}(\mathbf{X})$ at each point of the domain $\mathbf{X}$ is calculated through the equation (without considering distributed forces q_α):

$$N_{\gamma\theta}(\mathbf{X}') = \int_\Gamma U_{\gamma\theta\alpha}^{*}\left(\mathbf{X}',\mathbf{x}\right) t_\gamma\left(\mathbf{x}\right) d\Gamma - \int_\Gamma T_{\gamma\theta\alpha}^{*}\left(\mathbf{X}',\mathbf{x}\right) u_\gamma\left(\mathbf{x}\right) d\Gamma$$

$$- \int_\Gamma U_{\gamma\theta\alpha}^{*}\left(\mathbf{X}',\mathbf{x}\right) B\left[\kappa_{\alpha\beta}(1-\nu) + \nu\delta_{\alpha\beta}\delta_{\phi\phi}\right] n_\beta(\mathbf{x}) d\Gamma$$

$$+ \int_\Gamma U_{\gamma\theta\alpha}^{*}\left(\mathbf{X}',\mathbf{X}\right) B\left[\kappa_{\alpha\beta}(1-\nu) + \nu\delta_{\alpha\beta}\delta_{\phi\phi}\right] w_{3,\beta}(\mathbf{x}) d\Gamma$$

$$+ B\left[\kappa_{\gamma\theta}(1-\nu) + \nu\delta_{\gamma\theta}\delta_{\phi\phi}\right] w_3(\mathbf{X}'). \tag{12.36}$$

12.3.2 Domain Integral Transformations

Domain integrals are transformed into boundary integrals using DRBEM, following the procedure presented in the previous section. Thus:

$$c_{ij}(\mathbf{x}') w_j\left(\mathbf{x}'\right) + \int\!\!\!\!\!\!-_{\Gamma} P_{ij}^{*}\left(\mathbf{x}',\mathbf{x}\right) w_j\left(\mathbf{x}\right) d\Gamma = \int_\Gamma W_{ij}^{*}\left(\mathbf{x}',\mathbf{x}\right) p_j\left(\mathbf{x}\right) d\Gamma$$

$$- \int_\Omega B[\kappa_{11} + \nu\kappa_{22}] W_{i3}^{*}\left(\mathbf{x}',\mathbf{X}\right) u_{1,1} d\Omega - \int_\Omega B[\nu\kappa_{11} + \kappa_{22}] W_{i3}^{*}\left(\mathbf{x}',\mathbf{X}\right) u_{2,2}(\mathbf{X}) d\Omega$$

$$- \int_\Omega \kappa_{\alpha\beta} B[(1-\nu)\kappa_{\alpha\beta} + \nu\delta_{\alpha\beta}\kappa_{\gamma\gamma}] W_{i3}^{*}\left(\mathbf{x}',\mathbf{X}\right) w_3(\mathbf{X}) d\Omega + \lambda \int_\Omega W_{i3}^{*}\left(\mathbf{x}',\mathbf{X}\right) q_b\left(\mathbf{X}\right) d\Omega \tag{12.37}$$

where $q_b = q_3 + (N_{\alpha\beta} w_{3,\beta})_{,\alpha}$. The variables u_α are approximated by a radial basis function as

$$u_\alpha(\mathbf{X}) = \sum_{m=1}^{M} f(r)^m \psi_\alpha^m = \mathbf{f}\boldsymbol{\psi}_\alpha \tag{12.38}$$

where

$$\boldsymbol{\psi}_\alpha = \mathbf{F}^{-1}\mathbf{u}_\alpha \tag{12.39}$$

Similarly, the approximation for $u_{\alpha,\beta}$ is given by

$$u_{\alpha,\beta}(\mathbf{X}) = \sum_{m-1}^{M} f(r)_{,\beta}^m \psi_\alpha^m = \mathbf{f}(r)_{,\beta}\boldsymbol{\psi}_\alpha \tag{12.40}$$

and therefore replacing $\boldsymbol{\psi}_\alpha$ in this expression, we obtain $u_{\alpha,\beta}$ in terms of $\mathbf{u}_\alpha$:

$$u_{\alpha,\beta}(\mathbf{X}) = \mathbf{f}(r)_{,\beta}\mathbf{F}^{-1}\mathbf{u}_\alpha \tag{12.41}$$

The distributed force in the domain q_b, the displacements w_3, and the force resultants $N_{\alpha\beta}$ and their derivatives $N_{\alpha\beta,\beta}$ in the domain are approximated as presented in the previous section. In this way, the integral equation (12.37) is expressed as

$$\begin{aligned} c_{ij}(\mathbf{x}')w_j(\mathbf{x}') &+ \fint_\Gamma P_{ij}^*(\mathbf{x}',\mathbf{x})\, w_j(\mathbf{x})\, d\Gamma = \int_\Gamma W_{ij}^*(\mathbf{x}',\mathbf{x})\, p_j(\mathbf{x})\, d\Gamma \\ &- \int_\Omega B[\kappa_{11} + \nu\kappa_{22}]W_{i3}^*(\mathbf{x}',\mathbf{X})\,\mathbf{f}(r)_{,1}\mathbf{F}^{-1}\mathbf{u}_1 d\Omega - \int_\Omega B[\nu\kappa_{11} + \kappa_{22}]W_{i3}^*(\mathbf{x}',\mathbf{X})\,\mathbf{f}(r)_{,2}\mathbf{F}^{-1}\mathbf{u}_2 d\Omega \\ &- \int_\Omega \kappa_{\alpha\beta}B[(1-\nu)\kappa_{\alpha\beta} + \nu\delta_{\alpha\beta}\kappa_{\gamma\gamma}]W_{i3}^*(\mathbf{x}',\mathbf{X})\,\mathbf{f}(r)\mathbf{F}^{-1}\mathbf{w}_3 d\Omega + \lambda\int_\Omega W_{i3}^*(\mathbf{x}',\mathbf{X})\,\mathbf{f}(r)\mathbf{F}^{-1}\mathbf{q}_b d\Omega \end{aligned} \tag{12.42}$$

12.4 Solving the Plate/Shell Buckling Problem

Quadratic isoparametric boundary elements are used to describe the geometry and function along the boundary. At the same time, implementing the dual reciprocity method and evaluating the derivative terms require several points uniformly distributed across the domain. After discretizing equations and performing internal point placement, the following system of linear equations is obtained for each placement node and each placement node in the domain. The above equations can be written in matrix form as

$$\mathbf{Cw} + \mathbf{H}_c\mathbf{w} - \mathbf{G}_c\mathbf{q} = \lambda(\mathbf{H}_c\hat{\mathbf{w}} - \mathbf{G}_c\hat{\mathbf{q}})\boldsymbol{\psi} \tag{12.43}$$

and

$$\mathbf{I}\mathbf{w}_3 + \mathbf{H}_d\mathbf{w} - \mathbf{G}_d\mathbf{q} = \boldsymbol{\lambda}(\mathbf{H}_d\hat{\mathbf{w}} - \mathbf{G}_d\hat{\mathbf{q}})\boldsymbol{\psi} \tag{12.44}$$

where the subscripts c and d refer to the placement points on the boundary and in the domain, respectively; $\hat{\mathbf{w}}$ and $\hat{\mathbf{q}}$ are matrices related to the particular solutions $\hat{w}_{mk}$ and $\hat{p}_{mk}$, respectively. The vector $\boldsymbol{\psi}$ is formed by the expression $\mathbf{F}^{-1}\mathbf{q}_b$ evaluated at each placement point. Once the displacements and tractions for the reference state are known, the membrane stresses at the nodes of the domain $N_{\alpha\beta}$ are calculated as.

The vector $\mathbf{q}_b$ can be rewritten in terms of the displacement $\mathbf{w}_3$ as

$$\mathbf{q}_b = (N_{\alpha\beta,\alpha}\mathbf{f}(r)_{,\beta}\mathbf{F}^{-1} + N_{\alpha\beta}\mathbf{f}(r)_{,\alpha}\mathbf{f}(r)_{,\beta}\mathbf{F}^{-1})\mathbf{w}_3 = \mathbf{f}_b\mathbf{w}_3 \tag{12.45}$$

Considering then $\boldsymbol{\psi} = \mathbf{F}^{-1}\mathbf{f}_b\mathbf{w}_3$, the previous equations are written as

$$\mathbf{C}\mathbf{w} + \mathbf{H}_c\mathbf{w} - \mathbf{G}_c\mathbf{q} = \lambda\mathbf{E}_q\mathbf{w}_3 \tag{12.46}$$

and

$$\mathbf{I}\mathbf{w}_3 + \mathbf{H}_d\mathbf{w} - \mathbf{G}_d\mathbf{q} = \lambda\mathbf{D}_q\boldsymbol{w}_3 \tag{12.47}$$

where $\mathbf{E}_q = (\mathbf{H}_c\hat{\mathbf{w}} - \mathbf{G}_c\hat{\mathbf{q}})\mathbf{F}^{-1}\mathbf{f}_b$ and $\mathbf{D}_q = (\mathbf{H}_d\hat{\mathbf{w}} - \mathbf{G}_d\hat{\mathbf{q}})\mathbf{F}^{-1}\mathbf{f}_b$. Once the boundary conditions have been imposed on the above equations, they can be re-organized as

$$\mathbf{A}\mathbf{y} = \lambda\mathbf{E}_q\boldsymbol{w}_3 \tag{12.48}$$

$$\mathbf{I}\mathbf{w}_3 = \mathbf{B}\mathbf{y} + \lambda\mathbf{D}_q\boldsymbol{w}_3 \tag{12.49}$$

where matrices **A** and **B** contain the coefficient matrices **H** and **G**. From former equation, **y** vector can be expressed as

$$\mathbf{y} = \lambda\mathbf{A}^{-1}\mathbf{E}_q\boldsymbol{w}_3 \tag{12.50}$$

Substituting in the second equation:

$$\mathbf{I}\mathbf{w}_3 = \lambda\mathbf{B}\mathbf{A}^{-1}\mathbf{E}_q\boldsymbol{w}_3 + \lambda\mathbf{D}_q\boldsymbol{w}_3 \tag{12.51}$$

Rearranging terms:

$$\mathbf{I}\mathbf{w}_3 = \lambda(\mathbf{B}\mathbf{A}^{-1}\mathbf{E}_q\boldsymbol{w}_3 + \mathbf{D}_q\boldsymbol{w}_3) \tag{12.52}$$

This equation can be written as a standard eigenvalue problem equation as follows:

$$\left(\mathbf{K} - \frac{1}{\lambda}\mathbf{I}\right)\boldsymbol{w}_3 = \mathbf{0} \tag{12.53}$$

where $\mathbf{K} = \mathbf{B}\mathbf{A}^{-1}\mathbf{E}_q + \mathbf{D}_q$. In this way, the equation for the eigenvector problem is obtained. Buckling coefficients corresponding to the problem can be obtained by the solution of Eq. (12.53).

12.5 Summary

In this chapter, the boundary element formulation of the buckling problem was established as a standard eigenvalue and eigenvector problem based on the plate equations. This formulation is based on the plate and shell equations for isotropic and composite materials. The buckling equations were established from the initial deflection calculation, and the generalized forces acting on the plate were calculated in the linear elastic problem. Open research topics include the analysis of buckling in plates and shells under thermal stresses, as well as the analysis of buckling in fractured plates and shells. Likewise, the analysis of buckling in thick laminated plates/shells is an open problem with no research reported in the literature.

References

1. Baiz PM, Aliabadi MH (2007) Buckling analysis of shear deformable shallow shells by the boundary element method. Eng Anal Bound Elem 31(4):361–372
2. Baiz PM, Aliabadi MH (2008) Buckling analysis of cracked plates using the boundary element method. In: Computational and experimental methods in structures buckling and postbuckling structures, vol 23, pp 375–412
3. Baiz PM, Aliabadi MH (2008) The boundary element method for buckling and postbuckling analysis of plates and shells. In: Computational and experimental methods in structures: buckling and postbuckling structures, pp 375–412
4. Baiz PM, Aliabadi MH (2009) Local buckling of thin-walled structures by the boundary element method. Eng Anal Bound Elem 33(3):302–313
5. Baiz PM, Aliabadi MH (2010) Post buckling analysis of shear deformable shallow shells by the boundary element method. Eng Anal Bound Elem J 84:379–433
6. Chang Jun C, Wang R (1996) Boundary integral equations and the boundary element method for buckling analysis of perforated plates. Eng Anal Bound Elem 17(1):57–68
7. Costa Jr JA, Brebbia C (1985) Plate bending problems using BEM: in boundary elements VII, 7 edn. Springer
8. Lin J, Duffield RC, Hui-Ru S (1999) Buckling analysis of elastic plates by boundary element method. Int J Numer Methods Eng, Eng Anal Bound Elem J 23:131–137
9. Nunes-Rangel A, Palermo Jr L, Wrobel LC (2024) A simplified dual reciprocity boundary element method applied to plate buckling problems. Eng Anal Bound Elem J 166
10. Purbasukano J (2003) Buckling and post-buckling analysis of cracked plates by the boundary element method. PhD thesis-Queen Mary University of London, 2003
11. Purbolaksono J, Aliabadi MH (2005) Buckling analysis of shear deformable plates by boundary element method. Int J Numer Meth Eng 62(4):537–563
12. Reddy JN (2004) Mechanics of laminated composite plates and shells: theory and analysis, 2nd edn. CRC Press, New York
13. Reissner E (1955) On transverse vibration of thin, shallow elastic shells. Q Appl Math 13(2):169–176
14. Shi G (1990) Flexural vibration and buckling analysis of orthotropic plates by the boundary element method. Int J Solids Struct 26(12):1351–1370
15. Soares RA Jr, Palermo L Jr, Wrobel LC (2020) Application of the radial integration method for the buckling analysis of plates with shear deformation. Eng Anal Bound Elem J 118:250–264
16. Stern M (1979) A general boundary integral formulation for the numerical solution of plate bending problems. Int J Solids Struct 15(10):769–782

17. Syngellakis S, Elzein A (1994) Plate buckling loads by the boundary element method. Int J Numer Meth Eng 37:1763–1778
18. Tanaka M, Miyazaki K (1988) A boundary element method for elastic buckling analysis of assembled plate structures. Comput Mech 3:49–57
19. Tanaka M, Miyazaki K (1986) Elastic buckling analysis of assembled plate structures by boundary element method. In: Boundary elements VIII: proceedings of the 8th international conference-Tokyo

Chapter 13
Special Topics: Boundary Element Nonlinear Analysis of Plates and Shells

Abstract Nonlinear plate and shell analysis has significant application in the analysis of thin-walled structures, such as those found in the aeronautical industry. The nonlinear problems addressed in this chapter relate to deflection problems in elastic plates and shells with large deflections and small strains, which allows for the analysis of post-buckling behavior. Similarly, the analysis of plates with elastoplastic response considering small deflections and strains represents another large group of nonlinear problems in the study of plates and shells. This chapter is presented as an introduction to nonlinear problems in BEM. The topics presented here are advanced topics in BEM, and their study is left to the reader in the bibliographical sources provided.

13.1 Introduction

Nonlinear plate and shell analysis has significant application in the analysis of thin-walled structures, such as those found in the aeronautical industry. The nonlinear problems addressed in this chapter relate to deflection problems in elastic plates and shells with large deflections and small strains, which allows for the analysis of post-buckling behavior. Similarly, the analysis of plates with elastoplastic response considering small deflections and strains represents another large group of nonlinear problems in the study of plates and shells. This chapter is presented as an introduction to nonlinear problems in BEM. The topics presented here are advanced topics in BEM, and their study is left to the reader in the bibliographical sources provided.

13.2 Large Deflection Analysis of Reissner Plates

Historically the analysis of large deflections in plates and shells using boundary elements dates back to the work of Sabir and Lock [11]. Later, in Ye and Liu [16] an integral equation formulation for finite deflection analysis of thin elastic plates is presented, based on the von Kármán theory. Kamiya and Sawaki [5] where the

J. Useche Vivero, *The Boundary Element Method for Engineering Analysis of Plates and Shells*, Solid Mechanics and Its Applications 281,
https://doi.org/10.1007/978-3-031-93050-8_13

finite deformation analysis of thin plates subjected to thermal loads is presented. Subsequently, Lei et al. [6] where the boundary element method is used in the geometrically nonlinear analysis of Reissner plates. Sun et al. [12] present a variational principle to derive the exact boundary equation for the analysis of nonlinear Reissner plates. Purbasukano[10] presents a boundary element formulation for the buckling and postbuckling analysis of fractured thick plates. In Wen et al. [14] a boundary element formulation of large deflection for Reisser plate is presented where the bending and plane stress response are coupled through domain integrals. In Wen et al. [15] the postbuckling analysis of shear deformable plates is performed, using the dual reciprocity boundary element method to treat the domain integrals. Dirgantara and Aliabadi [3] presents the geometrically nonlinear analysis of shear deformable shells using a dual reciprocity boundary element method. In 2008, Baiz and Aliabadi [1] presents the postbuckling analysis of thick plates using a formulation based on the dual reciprocity boundary element method. In Baiz and Aliabadi [2] This paper presents four boundary element formulations for post buckling analysis of shear deformable shallow shells using a coupled formulation based on the in-plane elasticity and bending response where the Newton-Raphson and the Arc-Length method were used.

In Moshaiov and Vorus [7] the analysis of elasto-plastic thin plate bending using a boundary integral equation formulation is presented. A direct domain boundary element method is presented in Providakis and Beskos [9] for the dynamic response analysis of thin elasto-plastic flexural plates. Supriyono and Aliabadi [13] presented a boundary element formulation for shear deformable plates with combined large deflexion and elastic-plastic material, where the boundary integral equations is based on the von Kármán's nonlinear plates theory. Dirgantara and Aliabadi [4] developed a boundary element method formulation elastoplastic analysis of shear deformable shells where domain integrals are treated by Cell integration method. Oliveira and Karam [8] present a boundary element formulation for elasto-plastic analysis of Reissner's plates in which plastic strains due to bending are included. This chapter is presented as an introduction to nonlinear problems in BEM. The topics presented here are advanced topics in BEM, and their study is left to the reader in the bibliographical sources provided. This chapter is entirely based on the works of Wen et al. [14], Dirgantara and Aliabadi [3], Oliveira and Karam [8], Dirgantara and Aliabadi [4].

13.3 Equilibrium Equations

As presented in Chap. 3, the plate strain tensor for an isotropic shell can be approximated by

$$\epsilon_{\alpha\beta} = \frac{1}{2}\left(u_{\alpha,\beta} + u_{\beta,\alpha} + w_{3,\alpha}w_{3,\beta}\right) + \frac{1}{2}x_3\left(w_{\alpha,\beta} + w_{\beta,\alpha}\right) \tag{13.1}$$

Stress resultants and moments resolved in the coordinate directions are given by

$$N_{\alpha\beta} = \int_{-h/2}^{+h/2} \sigma_{\alpha\beta} \mathrm{d}x_3 \tag{13.2}$$

$$Q_\alpha = \int_{-h/2}^{+h/2} \left(\sigma_{\alpha 3} + \sigma_{\alpha\beta} w_{3,\beta}\right) dx_3 \tag{13.3}$$

$$M_{\alpha\beta} = \int_{-h/2}^{+h/2} \sigma_{\alpha\beta} x_3 dx_3 \tag{13.4}$$

Using Hooke's law, the resultants of stresses and moments can be expressed in terms of displacements and rotations Wen et al. [14]:

$$\begin{aligned} N_{\alpha\beta} =& \frac{B}{2}\left((1-\nu)\left(u_{\alpha,\beta} + u_{\beta,\alpha}\right) + 2\nu u_{\gamma,\gamma}\delta_{\alpha,\beta}\right) \\ &+ \frac{B}{2}\left((1-\nu) w_{3,\alpha} w_{3,\beta} + v w_{3,\gamma} w_{3,\gamma}\delta_{\alpha,\beta}\right) = N^l_{\alpha,\beta} + N^n_{\alpha,\beta} \end{aligned} \tag{13.5}$$

Based on the first-order shear deformation theory, the resultant stress-strain relationships of the shear-deformable plate, for large deflection, can be written as

$$Q_\alpha = \frac{5}{6} Dh \left(w_\alpha + w_{3,\alpha}\right) + N_{\alpha\beta} w_{3,\beta} = Q^l_\alpha + Q^n_\alpha \tag{13.6}$$

$$M_{\alpha\beta} = \frac{D}{2}\left((1-\nu)\left(w_{\alpha,\beta} + w_{\beta,\alpha}\right) + 2\nu w_{\gamma,\gamma}\delta_{\alpha,\beta}\right) \tag{13.7}$$

The nonlinear terms $N^n_{\alpha\beta}$ and Q^n_α incorporate the effects of the large deflection (w_3) of the plate and result in a coupling between the equations describing the loading and bending of the membrane. The equilibrium equations in terms of displacements u_α, as for two-dimensional plane tension elasticity, are:

$$L^s_{\alpha\beta} u_\alpha + f_\alpha = 0 \tag{13.8}$$

where $L^s_{\alpha\beta}$ is the linear operator:

$$L^s_{\alpha\beta} = \frac{B}{2}\left[(1-\nu)\delta_{\alpha\beta}\nabla^2 + (1+v)\frac{\partial}{\partial x_\alpha}\frac{\partial}{\partial x_\beta}\right] \tag{13.9}$$

The equivalent body forces are defined as $f_\alpha = f^0_\alpha(x_1, x_2) + N^n_{\alpha\beta,\beta}$. The equilibrium equations for the shear force resultants and bending moments as functions of rotation w_α and deflection w_3 are:

$$L^p_{ij} w_j + q_i = 0 \tag{13.10}$$

where L^p_{ij} is the differential operator for the plate bending problems given in Chap. 3, and the equivalent body force is defined as $q_i = q^0_3(x_1, x_2) + \delta_{i3}\left(N_{\alpha\beta} w_{3,\beta}\right)_{,\alpha}$.

13.4 Boundary Integral Equations

The displacements for the two-dimensional elasticity in $\mathbf{X}'$ in the domain Ω can be written as Wen et al. [14]:

$$c^s_{\alpha\beta}(\mathbf{x}')\,u_\beta(\mathbf{x}') - \int_\Gamma U^s_{\alpha\beta}(\mathbf{x}',\mathbf{x})\,t_\beta(\mathbf{x})\mathrm{d}\Gamma + \int_\Gamma T^s_{\alpha\beta}(\mathbf{x}',\mathbf{x})\,u_\beta(\mathbf{x})\mathrm{d}\Gamma = \int_\Omega U^s_{\alpha\beta}(\mathbf{x}',\mathbf{X})\,f^0_\beta\,\mathrm{d}\Omega + \int_\Gamma U^s_{\alpha\beta}(\mathbf{x}',\mathbf{x})\,N^n_{\beta\gamma}n_\gamma(\mathbf{x})\mathrm{d}\Gamma - \int_\Omega U^s_{\alpha\beta}(\mathbf{x}',\mathbf{X})\,N^n_{\beta\gamma,\gamma}(\mathbf{X})\mathrm{d}\Omega \tag{13.11}$$

The tensile boundary conditions correspond to the linear part are given by

$$t^l_\beta = N^l_{\alpha\beta}n_\beta = \left(N_{\alpha\beta} - N^n_{\alpha\beta}\right)n_\beta \tag{13.12}$$

where n_β denotes the normal vector components external to the boundary, $\mathbf{x}$ is a limit point. Similarly, the rotations and deflections w_i in $\mathbf{X}'$ in the domain Ω can be obtained as

$$c^p_{ik}(\mathbf{x}')\,w_k(\mathbf{x}') - \int_\Gamma U^p_{ik}(\mathbf{x}',\mathbf{x})\,p_k(\mathbf{x})\mathrm{d}\Gamma + \int_\Gamma T^p_{ik}(\mathbf{x}',\mathbf{x})\,w_k(\mathbf{x})\mathrm{d}\Gamma \tag{13.13}$$

$$= \int_\Omega U^p_{ik}(\mathbf{x}',\mathbf{X})\,q^0_k\,\mathrm{d}\Omega + \int_\Omega U^p_{ik}(\mathbf{x}',\mathbf{X})\left(N_{\alpha\beta}w_{3,\beta}\right)_{,\alpha}\delta_{3k}\,\mathrm{d}\Omega \tag{13.14}$$

The internal stress resultants are obtained by differentiating (13.11) concerning the source point coordinate $\mathbf{X}'$ followed by applying Eq. (13.5) to obtain

$$N^l_{\alpha\beta}(\mathbf{X}') = \int_\Gamma U^s_{\alpha\beta\gamma}(\mathbf{X}',\mathbf{x})\,t_\gamma(\mathbf{x})\mathrm{d}\Gamma - \int_\Gamma T^s_{\alpha\beta\gamma}(\mathbf{X}',\mathbf{x})\,u_\gamma(\mathbf{x})\mathrm{d}\Gamma + \int_\Omega U^s_{\alpha\beta\gamma}(\mathbf{X}',\mathbf{X})\,f^s_\gamma(\mathbf{X})\mathrm{d}\Omega - \int_\Gamma U^s_{\alpha\beta\gamma}(\mathbf{X}',\mathbf{x})\,N^n_{\beta\gamma}n_\gamma\mathrm{d}\Gamma - \int_\Omega U^s_{\alpha\beta\gamma}(\mathbf{x}',\mathbf{X})\,N^n_{\gamma\varsigma,\varsigma}(\mathbf{X})\mathrm{d}\Omega \tag{13.15}$$

The moments and shear force for the plate bending problem can be obtained as

$$M^l_{\alpha\beta}(\mathbf{X}') = \int_\Gamma U^p_{\alpha\beta k}(\mathbf{X}',\mathbf{x})\,p_k(\mathbf{x})\mathrm{d}\Gamma - \int_\Gamma T^p_{\alpha\beta k}(\mathbf{X}',\mathbf{x})\,w_k(\mathbf{x})\mathrm{d}\Gamma + \int_\Omega U^p_{\alpha\beta k}(\mathbf{X}',\mathbf{X})\,q^0_k\,\mathrm{d}\Omega + \int_\Omega U^p_{\alpha\beta k}(\mathbf{X}',\mathbf{X})\left(N_{\alpha\beta}w_{3,\beta}\right)_{,\alpha}\delta_{3k}\,\mathrm{d}\Omega \tag{13.16}$$

$$Q^{l}_{\beta}\left(\mathbf{X}'\right) = \int\limits_{\Gamma} U^{p}_{3\beta k}\left(\mathbf{X}', \mathbf{x}\right) p_k(\mathbf{x})\mathrm{d}\Gamma - \int\limits_{\Gamma} T^{p}_{3\beta k}\left(\mathbf{X}', \mathbf{x}\right) w_k(\mathbf{x})\mathrm{d}\Gamma$$
$$+ \int\limits_{\Omega} U^{p}_{3\beta k}\left(\mathbf{X}', \mathbf{X}\right) q^{0}_{k}\, \mathrm{d}\Omega - \int\limits_{\Omega} U^{p}_{3\beta k}\left(\mathbf{X}', \mathbf{X}\right) \left(N_{\alpha\beta} w_{3,\beta}\right)_{,\alpha} \delta_{3k}\, \mathrm{d}\Omega \tag{13.17}$$

The dual reciprocity method transforms the domain integrals (19) of plate problems into boundary integrals, as presented in reference Wen et al. [14]. In the numerical solution scheme described later, it is necessary to evaluate the nonlinear terms:

$$N^{n}_{\alpha\beta} = \frac{B}{2}\left((1-\nu) w_{3,\alpha} w_{3,\beta} + v w_{3,\gamma} w_{3,\gamma} \delta_{\alpha,\beta}\right) \tag{13.18}$$
$$Q^{n}_{\alpha} = \left(N^{l}_{\alpha\beta} + N^{n}_{\alpha\beta}\right) w_{3,\beta} \tag{13.19}$$

The derivatives $w_{3,\alpha}$ can be obtained from $w_{3,\beta} = 2Q^{l}_{\beta}/D\left(1-\nu^2\right)\lambda^2 - w_{\beta}$, which allows evaluating the terms $N^{n}_{\alpha\beta}$ and Q^{n}_{α} in (29). To evaluate the derivatives $N^{n}_{\alpha,\beta,\beta}$ and $Q^{n}_{\alpha,\alpha}$, the nonlinear terms are approximated as follows Wen et al. [14]:

$$S(\mathbf{X}) = \sum_{m=1}^{M} f^{m}\left(r^{m}\right)\alpha^{m} \tag{13.20}$$

where the radial basis function is defined as $f(r) = \sqrt{1+r^2}$. Then, the derivatives of $S(\mathbf{X})$ are calculated from the formula:

$$S_{,\alpha} = \sum_{m=1}^{M} \frac{\left(x_{\alpha} - x^{m}_{\alpha}\right)\alpha^{m}}{\sqrt{c^2 + r^2}} \tag{13.21}$$

To solve the system of nonlinear equations, the load is applied incrementally according to the following procedure as proposed by Wen et al. [14]:

Incremental load solution procedure

i Let $n = 1$, $f^{1}_{\alpha} = \Delta f^{0}_{\alpha}$; $q^{1}_{\alpha} = \Delta q^{0}_{\alpha}$ and nonlinear terms $N^{n}_{\alpha\beta} = Q^{n}_{\beta} = 0$, so that the equations to be solved are linear.
ii Solving the linear system gives boundary values of u_{α} and w_k from which membrane forces $N^{n}_{\alpha\beta}$ and derivatives $w^{n}_{3,\alpha}$ in the domain are evaluated from Equations.
iii The nonlinear terms $\left[N^{n}_{\alpha\beta}\right]^{(n)}_{,\beta}$ and $\left[N_{\alpha\beta} w_{3,\beta}\right]^{(n)}_{,\alpha}$ are evaluated for use in the next step.
iv If the final load $f^{n}_{\alpha} = f^{0}_{\alpha}$ and $q^{n}_{\alpha} = q^{0}_{\alpha}$ Goto Step 5, otherwise let $n = n+1$ and $f^{n}_{\alpha} = n\Delta f^{0}_{\alpha}$; $q^{n}_{\alpha} = n\Delta q^{0}_{\alpha}$ Go to Step 2.
v print results for each step.

13.5 Geometrically Nonlinear Analysis of Thick Shells

13.5.1 Displacement Integral Equations

The integral displacement equations for shear-deformable surface shell bending problems have been derived to obtain five integral equations in terms of displacements: two rotations, one out-of-plane displacement, and two in-plane displacements in Dirgantara and Aliabadi [3]:

$$
\begin{aligned}
& c_{ij}\left(\mathbf{x}'\right) w_j\left(\mathbf{x}'\right) + \int_\Gamma P_{ij}^*\left(\mathbf{x}', \mathbf{x}\right) w_j(\mathbf{x}) \mathrm{d}\Gamma = \int_\Gamma W_{ij}^*\left(\mathbf{x}', \mathbf{x}\right) p_j(\mathbf{x}) \mathrm{d}\Gamma - \\
& \int_\Omega W_{i3}^*\left(\mathbf{x}', \mathbf{X}\right) k_{\alpha\beta} B \times \frac{1-\nu}{2}\left[u_{\alpha,\beta}(\mathbf{X}) + u_{\beta,\alpha}(\mathbf{X}) + \frac{2\nu}{1-\nu} u_{\phi,\phi}(\mathbf{X}) \delta_{\alpha\beta}\right] \mathrm{d}\Omega \\
& - \int_\Omega W_{i3}^*\left(\mathbf{x}', \mathbf{X}\right) k_{\alpha\beta} B\left((1-\nu) k_{\alpha\beta} + v \delta_{\alpha\beta} k_{\phi\phi}\right) w_3(\mathbf{X}) \mathrm{d}\Omega \\
& - \int_\Omega W_{i3}^*\left(\mathbf{x}', \mathbf{X}\right) k_{\alpha\beta} N_{\alpha\beta}^{(n)}(\mathbf{X}) \mathrm{d}\Omega + \int_\Omega W_{i3}^*\left(\mathbf{x}', \mathbf{X}\right)\left[N_{\alpha\beta}(\mathbf{X}) w_{3,\beta}(\mathbf{X})\right]_{,\alpha} \mathrm{d}\Omega \\
& + \int_\Omega W_{i3}^*\left(\mathbf{x}', \mathbf{X}\right) q_3(\mathbf{X}) \mathrm{d}\Omega
\end{aligned} \tag{13.22}
$$

The integral representations of displacement in the plane stress are given by

$$
\begin{aligned}
& c_{\alpha\beta}\left(\mathbf{x}'\right) u_\beta\left(\mathbf{x}'\right) + f_\Gamma T_{\alpha\beta}^{(i)*}\left(\mathbf{x}', \mathbf{x}\right) u_\beta(\mathbf{x}) \mathrm{d}\Gamma \\
& \quad + \int_\Gamma U_{\alpha\beta}^*\left(\mathbf{x}', \mathbf{x}\right) B\left[k_{\beta\gamma}(1-\nu) + v \delta_{\beta\gamma} k_{\phi\phi}\right] w_3(\mathbf{x}) n_\gamma(\mathbf{x}) \mathrm{d}\Gamma \\
& \quad - \int_\Omega U_{\alpha\beta}^*\left(\mathbf{x}', \mathbf{X}\right) B\left[k_{\beta\gamma}(1-\nu) + v \delta_{\beta\gamma} k_{\phi\phi}\right] w_{3,\gamma}(\mathbf{X}) \mathrm{d}\Omega \\
& \quad = \int_\Gamma U_{\alpha\beta}^*\left(\mathbf{x}', \mathbf{x}\right) t_\beta(\mathbf{x}) \mathrm{d}\Gamma - \int_\Gamma U_{\alpha\beta}^*\left(\mathbf{x}', \mathbf{x}\right) N_{\beta\gamma}^{(n)}(\mathbf{x}) n_\gamma(\mathbf{x}) \mathrm{d}\Gamma + \\
& \int_\Omega U_{\alpha\beta}^*\left(\mathbf{x}', \mathbf{X}\right) N_{\beta\gamma,\gamma}^{(n)}(\mathbf{X}) \mathrm{d}\Omega + \int_\Omega U_{\alpha\beta}^*\left(\mathbf{x}', \mathbf{X}\right) q_\beta \mathrm{d}\Omega,
\end{aligned} \tag{13.23}
$$

Equations (13.22) and (13.23) represent integral displacement equations for analyzing large deformations of a surface lamina. The out-of-plane displacement derivatives necessary for the calculation of nonlinear terms can be obtained by differentiating equation (13.22) to get

$$w_{3,\gamma}\left(\mathbf{X}'\right)+\int_{\Gamma}P^{*}_{3j,\gamma}\left(\mathbf{X}',\mathbf{x}\right)w_{j}(\mathbf{x})\mathrm{d}\Gamma=\int_{\Gamma}W^{*}_{3j,\gamma}\left(\mathbf{X}',\mathbf{x}\right)p_{j}\mathbf{x}\,\mathrm{d}\Gamma-$$
$$\int_{\Omega}W^{*}_{33,\gamma}\left(\mathbf{X}',\mathbf{X}\right)k_{\alpha\beta}B\frac{1-\nu}{2}\left(u_{\alpha,\beta}(\mathbf{X})+u_{\beta,\alpha}(\mathbf{X})+\frac{2\nu}{1-\nu}u_{\phi,\phi}(\mathbf{X})\delta_{\alpha\beta}\right)\mathrm{d}\Omega$$
$$-\int_{\Omega}W^{*}_{33,\gamma}\left(\mathbf{X}',\mathbf{X}\right)k_{\alpha\beta}B\left((1-\nu)k_{\alpha\beta}+\upsilon\delta_{\alpha\beta}k_{\gamma\gamma}\right)w_{3}(\mathbf{X})\mathrm{d}\Omega-$$
$$\int_{\Omega}W^{*}_{33,\gamma}\left(\mathbf{X}',\mathbf{X}\right)k_{\alpha\beta}N^{(n)}_{\alpha\beta}(\mathbf{X})\mathrm{d}\Omega+\int_{\Omega}W^{*}_{33,\gamma}\left(\mathbf{X}',\mathbf{X}\right)\left[N_{\alpha\beta}(\mathbf{X})w_{3,\beta}(\mathbf{X})\right]_{,\alpha}\mathrm{d}\Omega+$$
$$\int_{\Omega}W^{*}_{33,\gamma}\left(\mathbf{X}',\mathbf{X}\right)q_{3}(\mathbf{X})\mathrm{d}\Omega \tag{13.24}$$

13.5.2 Stress Resultant Integral Equations

The stress resultants at the domain point $\mathbf{X}'$ can be evaluated from the following equations (see Dirgantara and Aliabadi [3]):

$$M_{\alpha\beta}(\mathbf{X}')=\int_{\Gamma}W^{*}_{\alpha\beta k}\left(\mathbf{X}',\mathbf{x}\right)p_{k}(\mathbf{x})\mathrm{d}\Gamma-\int_{\Gamma}P^{*}_{\alpha\beta k}\left(\mathbf{X}',\mathbf{x}\right)w_{k}(\mathbf{x})\mathrm{d}\Gamma+$$
$$\int_{\Omega}W^{*}_{\alpha\beta 3}\left(\mathbf{X}',\mathbf{X}\right)\left[q_{3}-k_{\gamma\theta}N_{\gamma\theta}+\left(N_{\gamma\theta}w_{3,\theta}\right)_{,\gamma}\right]\mathrm{d}\Omega \tag{13.25}$$

$$Q_{\beta}(\mathbf{X}')=\int_{\Gamma}W^{*}_{3\beta k}\left(\mathbf{X}',\mathbf{x}\right)p_{k}(\mathbf{x})\mathrm{d}\Gamma-\int_{\Gamma}P^{*}_{3\beta k}\left(\mathbf{X}',\mathbf{x}\right)w_{k}(\mathbf{x})\mathrm{d}\Gamma+$$
$$\int_{\Omega}W^{*}_{3\beta 3}\left(\mathbf{X}',\mathbf{X}\right)\left[q_{3}-k_{\gamma\theta}N_{\gamma\theta}+\left(N_{\gamma\theta}w_{3,\theta}\right)_{,\gamma}\right]\mathrm{d}\Omega \tag{13.26}$$

and

$$N_{\alpha\beta}(\mathbf{X}')=\int_{\Gamma}U^{*}_{\alpha\beta\gamma}\left(\mathbf{X}',\mathbf{x}\right)t_{\gamma}(\mathbf{x})\mathrm{d}\Gamma-\int_{\Gamma}T^{*}_{\alpha\beta\gamma}\left(\mathbf{X}',\mathbf{x}\right)u_{\gamma}(\mathbf{x})\mathrm{d}\Gamma$$
$$-\int_{\Gamma}U^{*}_{\alpha\beta\gamma}\left(\mathbf{X}',\mathbf{x}\right)B\left[k_{\gamma\theta}(1-\nu)+\upsilon\delta_{\gamma\theta}k_{\phi\phi}\right]w_{3}(\mathbf{x})n_{\theta}(\mathbf{x})\mathrm{d}\Gamma$$
$$-\int_{\Gamma}U^{*}_{\alpha\beta\gamma}\left(\mathbf{X}',\mathbf{x}\right)N^{(n)}_{\gamma\theta}(\mathbf{x})n_{\theta}(\mathbf{x})\mathrm{d}\Gamma+\int_{\Omega}U^{*}_{\alpha\beta\gamma}\left(\mathbf{X}',\mathbf{X}\right)\left[q_{\gamma}+N^{(ii)}_{\gamma\theta,\theta}+N^{(n)}_{\gamma\theta,\theta}\right]\mathrm{d}\Omega+$$
$$B\left[(1-\nu)k_{\alpha\beta}+\upsilon\delta_{\alpha\beta}k_{\phi\phi}\right]w_{3} \tag{13.27}$$

The above domain integrals are transformed to boundary integral equations using the dual reciprocity boundary element method.

The total membrane tension resultants $N_{\alpha\beta}$ can be obtained from:

$$N^n_{\alpha\beta} = \frac{1-\nu}{2} B \left(w_{3,\alpha} w_{3,\beta} + \frac{\upsilon}{1-\nu} w_{3,\gamma} w_{3,\gamma} \delta_{\alpha,\beta} \right) \tag{13.28}$$

and

$$P_\alpha = N_{\alpha\beta} w_{3,\beta} \tag{13.29}$$

The nonlinear terms can be approximated by using expression (13.20) and (13.21). After discretization and placement of points at the nodes of the Γ boundary and in the Ω domain, the system matrix can be written as

$$\mathbf{Hw} = \mathbf{Gp} + \mathbf{Q} + \mathbf{C}_{nl}$$

where $\mathbf{H}$ and $\mathbf{G}$ are the well-known boundary element influence matrices $\mathbf{C}_{nl}$ is influence matrix for nonlinear terms, $\mathbf{w}$ is the boundary and domain displacement vector, $\mathbf{p}$ is the boundary traction vector, $\mathbf{n}_l$ is the nonlinear terms and $\mathbf{Q}$ is the domain load vectors.

13.5.3 Load Increment Technique

To solve the nonlinear system, a load increment technique is applied as follows Dirgantara and Aliabadi [3]:

Load increment technique

- Step 1: Let $n = 1$, $f^1_\alpha = \Delta f^0_\alpha$; $q^1_\alpha = \Delta q^0_\alpha$ and the nonlinear terms $N^n_{\alpha\beta} = P_\alpha = 0$.
- Step 2: Solve the linear system of equations to obtain values at the boundary.
- Step 3: Obtain values in the domain: derivatives of the out-of-plane displacement $w^n_{3,\alpha}$, Eq. (13.24); nonlinear membrane tension $N^n_{\alpha\beta}$ and total membrane tension $N_{\alpha\beta}$.
- Step 4: Calculate the nonlinear terms $N^n_{\alpha\beta,\beta}$ and $P_{\alpha,\alpha}$.
- Step 5: If the final charge $q^n_i = q^0_i$, go to Step 7; otherwise, set $n = n + 1$.
- Step 6: Set $q^n_i = q^{n-1}_i + \Delta q^0_i$, go to Step 2.
- Step 7: End the increment.

13.6 Elastoplastic Analysis of Thick Plates

Since only bending deformations are considered in this work, the total bending deformations $\chi_{\alpha\beta}$ are expressed as Oliveira and Karam [8]:

$$\chi_{\alpha\beta} = \chi^{e}_{\alpha\beta} + \chi^{p}_{\alpha\beta} \tag{13.30}$$

and the total shear strains ψ_{α}, by

$$\psi_{\alpha} = \psi^{e}_{\alpha} \tag{13.31}$$

where $\chi^{e}_{\alpha\beta}$ and ψ^{e}_{α} represent the elastic parcels and $\chi^{p}_{\alpha\beta}$ is the plastic parcel. The total generalized strains, $\chi_{\alpha\beta}$ and ψ_{α}, can also be expressed in terms of the generalized displacements, as

$$\chi_{\alpha\beta} = \frac{1}{2}\left(u_{\alpha,\beta} + u_{\beta,\alpha}\right) \tag{13.32}$$

$$\psi_{\alpha} = u_{\alpha} + u_{3,\alpha} \tag{13.33}$$

The plate equilibrium equations can be written as

$$M_{\alpha\beta,\beta} - Q_{\alpha} = 0 \tag{13.34}$$

$$Q_{\alpha,\alpha} + q = 0 \tag{13.35}$$

where $M_{\alpha\beta}$ and Q_{α} are, respectively, moments and shear resultants, which can be expressed as Oliveira and Karam [8]:

$$M_{\alpha\beta} = \frac{D(1-\nu)}{2}\left[2\chi_{\alpha\beta} + \frac{2\nu}{1-\nu}\chi_{\gamma\gamma}\delta_{\alpha\beta}\right] + \frac{\nu q}{(1-\nu)\lambda^2}\delta_{\alpha\beta} - M^{p}_{\alpha\beta} \tag{13.36}$$

$$Q_{\alpha} = \frac{D(1-\nu)\lambda^2}{2}\psi_{\alpha} \tag{13.37}$$

The plot $M^{p}_{\alpha\beta}$ in equation (13.36) represents the components of the plastic moments, defined by

$$M^{p}_{\alpha\beta} = \frac{D(1-\nu)}{2}\left[2\chi^{p}_{\alpha\beta} + \frac{2\nu}{1-\nu}\chi^{p}_{\gamma\gamma}\delta_{\alpha\beta}\right] \tag{13.38}$$

To facilitate the formulation of the initial deformation, the elastic bending deformations can be written in terms of the elastic moments, and thus the following expression is obtained:

$$\chi^{e}_{\alpha\beta} = \frac{1}{D(1-\nu)}\left[M_{\alpha\beta} - \frac{\nu}{1+\nu}M_{\gamma\gamma}\delta_{\alpha\beta}\right] - \frac{\nu q}{D\left(1-\nu^2\right)\lambda^2}\delta_{\alpha\beta} \tag{13.39}$$

Furthermore, plastic deformations due to bending can be written as

$$\chi^{p}_{\alpha\beta} = \frac{1}{D(1-\nu)}\left[M^{p}_{\alpha\beta} - \frac{\nu}{1+\nu}M^{p}_{\gamma\gamma}\delta_{\alpha\beta}\right] \tag{13.40}$$

Using Eq. (13.30), the total bending strains can also be written as Oliveira and Karam [8]:

$$\chi_{\alpha\beta} = \frac{1}{D(1-\nu)}\left[\left(M_{\alpha\beta} + M^{p}_{\alpha\beta}\right) - \frac{\upsilon}{1+\upsilon}\left(M_{\gamma\gamma} + M^{p}_{\gamma\gamma}\right)\delta_{\alpha\beta}\right] - \frac{\nu q}{D\left(1-\nu^{2}\right)\lambda^{2}}\delta_{\alpha\beta} \tag{13.41}$$

Let Ω be the problem domain, which consists of the midsurface of the plate, and Γ be the corresponding boundary. Representing the direction cosines of the external normal to the boundary by n_β, the generalized tractions can be written as $p_\alpha = M_{\alpha\beta}n_\beta$ and $p_3 = Q_\beta n_\beta$.

13.6.1 Constitutive Equations

The existence of a yield function F expressed in terms of moments M_{ij} and a hardening parameter κ is assumed. During the loading that produces yield, the moments M_{ij} must remain on the yield surface, such that the following equation is fulfilled:

$$F\left(M_{ij}, \kappa\right) = f\left(M_{ij}\right) - \Psi(\kappa) = M_e - M_o = 0 \tag{13.42}$$

where M_e is the equivalent moment (or effective moment), calculated here following the von Mises yield criterion, and M_o is the uniaxial yield moment, calculated considering that whenever the equivalent moment, at any point, reaches the yield moment, the entire cross-section yields simultaneously, as

$$M_o = \frac{\sigma_o h^2}{4} \tag{13.43}$$

where σ_o is the uniaxial yield stress of the material. From this criterion, the equivalent moment (or effective moment) M_e is calculated using the following equation:

$$M_e = \sqrt{\frac{3}{2}M'_{ij}M'_{ij}} \tag{13.44}$$

where M'_{ij} are the "deviator moments" calculated as

$$M'_{ij} = M_{ij} - \frac{1}{3}\delta_{ij}M_{kk} \tag{13.45}$$

Now, considering the Prandtl-Reuss equations, the appropriate relations for determining the plastic strain increments can be expressed as

$$d\chi_{ij}^{p} = M'_{ij} d\gamma \tag{13.46}$$

where $d\gamma$ is a positive proportionality factor that can vary during the plastic deformation process, it is also defined as an equivalent or effective plastic strain increment, as follows:

$$d\chi_{e}^{p} = \sqrt{\frac{2}{3} d\chi_{ij}^{p} d\chi_{ij}^{p}} \tag{13.47}$$

The proportionality factor $d\gamma$ can be expressed in terms of M_e and $d\chi_e^p$, as follows Oliveira and Karam [8]:

$$d\gamma = \frac{3}{2}\frac{d\chi_e^p}{M_e} \tag{13.48}$$

Applying the BEM procedure with initial deformation, the plastic strain increments are calculated using the expressions presented above, assuming that a loading path is defined to reach a given state of accumulated plastic stresses and strains χ_{ij}^{p}.

A modified total strain can be easily defined as follows Oliveira and Karam [8]:

$$\hat{\chi}_{\mathrm{ij}} = \frac{1}{D(1-\nu)}\left[M_{ij} - \frac{\nu}{1+\nu} M_{kk}\delta_{ij}\right] - \frac{\nu q}{D\left(1-\nu^2\right)\lambda^2}\delta_{ij} + \Delta\chi_{ij}^{p} \tag{13.49}$$

In deviatoric form, this equation can be written as

$$\chi'_{ij} = \frac{M^{i}_{ij}}{D(1-\nu)} + \Delta\chi_{ij}^{p} \tag{13.50}$$

Defining χ_{et} as the equivalent strain as, the plastic bending strain increment is given by

$$\Delta\chi_{ij}^{p} = \frac{\Delta\chi_e^p}{\chi_{et}}\chi'_{ij} \tag{13.51}$$

where

$$\chi_{et} = \sqrt{\frac{2}{3}\chi'_{ij}\chi'_{ij}} \tag{13.52}$$

The equivalent plastic strain increment is defined as

$$\Delta\chi_e^p = \chi_{et} - \frac{2M_o}{3D(1-\nu)} \tag{13.53}$$

M_o corresponds to the uniaxial yield moment acting after applying the current load increment and can be approximated by a Taylor series, truncated around the previous value of M_o (i.e., before the application of the load increment), as ${}^{k}M_o = {}^{k-1}M_o + {}^{k-1}H'\Delta\chi_e^p + \ldots$, where H' is the slope of the effective moment vs. effective plastic strain curve, defined as

$$H' = \frac{\Delta M_e}{\Delta \chi_e^p} \tag{13.54}$$

Finally, it can be seen that this increment can be obtained from the uniaxial moment-strain curve, as follows Oliveira and Karam [8]:

$$\Delta\chi_e^p = \frac{3D(1-\nu)\chi_{et} - 2M_o}{3D(1-\nu) + 2H'} \tag{13.55}$$

13.7 Boundary Integral Equations

The integral equations for the generalized displacements at an origin point ξ of the domain Ω can be obtained by a weighted residue procedure or Betti's second theorem. Therefore, three equations can be written as follows Oliveira and Karam [8]:

$$\begin{aligned} c_{ij}w_j(\xi) &= \int_\Gamma u_{ij}^*(\mathbf{x}', \mathbf{x})p_j(\mathbf{x})d\Gamma - \int_\Gamma p_{ij}^*(\mathbf{x}', \mathbf{x})w_j(\mathbf{x})d\Gamma \\ &+ \int_\Omega \left[u_{i3}^*(\mathbf{X}', \mathbf{x}) - \frac{\nu}{(1-\nu)\lambda^2}u_{i\alpha,\alpha}^*(\mathbf{X}', \mathbf{x})\right]q(\mathbf{x})d\Omega \\ &+ \int_\Omega M_{\alpha\beta i}^*(\mathbf{X}', \mathbf{x})\chi_{\alpha\beta}^p(\mathbf{x})d\Omega \end{aligned} \tag{13.56}$$

This equation is the basic integral equation for the elastoplastic analysis of plate bending in a formulation that considers the initial plastic strains. The kernel $M_{\alpha\beta i}^*$ is defined as

$$\begin{aligned} M_{\alpha\beta\gamma}^* &= -\frac{1}{4\pi r}\left[(4A + 2zK_1 + 1 - \nu)\left(\delta_{\beta\gamma}r_{,\alpha} + \delta_{\alpha\gamma}r_{,\beta}\right)\right. \\ &-2\left(8A + 2zK_1 + 1 - \nu\right)r_{,\alpha}r_{,\beta}r_{,\gamma} + (4A + 1 + \nu)\delta_{\alpha\beta}r_{,\gamma}\big] \\ &-2\left(8A + 2zK_1 + 1 - \nu\right)r_{,\alpha}r_{,\beta}r_{,\gamma} + (4A + 1 + \nu)\delta_{\alpha\beta}r_{,\gamma}\big] \\ &-2\left(8A + 2zK_1 + 1 - \nu\right)r_{,\alpha}r_{,\beta}r_{,\gamma} + (4A + 1 + \nu)\delta_{\alpha\beta}r_{,\gamma}\big] \\ M_{\alpha\beta3}^* &= -\frac{(1-\nu)}{8\pi}\left[\left(2\frac{(1+\nu)}{(1-\nu)}\ln z - 1\right)\delta_{\alpha\beta} + 2r_{,\alpha}r_{,\beta}\right] \end{aligned} \tag{13.57}$$

in which: $A = A(z) = K_0(z) + \frac{2}{z}\left[K_1(z) - \frac{1}{z}\right]$, with $z = \lambda r$ and with $K_0(z)$ and $K_1(z)$ representing modified Bessel functions of integer order.

13.7.1 Integral Equations for Moments and Shear Forces at Internal Points

The moment and shear resultants at the internal points can be calculated, respectively, as follows:

$$M_{\alpha\beta}(\mathbf{X}') = \int_{\Gamma} W^*_{\alpha\beta k} p_k d\Gamma - \int_{\Gamma} P^*_{\alpha\beta k} u_k d\Gamma + q \int_{\Gamma} W^*_{\alpha\beta} d\Gamma$$
$$+ \int_{\Omega} M^*_{\alpha\beta\gamma} \chi^p_{\gamma\theta} d\Omega + \frac{\nu q}{(1-\nu)\lambda^2}\delta_{\alpha\beta} - \frac{D\left(1-\nu^2\right)}{8}\left(2\chi^p_{\alpha\beta} + \delta_{\alpha\beta}\chi^p_{\theta\theta}\right) \tag{13.58}$$

$$Q_{\beta}(\mathbf{X}') = \int_{\Gamma} W^*_{3\beta k} p_k d\Gamma - \int_{\Gamma} P^*_{3\beta k} u_k d\Gamma + q \int_{\Gamma} W^*_{3\beta} d\Gamma + \int_{\Omega} M^*_{3\beta\gamma}\theta \chi^p_{\gamma\theta} d\Omega \tag{13.59}$$

The expressions for the tensor $M^*_{i\beta\gamma\theta}$ are given by the following expressions:

$$\begin{aligned}
M^*_{\alpha\beta\gamma\theta} =& \frac{D(1-\nu)}{8\pi r^2}\left[2\left(4A + 2zK_1 + 1 - \nu\right)\left(\delta_{\alpha\theta}\delta_{\beta\gamma} + \delta_{\alpha\gamma}\delta_{\theta\beta}\right)\right. \\
&+2(4A + 1 + 3\nu)\delta_{\alpha\beta}\delta_{\gamma\theta} - 4\left(8A + 2zK_1 + 1 + \nu\right) \\
&\times\left(\delta_{\gamma\theta}r_{,\alpha}r_{,\beta} + \delta_{\alpha\beta}r_{,\theta}r_{,\gamma}\right) - 2\left(16A + 6zK_1 + z^2K_0 + 2 - 2\nu\right) \\
&\times\times\left(\delta_{\alpha\theta}r_{,\gamma}r_{,\beta} + \delta_{\alpha\gamma}r_{,\theta}r_{,\beta} + \delta_{\gamma\beta}r_{,\alpha}r_{,\theta} + \delta_{\theta\beta}r_{,\alpha}r_{,\gamma}\right) \\
&\left. + 8\left(24A + 8zK_1 + z^2K_0 + 2 - 2\nu\right)r_{,\alpha}r_{,\beta}r_{,\gamma}r_{,\theta}\right] \\
M^*_{3\beta\gamma\theta} =& -\frac{D(1-\nu)\lambda^2}{8\pi r}\left[\left(4A + 2zK_1\right)\left(\delta_{\theta\beta}r_{,\gamma} + \delta_{\gamma\beta}r_{,\theta}\right)\right. \\
&\left. +4Ar_{,\beta}\delta_{\gamma\theta} - 4\left(4A + zK_1\right)r_{,\gamma}r_{,\theta}r_{,\beta}\right].
\end{aligned}$$

13.7.2 Solution Procedure

Writing Eq. (13.56) for all nodal boundary points, a system is obtained with a number of equations equal to three times the number of nodes, represented as

$$\mathbf{Hw} = \mathbf{Gp} + \mathbf{b} + \mathbf{D}\chi^p \tag{13.60}$$

As usual, this system of equations can be rearranged, placing all unknowns in the vector $\mathbf{y}$ and all known values:

$$\mathbf{Ay} = \mathbf{f} + \mathbf{D}\chi^p \tag{13.61}$$

where $\mathbf{A}$ is a complete and non-symmetric matrix. Solving for $\mathbf{y}$:

$$\mathbf{y} = \mathbf{K}\chi^p + \mathbf{m} \tag{13.62}$$

where $\mathbf{K} = \mathbf{A}^{-1}\mathbf{D}$ and $\mathbf{m} = \mathbf{A}^{-1}\mathbf{f}$.

Now, applying Eqs. (13.58) and (13.59) to all internal points located at the geometric center of the cells, we obtain:

$$\mathbf{M} = \mathbf{G}'\mathbf{p} - \mathbf{H}'\mathbf{w} + (\mathbf{w}' + \mathbf{v}') + (\mathbf{D}' + \mathbf{C}')\chi^p \tag{13.63}$$

$$\mathbf{Q} = \mathbf{G}''\mathbf{p} - \mathbf{H}''\mathbf{w} + \mathbf{w}'' + \mathbf{D}''\chi^p \tag{13.64}$$

In which the matrices $\mathbf{G}'$, $\mathbf{G}''$, $\mathbf{H}'$ and $\mathbf{H}''$, the vectors $\mathbf{w}'$ and $\mathbf{w}''$ contain contour integrals related to the fundamental solution; $\mathbf{v}'$ contains the free term related to the transverse load; $\mathbf{D}'$ and $\mathbf{D}''$ contain domain integrals that multiply the plastic strains, and $\mathbf{C}'$ represents the free term related to the plastic strains. Similarly, his system of equations can be rearranged, placing all unknowns in the vector $\mathbf{y}$ and all known values:

$$\mathbf{M} = -\mathbf{A}'\mathbf{y} + \mathbf{f}' + \mathbf{D}^*\chi^p \tag{13.65}$$

$$\mathbf{Q} = -\mathbf{A}''\mathbf{y} + \mathbf{f}'' + \mathbf{D}''\chi^p \tag{13.66}$$

with $\mathbf{D}^* = \mathbf{D}' + \mathbf{C}'$. Vectors $\mathbf{f}$, $\mathbf{f}'$, and $\mathbf{f}''$ containing the prescribed values, including the transverse load. Substituting Eq. (13.62) into Eqs. (13.63) and (13.64) results in

$$\mathbf{M} = \mathbf{S}'\chi^p + \mathbf{n}' \tag{13.67}$$

$$\mathbf{Q} = \mathbf{S}''\chi^p + \boldsymbol{n}'' \tag{13.68}$$

in which:

$$\mathbf{S}' = \mathbf{D}^* - \mathbf{A}'\mathbf{K}, \quad \mathbf{S}'' = \mathbf{D}'' - \mathbf{A}''\mathbf{K} \tag{13.69}$$

$$\mathbf{n}' = \mathbf{f}' - \mathbf{A}'\mathbf{m}, \quad \mathbf{n}'' = \mathbf{f}'' - \mathbf{A}''\mathbf{m} \tag{13.70}$$

13.7.3 Incremental Solution Procedure

The incremental process begins by reducing the maximum equivalent moment calculated at the cell points, represented as $M_e^{\max}$, to the initial yield moment M_o. The initial load factor is then calculated as

$$\mu_o = \frac{M_o}{M_e^{\max}} \tag{13.71}$$

Subsequent load factor values for the incremental process are calculated using the following recursive equation: $\mu_i = \mu_{i-1} + \Delta\mu_i$, where $\Delta\mu_i$ represents the increment defined as a percentage β_i in terms of the load at first yield: $\Delta\mu_i = \beta_i\mu_o$. Equations (13.62), (13.63), and (13.64) can be written, for an incremental procedure, as

$$\mathbf{y} = \mathbf{K}(\boldsymbol{\chi}^p + \boldsymbol{\Delta\chi}^p) + \mu_i \mathbf{m} \tag{13.72}$$

$$\mathbf{M} = \mathbf{S}'(\boldsymbol{\chi}^p + \boldsymbol{\Delta\chi}^p) + \mu_i \mathbf{n}'' \tag{13.73}$$

$$\mathbf{Q} = \mathbf{S}''(\boldsymbol{\chi}^p + \boldsymbol{\Delta\chi}^p) + \mu_i \mathbf{n}'' \tag{13.74}$$

For each value of μ_i, the plastic strain increment is calculated iteratively at each internal cell point considered, following a solution technique for an initial strain formulation Oliveira and Karam [8].

13.8 Elastoplastic Boundary Element Method for Shear Deformable Shells

In the elastoplastic response of thick shells, it is considered that plastic strains are only due to bending and membrane, hence a total strain rates can be defined as Dirgantara and Aliabadi [4]:

$$\dot{\varepsilon}_{\alpha\beta} = \frac{1}{2}\left(\dot{u}_{\alpha,\beta} + \dot{u}_{\beta,\alpha}\right) + \frac{\dot{w}_3}{R_\alpha}\delta_{\alpha\beta} \tag{13.75}$$

$$\dot{\chi}_{\alpha\beta} = \dot{\chi}^e_{\alpha\beta} + \dot{\chi}^p_{\alpha\beta} \tag{13.76}$$

$$\varepsilon_{\alpha\beta} = \dot{\varepsilon}^e_{\alpha\beta} + \dot{\varepsilon}^p_{\alpha\beta} \tag{13.77}$$

$$\dot{\gamma}_{\alpha 3} = \dot{\gamma}^e_{\alpha 3} \tag{13.78}$$

The normal strain tensor in the plane can be expressed in terms of generalized displacement rates, as

$$\dot{\varepsilon}_{\alpha\beta} = \frac{1}{2}\left(\dot{u}_{\alpha,\beta} + \dot{u}_{\beta,\alpha}\right) + \frac{\dot{w}_3}{R_\alpha}\delta_{\alpha\beta} \tag{13.79}$$

and the shear deformation in the plane

$$\dot{\gamma}_{\alpha\beta} = 2\dot{\varepsilon}_{\alpha\beta}; \quad \alpha \neq \beta \tag{13.80}$$

Transverse deformations are:

$$\dot{\gamma}_{\alpha 3} = \dot{w}_\alpha + \dot{w}_{3,\alpha} \tag{13.81}$$

where the tangential displacements were neglected. The curvature relations with the derivatives of generalized displacements can be written as

$$\dot{\kappa}_{\alpha\beta} = 2\dot{\chi}_{\alpha\beta} = \dot{w}_{\alpha,\beta} + \dot{w}_{\beta,\alpha} \tag{13.82}$$

The equilibrium equations of the layer are given as

$$\dot{M}_{\alpha\beta,\beta} - \dot{Q}_\alpha = 0 \tag{13.83}$$

$$\dot{Q}_{\alpha,\alpha} - k_{\alpha\beta}\dot{N}_{\alpha\beta} + \dot{q}_3 = 0 \tag{13.84}$$

$$\dot{N}_{\alpha\beta,\beta} + k_{\alpha\beta} + \dot{q}_\alpha = 0. \tag{13.85}$$

The relationships between the resulting stress and strain are as follows:

$$\dot{M}_{\alpha\beta} = D\frac{1-\nu}{2}\left(2\dot{\chi}_{\alpha\beta} + \frac{2\nu}{1-\nu}\dot{\chi}_{\gamma\gamma}\delta_{\alpha\beta}\right) - \dot{M}^p_{\alpha\beta} = \dot{M}^e_{\alpha\beta} - \dot{M}^p_{\alpha\beta} \tag{13.86}$$

$$\dot{Q}_\alpha = C\dot{\gamma}_{\alpha 3} \tag{13.87}$$

$$\dot{N}_{\alpha\beta} = B\frac{1-\nu}{2}\left(\dot{\varepsilon}_{\alpha\beta} + \dot{\varepsilon}_{\beta\alpha} + \frac{2\nu}{1-\nu}\dot{\varepsilon}_{\gamma\gamma}\delta_{\alpha\beta}\right) - \dot{N}^p_{\alpha\beta} = \dot{N}^e_{\alpha\beta} - \dot{N}^p_{\alpha\beta} \tag{13.88}$$

The resulting rates of plastic bending stress, $\dot{M}^p_{\alpha\beta}$, and the resulting rates of plastic membrane stress, $\dot{N}^p_{\alpha\beta}$, defined in an initial stress formulation are given by

$$\dot{M}^p_{\alpha\beta} = D\frac{1-\nu}{2}\left(2\dot{\chi}^p_{\alpha\beta} + \frac{2\nu}{1-\nu}\dot{\chi}^p_{\gamma\gamma}\delta_{\alpha\beta}\right) \tag{13.89}$$

$$\dot{N}^p_{\alpha\beta} = D\frac{1-\nu}{2}\left(2\dot{\varepsilon}^p_{\alpha\beta} + \frac{2\nu}{1-\nu}\dot{\varepsilon}^p_{\gamma\gamma}\delta_{\alpha\beta}\right) \tag{13.90}$$

By substituting the above expressions into the resulting stress and strain and then into the stress resultant-displacement relationships we obtain [4]:

$$\dot{M}_{\alpha\beta} = D\frac{1-\nu}{2}\left(\dot{w}_{\alpha,\beta} + \dot{w}_{\beta,\alpha} + \frac{2\nu}{1-\nu}\dot{w}_{\gamma,\gamma}\delta_{\alpha\beta}\right) - \dot{M}^p_{\alpha\beta} \tag{13.91}$$

$$\dot{Q}_\alpha = C\left(\dot{w}_\alpha + \dot{w}_{3,\alpha}\right) \tag{13.92}$$

$$\begin{aligned}\dot{N}_{\alpha\beta} = B\frac{1-\nu}{2}\left(\dot{u}_{\alpha,\beta} + \dot{u}_{\beta,\alpha} + \frac{2\nu}{1-\nu}\dot{u}_{\gamma,\gamma}\delta_{\alpha,\beta}\right) + B\left[(1-\nu)k_{\alpha\beta}\right.\\ \left.+\upsilon\delta_{\alpha\beta}k_{\phi\phi}\right]\dot{W}_3 - \dot{N}^p_{\alpha,\beta} = \dot{N}^{(i)}_{\alpha\beta} + \dot{N}^{(i)}_{\alpha\beta} - \dot{N}^p_{\alpha\beta}\end{aligned} \tag{13.93}$$

Finally, substituting above equations into the equilibrium equations we obtain:

$$L^b_{\alpha k}\dot{W}_k - \dot{M}^p_{\alpha\beta} = 0 \tag{13.94}$$

$$L^b_{3k}\dot{W}_k + \dot{q}_3 - k_{\alpha\beta}\dot{N}_{\alpha\beta}\left(\dot{N}^e_{\alpha\beta} - \dot{N}^p_{\alpha\beta}\right) = 0 \tag{13.95}$$

$$L^m_{\alpha\beta}\dot{u}_\beta + \dot{q}_\alpha + \dot{N}^{(i)}_{\alpha\beta,\beta} - \dot{N}^p_{\alpha\beta} = 0 \tag{13.96}$$

$$L^b_{\alpha,\beta} = \frac{D(1-\nu)}{2}\left[\left(\nabla^2 - \lambda^2\right)\delta_{\alpha\beta} + \frac{(1+\upsilon)}{(1-\nu)}\frac{\partial}{\partial x_\alpha}\frac{\partial}{\partial x_\beta}\right] \tag{13.97}$$

$$L^b_{\alpha 3} = -\frac{(1-\nu)D}{2}\lambda^2\frac{\partial}{\partial x_\alpha} \tag{13.98}$$

$$L^b_{3\alpha} = -L^b_{\alpha 3} \tag{13.99}$$

$$L^b_{33} = \frac{(1-\nu)D}{2}\lambda^2\nabla^2. \tag{13.100}$$

13.9 Elastoplastic Displacement Integral Equations

From the previous equilibrium equations and using the usual procedure, the contour integral equations for the bending problem are obtained Dirgantara and Aliabadi [4]:

$$c_{ij}\dot{w}_j(\mathbf{x}') + \int_\Gamma P^*_{ij}(\mathbf{x}',\mathbf{x})\dot{w}_j(\mathbf{x})d\Gamma = \int_\Gamma W^*_{ij}(\mathbf{x}',\mathbf{x})\dot{p}_j(\mathbf{x})d\Gamma - \int_\Omega W^*_{i3}(\mathbf{x}',\mathbf{x})k_{\alpha\beta}B$$
$$\times\frac{1-\nu}{2}\left[\dot{u}_{\alpha,\beta} + \dot{u}_{\beta,\alpha} + \frac{2\nu}{1-\nu}\dot{u}_{\phi,\phi}\delta_{\alpha\beta}\right]d\Omega - \int_\Omega W^*_{i3}(\mathbf{x}',\mathbf{x})k_{\alpha\beta}B\left((1-\nu)k_{\alpha\beta}\right.$$
$$\left.+\nu\delta_{\alpha\beta}k_{\phi\phi}\right)\dot{w}_3(\mathbf{X})d\Omega + \int_\Omega W^*_{i3}(\mathbf{x}',\mathbf{x})\dot{q}_3(\mathbf{X})d\Omega + \int_\Omega W^*_{i3}(\mathbf{x}',\mathbf{x})k_{\alpha\beta}\dot{N}^p_{\alpha\beta}(\mathbf{X})d\Omega +$$
$$\int_\Omega \chi^*_{i\alpha\beta}\dot{M}^p_{\alpha\beta}(\mathbf{X})d\Omega$$

The integral representations of displacement in the plane can be derived in the usual form to obtain:

$$c_{\alpha\beta}u_\beta(\mathbf{x}') + \int_\Gamma T^{(i)*}_{\alpha\beta}(\mathbf{x}',\mathbf{x})u_\beta(\mathbf{x})d\Gamma + \int_\Gamma U^*_{\alpha\beta}B\left[k_{\beta\gamma}(1-\nu) + \nu\delta_{\beta\gamma}k_{\phi\phi}\right]w_3(\mathbf{x})n_\gamma d\Gamma$$
$$-\int_\Omega U^*_{\alpha\beta}(\mathbf{x}',\mathbf{x})B\left[k_{\beta\gamma}(1-\nu) + \nu\delta_{\beta\gamma}k_{\phi\phi}\right]w_{3,\gamma}(\mathbf{x})d\Omega$$
$$= \int_\Gamma U^*_{\alpha\beta}(\mathbf{x}',\mathbf{x})\dot{N}_{\beta\gamma}n_\gamma(\mathbf{x})d\Gamma + \int_\Omega U^*_{\alpha\beta}(\mathbf{x}',\mathbf{x})q_\beta(\mathbf{x})d\Omega + \int_\Omega \varepsilon^*_{\alpha\beta\gamma}(\mathbf{x}',\mathbf{x})\dot{N}^{p(i)}_{\beta\gamma}(\mathbf{x})d\Omega$$

where the kernel $\chi^*_{\gamma\alpha\beta}$ is given by

$$\chi^*_{\gamma\alpha\beta} = \frac{-1}{8\pi D(1-\nu)r}\left[(8A(z) + 4zK_1(z) + 2 - 2\nu)\left(\delta_{\beta\gamma}r_{,\alpha} + \delta_{\alpha\gamma}r_{,\beta}\right)\right.$$
$$-4\left(8A(z) + 2zK_1(z) + 1 - \nu\right)r_{,\alpha}r_{,\beta}r_{,\gamma}$$
$$\left.+2(4A(z) + 1 - \nu)\delta_{\alpha\beta}r_{,\gamma}\right]$$
$$\chi^*_{3\alpha\beta} = \frac{-1}{8\pi D}\left[(2\ln z - 1)\delta_{\alpha\beta} + 2r_{,\alpha}r_{,\beta}\right]$$

Finally, the stress resultants at the domain point $\mathbf{X}'$ can be evaluated from the following integral equations for the stress resultants [4]:

$$\dot{M}_{\alpha\beta}(\mathbf{X}') = \int_\Gamma W^*_{\alpha\beta k}(\mathbf{X}',\mathbf{x})\dot{p}_k(\mathbf{x})d\Gamma - \int_\Gamma P^*_{\alpha\beta k}(\mathbf{X}',\mathbf{x})\dot{w}_k(\mathbf{x})d\Gamma$$
$$+\int_\Omega W^*_{\alpha\beta 3}(\mathbf{X}',\mathbf{X})\left[\dot{q}_3(\mathbf{X}) - k_{\gamma\theta}\dot{N}^e_{\gamma\theta}(\mathbf{X}) + k_{\gamma\theta}\dot{N}^p_{\gamma\theta}(\mathbf{X})\right]d\Omega +$$
$$\int_\Omega \chi^*_{\alpha\beta\gamma\theta}(\mathbf{X}',\mathbf{X})\dot{M}^p_{\gamma\theta}(\mathbf{X})d\Omega - \frac{1}{8}\left[2(1+\nu)\dot{M}^p_{\alpha\beta}(\mathbf{X}) + (1-3\nu)\dot{M}^p_{\theta\theta}(\mathbf{X})\delta_{\alpha\beta}\right] \tag{13.101}$$

$$\dot{Q}_\beta(\mathbf{X}') = \int_\Gamma W^*_{3\beta k}(\mathbf{X}',\mathbf{x})\dot{k}_k(\mathbf{X})d\Gamma - \int_\Gamma P^*_{3\beta k}(\mathbf{X}',\mathbf{x})\dot{w}_k(\mathbf{X})d\Gamma$$
$$+\int_\Omega W^*_{3\beta 3}(\mathbf{X}',\mathbf{X})\left[\dot{q}_3 - k_{\gamma\theta}\dot{N}^e_{\gamma\theta}(\mathbf{X}) + k_{\gamma\theta}\dot{N}_{\gamma\theta}(\mathbf{X})^p\right]d\Omega + \int_\Omega \chi^*_{3\beta\gamma\theta}(\mathbf{X}',\mathbf{X})\dot{M}^p_{\gamma\theta}(\mathbf{X}',\mathbf{X})d\Omega \tag{13.102}$$

and

$$\dot{N}_{\alpha\beta}(\mathbf{X}') = \int_\Gamma U^*_{\alpha\beta\gamma}(\mathbf{X}',\mathbf{x})\dot{t}_\gamma(\mathbf{x})d\Gamma - \int_\Gamma T^*_{\alpha\beta\gamma}(\mathbf{X}',\mathbf{x})\dot{u}_\gamma(\mathbf{x})d\Gamma - \int_\Gamma U^*_{\alpha\beta\gamma}(\mathbf{X}',\mathbf{x})B\left[k_{\gamma\theta}(1-\nu)\right.$$
$$\left.+v\delta_{\gamma\theta}k_{\phi\phi}\right]\dot{w}_3(\mathbf{x})n_\theta(\mathbf{x})d\Gamma + \int_\Omega U^*_{\alpha\beta\gamma}(\mathbf{X}',\mathbf{X})\left[\dot{q}_\gamma(\mathbf{X}) + \dot{N}^{(ii)}_{\gamma\theta,\theta}(\mathbf{X})\right]d\Omega$$
$$+\int_\Omega \varepsilon^*_{\alpha\beta\gamma\theta}(\mathbf{X}',\mathbf{X})\dot{N}^p_{\gamma\theta}(\mathbf{X})d\Omega + B\left[(1-\nu)k_{\alpha\beta} + \nu\delta_{\alpha\beta}k_{\phi\phi}\right]\dot{w}_3(\mathbf{X})d\Omega$$
$$-\frac{1}{8}\left[2(1+\nu)\dot{N}^p_{\alpha\beta}(\mathbf{X}) + (1-3\nu)\dot{N}^p_{\theta\theta}(\mathbf{X})\delta_{\alpha\beta}\right]$$

where the kernels $\varepsilon^*_{\alpha\beta\gamma\theta}$, $\varepsilon^*_{\alpha\beta\gamma\theta}$ y $\chi^*_{i\beta\gamma\theta}$ are defined as

$$\varepsilon^*_{\gamma\alpha\beta} = \frac{-1}{4\pi B(1-\nu)r}\left[(1-\nu)\left(\delta_{\gamma\alpha}r_{,\beta} + \delta_{\gamma\beta}r_{,\alpha}\right)\right.$$
$$\left.+(1+v)\left(-\delta_{\alpha\beta}r_{,\gamma} + 2r_{,\alpha}r_{,\beta}r_{,\gamma}\right)\right] \tag{13.103}$$

$$\varepsilon^*_{\alpha\beta\gamma\theta} = \frac{1}{4\pi r^2}\left\{2\nu\left[\delta_{\theta\alpha}r_{,\beta}r_{,\gamma} + \delta_{\theta\beta}r_{,\alpha}r_{,\gamma} + \delta_{\gamma\alpha}r_{,\beta}r_{,\theta} + \delta_{\gamma\beta}r_{,\alpha}r_{,\theta}\right]\right.$$
$$+(1-\nu)\left[\delta_{\beta\gamma}\delta_{\alpha\theta} + \delta_{\alpha\gamma}\delta_{\beta\theta} - \delta_{\alpha\beta}\delta_{\gamma\theta} + 2\delta_{\alpha\beta}r_{,\gamma}r_{,\theta}\right]$$
$$\left.+(1+v)\left[2\delta_{\gamma\theta}r_{,\alpha}r_{,\beta} - 8r_{,\alpha}r_{,\beta}r_{,\gamma}r_{,\theta}\right]\right\} \tag{13.104}$$

$$\chi^*_{\alpha\beta\gamma\theta} = \frac{1}{16\pi r^2}\left\{4\left(4A(z) + 2zK_1(z) + 1 - \nu\right)\left(\delta_{\gamma\alpha}\delta_{\theta\beta} + \delta_{\gamma\beta}\delta_{\theta\alpha}\right)\right.$$
$$+4(4A(z) + 1 + v)\delta_{\alpha\beta}\delta_{\gamma\theta} - 4\left(16A(z) + 6zK_1(z) + z^2K_0(z) + 2 - 2\nu\right)$$
$$\times\left[\delta_{\theta\alpha}r_{,\beta}r_{,\gamma} + \delta_{\theta\beta}r_{,\alpha}r_{,\gamma} + \delta_{\gamma\alpha}r_{,\beta}r_{,\theta} + \delta_{\gamma\beta}r_{,\alpha}r_{,\theta}\right]$$
$$-8\left(8A(z) + 2zK_1(z) + 1 + v\right)\delta_{\alpha\beta}r_{,\gamma}r_{,\theta} - 8\left(8A(z) + 2zK_1(z) + 1 - \nu\right)\delta_{\gamma\theta}r_{,\alpha}r_{,\beta}$$
$$\left.+16\left(24A(z) + 8zK_1(z) + z^2K_0(z) + 2 - 2\nu\right)r_{,\alpha}r_{,\beta}r_{,\gamma}r_{,\theta}\right\} \tag{13.105}$$

$$\chi^*_{3\beta\gamma\theta} = \frac{-\lambda^2}{16\pi r}\left[\left(8A(z) + 4zK_1(z)\right)\left(\delta_{\gamma\beta}r_{,\theta} + \delta_{\theta\beta}r_{,\gamma}\right)\right.$$
$$\left.-4\left(8A(z) + 2zK_1(z)\right)r_{,\beta}r_{,\gamma}r_{,\theta} + 8A(z)\delta_{\gamma\theta}r_{,\beta}\right] \tag{13.106}$$

13.10 Numerical Implementation

After discretization and imposition of boundary conditions, we obtain:

$$\mathbf{A}\dot{\mathbf{x}} = \dot{\mathbf{f}} + \begin{bmatrix} \mathbf{T}^b & 0 \\ 0 & \mathbf{T}^m \end{bmatrix} \begin{Bmatrix} \dot{\mathbf{M}}^p \\ \dot{\mathbf{N}}^p \end{Bmatrix} \tag{13.107}$$

$$\begin{Bmatrix} \dot{\mathbf{M}}^e \\ \dot{\mathbf{Q}}^e \\ \dot{\mathbf{N}}^e \end{Bmatrix} = -\left[\mathbf{A}'\right]\{\dot{\mathbf{X}}\} + \left\{\dot{\mathbf{f}}'\right\} + \begin{bmatrix} \mathbf{T}^{p*} & 0 \\ \mathbf{T}^{p*} & \mathbf{0} \\ \mathbf{0} & \mathbf{T}^{e^*} \end{bmatrix} \begin{Bmatrix} \dot{\mathbf{M}}^p \\ \dot{\mathbf{N}}^p \end{Bmatrix} \tag{13.108}$$

where $\mathbf{A}$ is the system matrix, $\dot{\mathbf{x}}$ is the unknown vector, $\mathbf{T}$ is the influence matrix for the plastic term, $\dot{\mathbf{f}}$ and $\left\{\dot{\mathbf{f}}''\right\}$ are the vector of prescribed boundary values, and:

$$\mathbf{T}^{p*} = \mathbf{T}^{p'} + \mathbf{E}^{p'} + \mathbf{I} \tag{13.109}$$

$$\mathbf{T}^{e*} = \mathbf{T}^{T'} + \mathbf{E}^{e'} + \mathbf{I} \tag{13.110}$$

The initial yield load is calculated by taking the most stressed node (boundary or internal) and comparing its equivalent stress, $\sigma_e^{\max}$, with the uniaxial yield stress of the material. The incremental process starts by reducing this stress value by a load factor, defined as $\lambda_o = \sigma_Y/\sigma_e^{max}$. Then, $\sigma_e^{\max}$ is set equal to the uniaxial yield stress σ_Y, and the initial load factor scales the elastic response. The new load increment is then calculated, and the additional load factor values are given by

$$\lambda_i = \lambda_{i-1} + \Delta\lambda_0 \quad \lambda_i \leqslant 1 \tag{13.111}$$

where $\Delta\lambda_0$ is the given value of the load increment with respect to the load at first yield. For a given value λ_i, the increment of the plastic stress resultants, $\Delta\dot{\mathbf{M}}^p$ and $\Delta\dot{\mathbf{N}}^p$, is determined iteratively at each selected node as follows Dirgantara and Aliabadi [4]:

Load increment technique

- Step 1. Estimate $\Delta\dot{\mathbf{M}}^p$ y $\Delta\dot{\mathbf{N}}^p$; $\Delta\dot{\mathbf{M}}^p = \Delta\dot{\mathbf{N}}^p = 0$ for the first iteration:
  $\left(\Delta\dot{\mathbf{M}}^p\right)_k = \left(\Delta\dot{\mathbf{M}}^p\right)_{k-1}, \quad \left(\Delta\dot{\mathbf{N}}^p\right)_k = \left(\Delta\dot{\mathbf{N}}^p\right)_{k-1}$
  for subsequent iterations.
- Step 2. Calculate $\dot{\mathbf{x}}$ y $\dot{\mathbf{M}}^e$, $\dot{\mathbf{Q}}^e$, $\dot{\mathbf{N}}^e$ and calculate the actual resultant voltage increment $\dot{\mathbf{M}}$, $\dot{\mathbf{Q}}$, $\dot{\mathbf{N}}$.
- Step 3. Check convergence by comparing the calculated $\Delta\varepsilon_{eq}^p$ with its accumulated value obtained during the current load increment if it can be neglected and calculate a new estimate of $\Delta\dot{\mathbf{M}}^p$ y $\Delta\dot{\mathbf{N}}^p$.
- Step 4. Accumulate initial stress resultants and actual stress resultants $\dot{\mathbf{M}}_{\alpha\beta}^p = \dot{\mathbf{M}}_{\alpha\beta}^p + \Delta\dot{\mathbf{M}}_{\alpha\beta}^p$.
- Step 5. Repeat Step 1 for a new iteration.

(13.112)

13.11 Summary

This chapter presented the application of the boundary element method to the analysis of plates and shells with large deflections and elastoplastic responses. These types of problems represent some of the most complex in solid mechanics, where they are considered advanced topics. As noted throughout the chapter, these formulations are not simple, and by their nature, their BEM formulation is complex. Their computational implementation is equally complex. The formulations presented here have been developed over the last 15 years, which also demonstrates the progress made in the application of BEM and demonstrates that they remain active areas of research.

References

1. Baiz PM, Aliabadi MH (2008) The boundary element method for buckling and postbuckling analysis of plates and shells. In: Computational and experimental methods in structures: buckling and postbuckling structures, pp. 375–412
2. Baiz PM, Aliabadi MH (2010) Post buckling analysis of shear deformable shallow shells by the boundary element method. Eng Anal Bound Elem J 84:379–433
3. Dirgantara T, Aliabadi MH (2006) A boundary element formulation for geometrically nonlinear analysis of shear deformable shells. Comput Methods Appl Mech Eng 195:4635–4654
4. Dirgantara T, Aliabadi MH (2012) Elastoplastic boundary element method for shear deformable shells. Eng Struct 45:62–67
5. Kamiya N, Sawaki Y (1988) Plate bending analysis by the dual reciprocity boundary elements. Eng Anal 5(1):36–40
6. Lei XY, Huang MK, Wang XX (1990) Geometrically nonlinear analysis of a reissner type plate by the boundary element method. Comput Struct J 37:911–916
7. Moshaiov A, Vorus WS (1986) Elasto-plastic plate bending analysis by a boundary element method with initial plastic moments. Int J Solids Struct 22:1213–1229
8. Oliveira SRC, Karam VJ (2016) Elastoplastic analysis of reissner's plates by the boundary element method. Eng Anal Bound Elem J 64:247–254
9. Providakis CP, Beskos DE (1994) Dynamic analysis of elasto-plastic flexural plates by the d/bem. Eng Anal Bound Elem 14:75–80
10. Purbasukano J (2003) Buckling and post-buckling analysis of cracked plates by the boundary element method. PhD thesis-Queen Mary University of London, 2003
11. Sabir AB, Lock AC (1972) The application of finite elements to the larege deflection geometrically nonlinear behaviour of cylindrical shells. In: Brebbia CA, Tottenham H (eds) Variational methods in engineering. Southampton Press
12. Sun YB, Xq HE, Qh QIN (1994) A new procedure for the nonlinear-analysis of reissner plate by boundary-element method. Comput Struct 53:649–652
13. Supriyono S, Aliabadi MH (2006) Boundary element method for shear deformable plates with combined geometric and material nonlinearities. Eng Anal Bound Elem J 30:31–43
14. Wen PH, Aliabadi MH, Young A (2005) Large deflection analysis of reissner plate by boundary element method. Comput Struct 83:870–879
15. Wen PH, Aliabadi MH, Young A (2006) Post buckling analysis of reissner plates by the boundary element method. J Strain Anal Eng Des 41:239–252
16. Ye TQ, Liu Y (1985) Finite deflection analysis of elastic plate by the boundary element method author links open overlay panel. Appl Math Modell J 9:183–188

MIX
Papier aus verantwortungsvollen Quellen
Paper from responsible sources
FSC® C105338

If you have any concerns about our products,
you can contact us on
ProductSafety@springernature.com

In case Publisher is established outside the EU,
the EU authorized representative is:
Springer Nature Customer Service Center GmbH
Europaplatz 3, 69115 Heidelberg, Germany

Printed by Libri Plureos GmbH
in Hamburg, Germany